METHODS IN MOLECULAR BIOLOGY

Series Editor
John M. Walker
School of Life and Medical Sciences
University of Hertfordshire
Hatfield, Hertfordshire, AL10 9AB, UK

For further volumes:
http://www.springer.com/series/7651

Matrix Metalloproteases

Methods and Protocols

Edited by

Charles A. Galea

Drug Delivery, Disposition and Dynamics, Monash Institute of Pharmaceutical Sciences, Monash University, Parkville, VIC, Australia

Humana Press

Editor
Charles A. Galea
Drug Delivery, Disposition and Dynamics
Monash Institute of Pharmaceutical Sciences
Monash University
Parkville, VIC, Australia

ISSN 1064-3745 ISSN 1940-6029 (electronic)
Methods in Molecular Biology
ISBN 978-1-4939-6861-9 ISBN 978-1-4939-6863-3 (eBook)
DOI 10.1007/978-1-4939-6863-3

Library of Congress Control Number: 2017930809

Printed on acid-free paper

This Humana Press imprint is published by Springer Nature
The registered company is Springer Science+Business Media LLC
The registered company address is: 233 Spring Street, New York, NY 10013, U.S.A.

Preface

The matrix metalloprotease (MMP) field has witnessed enormous advances since Jerome Gross and Charles Lapière first observed in 1962 an enzymatic activity (collagen degradation) associated with tadpole tail metamorphosis. Since the identification of this enzyme (interstitial collagenase or MMP-1), more than 20 closely related and evolutionarily conserved vertebrate MMPs have been discovered. These MMPs and their endogenous inhibitors (TIMPS) are involved in a diverse range of functions including tissue remodeling, immunity, inflammation, and angiogenesis. The first part of this book outlines recent advances in the expression and purification of MMPs in various expression systems, highlighting the advantages and disadvantages of each. In Part II we highlight how various biophysical methods such as X-ray crystallography, NMR spectroscopy, and small angle X-ray scattering, in combination with computational analyses (Part III), can provide a detailed understanding of the molecular mechanism of action of these enzymes. Part IV details how experimental and bioinformatics approaches can be used to define the substrate specificity of MMPs while Part V discusses methods for detecting MMP activity in vitro and in vivo. In Part VI we present various methods for the development and characterization of MMP-based inhibitors as potential therapeutics for the treatment of various diseases. The final part presents an overview of the involvement of MMPs in various diseases and their potential as diagnostic biomarkers.

Parkville, VIC, Australia *Charles A. Galea*

Contents

Contributors

OLAYIWOLA ADEDOTUN ADEKOYA • *Department of Pharmacy, University of Tromsø, Tromso, Norway*
BEATRIZ DA COSTA AGUIAR ALVES • *Laboratório de Análises Clínicas—Anexo 3, Faculdade de Medicina do ABC, Santo André, SP, Brazil*
LIGIA AJAIME AZZALIS • *Departamento de Ciências Biológicas, Instituto de Ciências Químicas, Ambientais e Farmacêuticas, Universidade Federal de São Paulo, Diadema, SP, Brazil*
CHRISTINE BEETON • *Department of Molecular Physiology and Biophysics, Baylor College of Medicine, Houston, TX, USA*
THAÍS MOURA GÁSCON BELARDO • *Laboratório de Análises Clínicas—Anexo 3, Faculdade de Medicina do ABC, Santo André, SP, Brazil*
LOUISE E. BUTT • *Institute and Biomedical and Biomolecular Science (IBBS) and School of Biological Sciences, University of Portsmouth, Portsmouth, UK*
WEN-CHI L. CHANG • *Cancer Prevention and Control Program, Fox Chase Cancer Center, Philadelphia, PA, USA*
JUN YONG CHOI • *Department of Chemistry, The Scripps Research Institute, Jupiter, FL, USA*
MARGIE L. CLAPPER • *Cancer Prevention and Control Program, Fox Chase Cancer Center, Philadelphia, PA, USA*
HARRY S. COOPER • *Department of Pathology, Fox Chase Cancer Center, Philadelphia, PA, USA; Cancer Prevention and Control Program, Fox Chase Cancer Center, Philadelphia, PA, USA*
ELENA DECANETO • *Max Planck Institute for Chemical Energy Conversion, Mülheim an der Ruhr, Germany*
NATALIA DÍAZ • *Dpto. Química Física y Analítica, Universidad de Oviedo, Oviedo, Asturias, Spain*
STEVEN R. VAN DOREN • *Department of Biochemistry, University of Missouri, Columbia, MO, USA*
FABIAN E. EGLI • *ETH Zurich, Department of Biology, Institute of Molecular Health Sciences, Zurich, Switzerland*
GREGG B. FIELDS • *Department of Chemistry and Biochemistry, Florida Atlantic University, Jupiter, FL, USA; Department of Chemistry, The Scripps Research Institute/Scripps Florida, Jupiter, FL, USA; Departments of Chemistry and Biology, Torrey Pines Institute for Molecular Studies, Port St. Lucie, FL, USA*
FERNANDO LUIZ AFFONSO FONSECA • *Departamento de Ciências Biológicas, Instituto de Ciências Químicas, Ambientais e Farmacêuticas, Universidade Federal de São Paulo, Diadema, SP, Brazil; Laboratório de Análises Clínicas—Anexo 3, Faculdade de Medicina do ABC, Santo André, SP, Brazil*
RITA FUERST • *Department of Chemistry, The Scripps Research Institute, Jupiter, FL, USA*
YAN G. FULCHER • *Department of Biochemistry, University of Missouri, Columbia, MO, USA*
CHARLES A. GALEA • *Drug Delivery, Disposition and Dynamics, Monash Institute of Pharmaceutical Sciences, Monash University, Parkville, VIC, Australia*

JOAN HARE • *Institute of Molecular Biophysics, Florida State University, Tallahassee, FL, USA*
HARVEY HENSLEY • *Biological Imaging Facility, Fox Chase Cancer Center, Philadelphia, PA, USA*
ROBERT A. HOLLAND • *Institute and Biomedical and Biomolecular Science (IBBS) and School of Biological Sciences, University of Portsmouth, Portsmouth, UK*
RUCHI JAIN • *Department of Molecular Reproduction, Development and Genetics, Indian Institute of Science, Bangalore, India*
B. JAYARAM • *Department of Chemistry, Indian Institute of Technology, HauzKhas, New Delhi, India; Supercomputing Facility for Bioinformatics & Computational Biology, Indian Institute of Technology, HauzKhas, New Delhi, India; Kusuma School of Biological Sciences, Indian Institute of Technology, HauzKhas, New Delhi, India*
ULRICH AUF DEM KELLER • *ETH Zurich, Department of Biology, Institute of Molecular Health Sciences, Zurich, Switzerland*
NIKUL S. KHUNTI • *Institute and Biomedical and Biomolecular Science (IBBS) and School of Biological Sciences, University of Portsmouth, Portsmouth, UK; Diamond Light Source, Diamond House, Harwell Science and Innovation Campus, Didcot, Oxfordshire, UK*
ANNA M. KNAPINSKA • *Department of Chemistry and Biochemistry, Florida Atlantic University, Jupiter, FL, USA*
RAMA K. KOPPISETTI • *Department of Biochemistry, University of Missouri, Columbia, MO, USA; Department of Medical Microbiology and Immunology, Life Sciences Center, University of Missouri, Columbia, MO, USA*
COLIN A. KRETZ • *Thrombosis and Atherosclerosis Research Institute and Department of Medicine, McMaster University, Hamilton, ON, Canada*
ZON W. LAI • *Institute of Molecular Medicine and Cell Research, University of Freiburg, Freiburg, Germany; Department of Genetics and Complex Diseases, Harvard T.H. Chan School of Public Health, Boston, MA, USA*
WOLFGANG LUBITZ • *Max Planck Institute for Chemical Energy Conversion, Mülheim an der Ruhr, Germany*
TARA C. MARCINK • *Department of Biochemistry, University of Missouri, Columbia, MO, USA*
JENNIFER S. MYERS • *Department of Chemistry and Biochemistry, Florida State University, Tallahassee, FL, USA*
HIDEAKI OGATA • *Max Planck Institute for Chemical Energy Conversion, Mülheim an der Ruhr, Germany*
RUTVIK S. PATEL • *Department of Molecular Physiology and Biophysics, Baylor College of Medicine, Houston, TX, USA*
ANDREW R. PICKFORD • *Institute of Biomedical and Biomolecular Science (IBBS) and School of Biological Sciences, University of Portsmouth, Portsmouth, UK*
DEBRA L. QUINN • *Institute of Biomedical and Biomolecular Science (IBBS) and School of Biological Sciences, University of Portsmouth, Portsmouth, UK*
HARINI RAMANAN • *Department of Molecular Reproduction, Development and Genetics, Indian Institute of Science, Bangalore, India*
DEEPAK K. SAINI • *Department of Molecular Reproduction, Development and Genetics, Indian Institute of Science, Bangalore, India*
QING-XIANG AMY SANG • *Department of Chemistry and Biochemistry, Florida State University, Tallahassee, FL, USA; Institute of Molecular Biophysics, Florida State University, Tallahassee, FL, USA*

Stephan Schilling • *Department of Drug Design and Target Validation (IZI-IMWT), Fraunhofer Institute for Cell Therapy and Immunology, Halle/Saale, Germany*
Oliver Schilling • *Institute of Molecular Medicine and Cell Research, University of Freiburg, Freiburg, Germany; BIOSS Centre of Biological Signaling Studies, University of Freiburg, Freiburg, Germany; German Cancer Consortium (DKTK) and German Cancer Research Center (DKFZ), Heidelberg, Germany*
Pascal Schlage • *ETH Zurich, Department of Biology, Institute of Molecular Health Sciences, Zurich, Switzerland*
Dagmar Schlenzig • *Department of Drug Design and Target Validation (IZI-IMWT), Fraunhofer Institute for Cell Therapy and Immunology, Halle/Saale, Germany*
Tanya Singh • *Department of Chemistry, Indian Institute of Technology, HauzKhas, New Delhi, India; Supercomputing Facility for Bioinformatics & Computational Biology, Indian Institute of Technology, HauzKhas, New Delhi, India*
Krishna K. Singh • *Department of Molecular Reproduction, Development and Genetics, Indian Institute of Science, Bangalore, India*
Maciej J. Stawikowski • *Department of Chemistry and Biochemistry, Florida Atlantic University, Jupiter, FL, USA*
Dimas Suárez • *Dpto. Química Física y Analítica, Universidad de Oviedo, Oviedo, Spain*
Rajeev B. Tajhya • *Department of Molecular Physiology and Biophysics, Baylor College of Medicine, Houston, TX, USA*

Part I

Expression and Purification of Matrix Metalloproteases

Chapter 1

Expression and Purification of Matrix Metalloproteinases in *Escherichia coli*

Krishna K. Singh, Ruchi Jain, Harini Ramanan, and Deepak K. Saini

Abstract

The MMP (matrix metalloproteinases) family of endopeptidases are involved in cleavage induced remodelling of the extracellular matrix including collagen, fibrinogen, elastin, and gelatin. Owing to their proteolytic activity which can cleave and degrade multiple intracellular substrates, the overexpression and purification of these proteins tends to be toxic. Here we describe a novel "matrix assisted refolding" protocol to overcome the technical challenges associated with overexpression and purification of full-length MMPs. The toxicity issue associated with MMP expression, is circumvented by expressing the recombinant protein in *Escherichia coli* in an inactive insoluble form. The methodology used for obtaining full-length MMP2 protein from these inclusion bodies, by its subsequent purification and refolding using affinity chromatography, through a single-step matrix based refolding protocol is presented here. The protocol described yields high concentrations of pure full-length and active MMP2 protein useful for downstream applications.

Key words MMP2, Inclusion bodies, Affinity chromatography, Refolding, Zymography

1 Introduction

MMPs (matrix metalloproteinases) comprise a family of 23 proteins which are metal ion dependent endopeptidases [1]. Owing to their proteolytic activity towards matrix proteins such as collagen, gelatin, and elastin, MMPs perform a number of physiological roles in normal tissues including tissue remodelling, wound healing, and bone morphogenesis. However, the activity of this class of proteins has been found to be highly upregulated under various pathological conditions including cancer [2–5]. Therefore, the expression and activity of MMP proteins can serve as a biomarker for malignancies associated with cancer progression [5, 6]. MMPs are also a promising target for cancer therapeutics, albeit with some caveats on account of the physiologically relevant roles played by MMPs in normal cellular physiology [7]. This arises due to a poor understanding of the functional roles of these proteins and their

Charles A. Galea (ed.), *Matrix Metalloproteases: Methods and Protocols*, Methods in Molecular Biology, vol. 1579,
DOI 10.1007/978-1-4939-6863-3_1, © Springer Science+Business Media LLC 2017

substrates in different biological conditions. It is hence imperative to further study the MMP family of proteins in greater detail. However, analysis of purified MMPs has been limited due to the toxic nature of these proteins, which poses a technical challenge in their overexpression, purification, and further characterization.

MMPs are secreted as zymogens in an inactive pro-MMP form, that are proteolytically processed and activated by other proteases (e.g., serine proteases) and can undergo an autocatalytic activation process [8]. This proteolytic processing is critical for the activation of pro-MMPs to yield the functional MMP protein. Under normal physiological conditions, the activity of the full-length protein is kept under regulatory check by members of another protein family, known as tissue inhibitor of metalloproteases (TIMPs). Under pathological conditions, such as cancer, the balance between TIMPs and MMPs is altered and is shifted towards activation of MMPs [9]. The activated MMPs lead to cleavage of a large variety of extracellular substrates facilitating cytoskeletal remodelling necessary for cancer metastasis and progression.

Traditional methods for MMP protein overexpression and purification include expressing recombinant proteins in *E. coli* or purification from human plasma [10] or from conditioned media of MMP-expressing mammalian cells [11]. However, MMPs purified using human plasma or cell culture media are often found to be contaminated with other associated cellular proteins including TIMPs, fibronectins etc. [10]. These co-purified contaminants interfere with the functional readouts used for MMP characterization, thereby limiting the use of eukaryotic cells as a source for MMPs. In this context, bacterial cells, specifically *E. coli* can serve as a viable alternate host for expression and purification of MMP proteins. However, due to the absence of inhibitory TIMPs, the expression of MMP proteins in bacterial cells causes proteolysis induced toxicity, thereby resulting in poor yield of the protein. These limitations can be overcome if the toxic protein is overexpressed as inactive inclusion bodies (IBs). However, the expression of proteins in IBs requires significant downstream processing, where they have to be solubilized and refolded to restore their bioactivity [12]. Refolding from IBs is generally tedious and depends on temperature, pH, salt concentration, as well as on the type and the concentration of the denaturants utilized. Conventional methods of protein refolding based on dialysis of denatured proteins against a large volume of refolding buffer or solvent-exchange chromatography, in general results in poor yield of active protein [13]. Keeping all this in mind, we developed a matrix assisted refolding protocol for the purification and refolding of MMP2 [14]. This is an efficient, single step protocol, which results in significantly higher yield of active protein from inclusion bodies compared to any other protocol reported to date. The steps, salient features, and advantages of the protocol are described below.

2 Materials

All the reagents utilized for protein work are from Sigma-Aldrich (MO, USA), of molecular biology grade, free of DNase, RNase, and proteases, unless otherwise mentioned.

Compositions of various media and buffers used in the protocols described herein are listed below:

2.1 Plasmids and *E. coli* Strains

1. MMP2 expression plasmid: cDNA for MMP2 was cloned into the pPROEx-HTc (Invitrogen, Carlsbad, CA) expression vector [14].
2. CFP-2RS expression plasmid: CFP-2RS CDS was cloned into the pPROEx-HTa expression vector [14].
3. *E. coli* C43: F $^-$ *ompT* *hsdS*$_B$ (rB $^-$ mB $^-$) *gal dcm* (DE3) pLysS (Cm^R) [15].
4. *E. coli* BL21 ArcticExpress™: *E. coli* B F$^-$ *ompT* *hsdS*(r$_B$ $^-$ mB $^-$) *dcm*$^+$ Tet^r *gal* λ(DE3) *endA* Hte [*cpn*10 *cpn*60 Gent^r] [16].

2.2 Protein Overexpression

1. Terrific broth media (pH ~ 7.4): Dissolve 12 g tryptone, 24 g yeast extract, 9.4 g potassium phosphate dibasic, 2.2 g potassium phosphate monobasic, and 4.0 ml glycerol in 1 l of deionized water. Autoclave to sterilize before use.
2. Antibiotics: Prepare a stock solution of 100 mg/ml of ampicillin or carbenicillin in water and filter-sterilize using a 0.22 μm filter before use. Stock solutions are kept frozen at −20 °C.
3. 1 M IPTG (isopropyl β-D-1-thiogalactopyranoside): 238 mg/ml stock solution in water. Dissolve 2.383 g of IPTG in a minimum volume of water (~5 ml) until it completely dissolves and then make up the volume to 10 ml. Filter-sterilize using a 0.22 μm filter before use. Sterile stock solutions are kept frozen at −20 °C.

2.3 MMP2 Purification and Refolding

1. 1 M sodium phosphate buffer, pH 7.4: Mix 77.4 ml of 1 M Na_2HPO_4 (monobasic) and 22.6 ml of 1 M NaH_2PO_4 (dibasic) to obtain 100 ml of 1 M sodium phosphate buffer, pH 7.4.
2. 5 M NaCl: Dissolve 29.25 g of NaCl in 100 ml of water.
3. 2 M imidazole: Dissolve 2.72 g imidazole in 20 ml of water. Store at 4 °C till use or prepare fresh.
4. 100 mM phenyl methyl sulfonylfluoride (PMSF): Dissolve 174 mg of PMSF in 10 ml of absolute ethanol and store at −20 °C in aliquots until use.
5. Native lysis buffer: 50 mM sodium phosphate buffer pH 7.4, 300 mM NaCl, 10 mM imidazole, 1.0 mM PMSF. (Tris–HCl pH 8.0 can also be used instead of 50 mM sodium phosphate buffer pH 7.4).

6. Inclusion bodies solubilization buffer: 20 mM Tris–HCl pH 8.0, 500 mM NaCl, 10% glycerol, and 8 M urea.
7. 100 mM benzamidine: Dissolve 120 mg of benzamidine hydrochloride in 10 ml of water. Store at 4 °C as aliquots till use.
8. 1 M dithiothreitol (DTT): Dissolve 154 mg of dithiothreitol in water to a final volume of 1 ml. Store as aliquots at −20 °C till use.
9. 0.5 M glutathione, oxidized (GSSG): Dissolve 3.06 g of GSSG in water to a final volume of 10 ml. Store at −20 °C till use.
10. 0.5 M glutathione, reduced (GSH): Dissolve 1.53 g of GSH in water to a final volume of 10 ml. Store at −20 °C till use.
11. Refolding buffer: 20 mM Tris–HCl pH 8.0, 500 mM NaCl, 10% glycerol, 20 mM imidazole, 0.5 mM oxidized glutathione (GSSG), and 5 mM reduced glutathione (GSH).
12. Elution buffer: 20 mM Tris–HCl, pH 8.0, 500 mM NaCl, 10% glycerol and 250 mM imidazole.
13. Dialysis/storage buffer: 50 mM Tris–HCl, pH 8.0, 50% glycerol, 50 mM NaCl and 1 mM DTT.

2.4 MMP2 Activity Analysis

1. 5× sample buffer: 250 mM Tris–HCl, pH 6.8, 5% SDS, 50% glycerol, 0.5% bromophenol blue, and 6.25% β-mercaptoethanol.
2. 2× sample buffer (for zymography): 100 mM Tris–HCl, pH 6.8, 2% SDS, 20% glycerol, and 0.2% bromophenol blue.
3. 10× activation buffer: 500 mM Tris–HCl, pH 7.4, 50 mM $CaCl_2$, and 10 μM $ZnCl_2$.
4. ARP100 (Cayman Chemical Co., USA): 5 mM ARP100 dissolved in DMSO. Aliquots stored at −20 °C till further use.
5. Staining solution: 0.5% CBB R_{250}, 40% methanol, 10% acetic acid, and 50% water.
6. Destaining solution: 40% methanol, 10% acetic acid, and 50% water.

3 Methods

3.1 Cloning MMP2 Gene for Prokaryotic Expression

1. PCR amplify cDNA template coding for MMP2, from total RNA extracted from HEK293 cells, using gene specific primers (*see* **Note 1**).
2. The amplified MMP2 ORF is cloned into the *Eco*RI and *Xho*I sites of the pPROEx-HT expression vector, which contains an N-terminal 6× His tag and a *trc* promoter upstream of the MMP2 ORF allowing for IPTG-dependent overexpression of the fusion protein.
3. Clones were verified by DNA sequencing to ensure the in-frame fusion of the purification tag with the MMP2 CDS.

3.2 Transformation and Propagation of the MMP Expression Strain

1. Chemically competent *E. coli* C43 (DE3) pLysS cells were transformed with the recombinant MMP2 expression plasmid DNA (*see* **Notes 2** and **3**).
2. The transformed cells were plated onto Luria Agar (LA) plates containing 100 μg/ml carbenicillin and incubate overnight at 37 °C. The colonies obtained were immediately used for protein overexpression (*see* **Note 4**).

3.3 Overexpression of Recombinant MMP2 Protein in Inclusion Bodies

Given that inclusion body formation is favoured when the expression of recombinant protein is high, conditions for MMP2 overexpression had to be optimized to maximize expression and inclusion body formation (*see* **Note 5**).

1. Inoculate a single colony of *E. coli* C43 cells into 10 ml of LB, containing 100 μg/ml carbenicillin (*see* **Note 4**) and incubate overnight at 37 °C.
2. Inoculate 1% of the overnight culture into 1 liter of terrific broth (*see* **Note 6**) containing 35 μg/ml chloramphenicol and 100 μg/ml ampicillin and grow at 37 °C to an OD_{600} of 0.8–1.0.
3. Add IPTG to the culture to a final concentration of 1 mM and incubate at 37 °C for a further 24 h at 180 rpm.
4. Pellet the induced culture by centrifuging at 6,000 × *g* for 10 min at room temperature.
5. Resuspend the pellet in 20 ml of native lysis buffer and sonicate on ice (Branson Sonifier model S-450D with 1/8″ tapered microtip) at 50% amplitude, six cycles of sonication with 10 s of ON and OFF pulse (*see* **Note 7**).
6. Centrifuge the sonicated lysate at 20,000 × *g* for 30 min at 4 °C to separate the soluble and insoluble fractions.
7. The soluble supernatant and insoluble pellet containing the inclusion bodies are analyzed by SDS-PAGE or western blotting using anti-His antibody (Fig. 1). If the yield of MMP2 in inclusion bodies is poor then it may be necessary to repeat **steps 1–7** using a different concentration of IPTG (*see* **Note 8**).

3.4 Matrix Assisted Purification and Refolding of Denatured MMP2

1. Resuspend the pellet containing the recombinant MMP2 inclusion bodies in solubilization buffer (5 ml buffer for inclusion bodies obtained from 1 l of culture) and incubate at 37 °C for 2 h at 180 rpm (*see* **Note 9**).
2. Centrifuged at 20,000 × *g* for 30 min at 15 °C to remove insoluble debris.
3. Load the supernatant containing the denatured MMP2 protein onto a column containing Ni^{2+}-NTA resin pre-equilibrated with solubilization buffer (*see* **Note 10**). In general, 1 ml of packed bead volume was used per litre of bacterial culture in a column of 12–20 ml of total capacity.

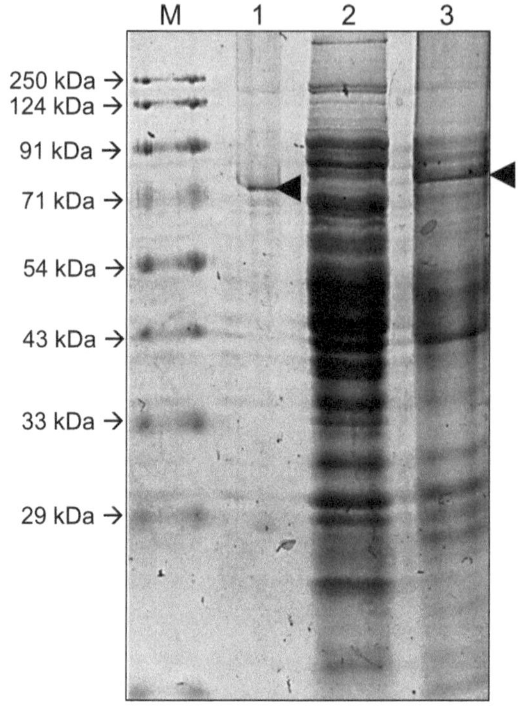

Fig. 1 SDS-PAGE analysis for MMP2 protein expression. Bacterial cell lysates were analyzed for MMP2 expression in soluble (supernatant) and insoluble (pellet) fraction. *Lane M*: Marker; *lane 1*, purified MMP2 protein; *lane 2*, soluble fraction; *lane 3*, insoluble fraction containing inclusion bodies

4. After loading, incubate the column for 2 h at room temperature (*see* **Note 11**).
5. After the incubation, the flow through containing the unbound proteins was collected and kept at room temperature until further analysis.
6. Wash the column with 50 bed volumes (50 ml) of solubilization buffer containing 20 mM imidazole under gravity flow (*see* **Note 12**).
7. To refold the protein the column was washed with 10 ml of a solution containing decreasing concentrations (from 8 to 0 M) of urea by mixing appropriate proportions of solubilization buffer and refolding buffer (*see* **Note 13**). The washing steps employed are described in Table 1.
8. Wash the column five times with 10 bed volumes of the refolding buffer to ensure proper refolding of MMP protein (*see* **Notes 14** and **15**).
9. Elute the refolded MMP2 protein in 5 bed volumes of elution buffer, in steps of 1 ml each on ice (*see* **Note 16**).

Table 1
Washing steps used for on-column refolding of the denatured MMP2 protein

Washing step	Final urea concentration (M)	Volume of solubilization buffer (ml)	Volume of refolding buffer (ml)
1	6	7.50	2.5
2	5	6.25	3.75
3	4	5.0	5.0
4	3	3.75	6.25
5	2	2.50	7.50
6	1	1.25	8.75

10. Fractions containing purified and refolded MMP2 protein are analyzed by microplate based Bradford's assay.
11. Pool peak elution fractions containing the refolded MMP2 protein (*see* **Note 17**) and dialyze against dialysis buffer at 4 °C for 4 h using a 12 kDa cutoff dialysis tube. Aliquot the dialyzed protein and store at −20 °C.
12. Confirm protein purity by SDS-PAGE. Mix 12 μl of purified protein with 3 μl of 5× loading dye. Heat the mixture at 95 °C for 10 min and resolve on a 12% SDS-PAGE gel (Fig. 2).
13. Determine total protein yield for the purified protein using the Bradford protein assay (typically approx. 2–4 mg/l of culture).

3.5 Activity Analysis of Purified and Refolded MMP2 Protein

The MMP2 protein activity was determined using a conventional "gelatin zymography" methodology as well as using an advanced "form invariant substrate cleavage assay" as described previously [14].

3.5.1 Gelatin Zymography

The biological activity of the purified MMP2 protein was assessed using gelatin based zymography as reported in various studies [17, 18], which is described below:

1. Add 2 μg of purified and refolded MMP2 protein to an equal volume of 2× sample buffer (*see* **Note 18**).
2. Resolve the sample on a 12% SDS-PAGE gel co-polymerized with 0.1% gelatin at 100 V for 2 h.
3. Carefully remove the gel and wash with activation buffer containing 2.5% Triton X-100 for 1 h at room temperature (*see* **Note 19**).
4. Briefly rinse the gel with deionized water and incubate in activation buffer (without Triton X-100) overnight at 37 °C.

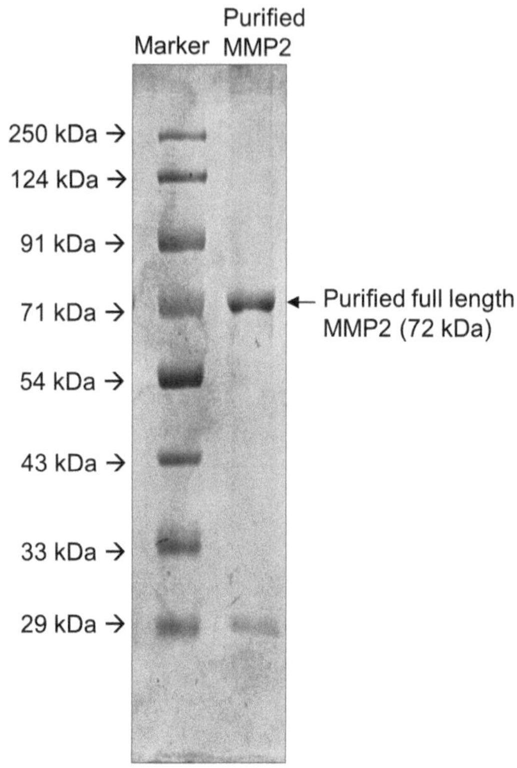

Fig. 2 SDS-PAGE analysis of refolded MMP2 protein. MMP2 protein purified and refolded from IBs using affinity chromatography were analyzed on 12% SDS-PAGE and stained with Coomassie brilliant blue

5. Stain the gel by submerging it in staining solution for 30 min at room temperature followed by destaining for 1–2 h at room temperature using the destaining solution.
6. Wash the gel with deionized water and scan using a image documentation system. The presence of light or unstained bands as a result of proteolytic cleavage of gelatin substrate on a dark background reveals the activity of the purified protein (*see* **Note 20**) (Fig. 3).

3.5.2 Form Invariant Substrate Cleavage Assay

The total activity of purified MMP2 protein can be further tested using an advanced substrate cleavage assay (*see* **Note 21**) [14].

1. In brief, a synthetic substrate site for MMP2 proteins composed of CFP fluorescent protein fused with MMP2 substrate recognition sequence IPVS↓LRSG (CFP-2RS), was overexpressed, purified and used for activity assessment as previously described [14].
2. Incubate 2 μg of purified CFP-2RS protein with 500 ng of purified MMP2 in the presence of 1× activation buffer with or without the MMP2 specific inhibitor, ARP-100 [19] for 16 h at 37 °C.

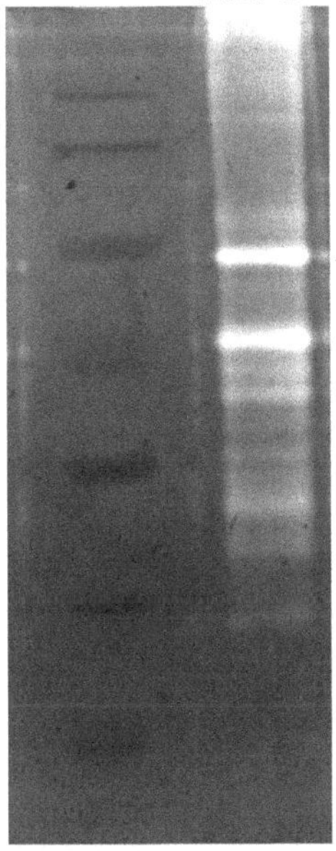

Fig. 3 Gelatin zymography for analysis of refolded MMP2 protein activity. Zymography was used to analyze the bioactivity of purified full-length MMP2 protein. The two bands corresponding to pro-MMP2 and MMP2 are as marked by *arrows*

3. Add 5× loading dye and heat denature at 95 °C for 10 min.
4. Resolve the samples on a 15% SDS-PAGE gel followed by staining for 30 min and destaining.
5. Rinse the destained gel with deionized water and image the gel using a gel image acquisition system (Fig. 4). In this assay, action of MMP2 protein on the CFP-2RS substrate leads to reduction in its size by approx. 3 kDa which can be recorded after resolving the proteins on SDS-PAGE. The refolded protein should show the presence of clear digested protein bands, which are abrogated when ARP100 is present in the reaction.

4 Notes

1. For amplification of MMP2 CDS, cDNA prepared using RNA isolated from HEK 293 cells was used. The RNA was reverse transcribed to cDNA in a single step RT-PCR protocol using oligo-dT primers and Superscript cDNA synthesis kit (Life Technologies, USA). The use of oligo-dT primer was preferred

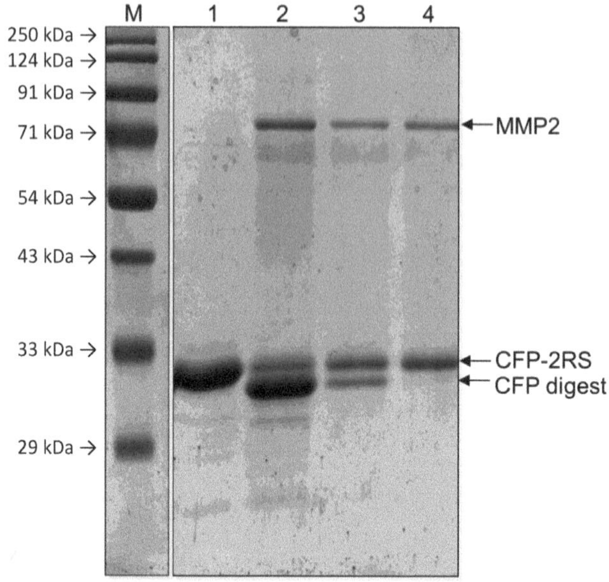

Fig. 4 Form invariant substrate cleavage assay. Activity of purified and refolded MMP2 was monitored using recombinant CFP-2RS protein containing a synthetic MMP2 substrate recognition sequence. SDS-PAGE analysis of CFP-2RS substrate protein incubated with purified and refolded MMP2 in the absence and presence of the MMP2 inhibitor, ARP100. *Lane 1*, CFP-2RS only; *lane 2*, CFP-2RS + MMP2; *lane 3*: CFP-2RS + MMP2 + 1 μM ARP 100; and *lane 4*, CFP-2RS + MMP2 + 10 μM ARP 100

over the random hexamer to ensure the amplification of full-length MMP2 CDS. One hundred nanogram of cDNA was used for PCR amplification of MMP2 CDS using gene specific primers and Pfu polymerase enzyme. The PCR conditions used had extension step done at 72 °C for 4 min, to ensure amplification larger amplicons corresponding to full-length MMP2. Details have been described previously [14].

2. The use of the *E. coli* C43 (DE3) pLysS strain is critical for efficient expression of toxic proteins like MMPs. It is a derivative of *E. coli* B strain containing the DE3 lysogen, a λ prophage carrying the T7 RNA polymerase gene and $lacI^q$ [15]. The pLysS plasmid encodes for T7 phage lysozyme, an inhibitor of T7 RNA polymerase (T7RNAP) that reduces the activity of T7 RNAP and confers resistance to cell death associated with toxic protein overexpression. The strain also contains additional mutations, which confer greater tolerance to the presence of toxic proteins. Alternatively, C43 (DE3) pLysS can be replaced by the BL21 ArcticExpress *E. coli* strain (*see* **Note 3**).
3. *E. coli* BL21 ArcticExpress (Agilent Inc., USA) is the preferred strain for expression of proteins in soluble form. The strain is genetically engineered to co-express the Cpn10 and Cpn60

chaperonins from *O. antarctica* which assists improved protein refolding at low temperature [16] .

4. Usage of carbenicillin was preferred over ampicillin because of improved stability of carbenicillin against the action of β-lactamase. This ensures that only cells containing the recombinant plasmid grow during overnight growth, thereby providing good primary cultures for protein overexpression in the next step.
5. The colonies obtained were immediately used for protein overexpression. The transformed plates stored beyond 4 days were not utilized for protein expression as the yield of MMP2 protein was consistently lower or absent from them.
6. Terrific broth (TB), nutrient rich media with glycerol as an extra carbon source is preferred as it confers robust growth of bacteria. Phosphate salts present in TB prevent changes in pH during bacterial growth which is critical for obtaining the high cell densities required for maximum expression and yield of the MMP2 protein.
7. The sonication protocol should be optimized for time and the power of the sonicator used for cell lysis. The process should be carefully optimized to maximize removal of contaminating soluble proteins, which may interfere with downstream processing. Additionally, sonication should be done on ice to avoid overheating, that can cause protein denaturation and thereby contamination of the inclusion bodies.
8. If MMP2 overexpression localizes the protein in the soluble fraction, its toxicity leads to extremely poor yield, in such case fresh induction is performed with higher concentration (2 mM) of IPTG to ensure higher expression of the recombinant protein, which facilitates its localization in inclusion bodies.
9. The following points can be considered if issues are faced with solubilization and recovery of MMP2 from the inclusion bodies:
   (a) Vortexing the pellet vigorously, in general, ensures better solubilization of inclusion bodies.
   (b) 6 M Guanidine hydrochloride (Gm.HCl) can be used as a denaturant in place of 8 M urea, however samples containing Gm.HCl cannot be analyzed on SDS-PAGE unlike urea containing fractions.
   (c) Do not centrifuge the solubilized protein at temperature less than 15 °C, as it can lead to the precipitation of the denaturant (urea/Gm.HCl), thereby reducing the recovery of soluble protein.
10. If Gm.HCl is used for solubilization of inclusion bodies instead of urea in the solubilization buffer, then the same buffer should be used for Ni^{2+}-NTA resin equilibration.

11. To facilitate efficient binding of the protein on the Ni^{+2}-NTA resin, resuspend the loaded protein and resin every 15 min by gentle mixing.
12. In our hands, the flow rate does not have significant bearing on the outcome of the folding, but faster flow rates always provide better yields.
13. In the refolding steps, the urea concentration is reduced concomitantly with an increase in the concentration of GSH/GSSH (contributed by the refolding buffer). This facilitates appropriate disulfide bond formation, which is critical for appropriate folding of the protein [20].
14. The final refolding step is performed at 4 °C to prevent protein degradation.
15. GSH and GSSG are added fresh in powder form to the refolding buffer. The gradual increase in the concentration of GSH during the refolding steps can be effectively visualized by the reduction of Ni^{+2} on the column, because the colour of the resin changes from blue to pink/white.
16. At this step since bound and refolded protein are eluted from the Ni^{+2}-NTA resin, the color of the reduced Ni^{+2} is restored and the resin becomes blue again. The restoration of the color is a good indicator of elution of the protein.
17. For rapid analysis of fractions containing high amounts of eluted MMP2 protein, we utilized a rapid Bradford's assay. Here, we use a 96-well microplate where 100 μl of Broadford's reagent is added to ten wells and kept ready. In one well marked as a control, 5 μl of elution buffer is added to determine the baseline colour. In the remaining wells, as the fractions are collected, 5 μl each of the eluate is added to rapidly identify fractions containing protein. Peak fractions (generally fractions 2–6) are then pooled for subsequent dialysis.
18. For SDS-PAGE based activity assays such as zymography, protein samples are processed in non-reducing conditions, without heat denaturation. For this, 2× sample buffer devoid of reducing agent is used. Also heating of samples before loading on SDS gel is avoided to prevent loss of MMP protein activity.
19. The complete exchange of SDS in the gel by Triton X-100 in the activation buffer is critical to restore the activity of MMP proteins. To ensure this the gel is pre-washed with dH_2O before treatment with activation buffer.
20. During the activation step, not only is the activity of denatured MMP2 protein restored but the inactive pro-MMP2 protein also undergoes activation. This leads to detection of both the pro- as well as the active form of MMP2 in the zymogram,

which is evident by the presence of two bands, one of high molecular weight (pro form, corresponding to the purified protein) and a smaller molecular weight band (corresponding to the processed form of MMP2, which is highly active).

21. The substrate cleavage assay utilizes a chimeric reporter protein (CFP-2RS) whose size reduces by approx. 3 kDa after cleavage by MMP2. The assay records activity from both proMMP2 as well as from processed and active MMP2 by a simple SDS-PAGE analysis, which cannot be recorded from conventional zymography. The design of the assay and construction of the chimeric reporter has been described previously [14]. The reporter protein consists of a fusion of ORF coding for CFP fluorescent protein, PCR amplified from pECFP vector (Clontech, USA) with a synthetic nucleotide sequence (5'GGATCCGGCGGAAGCATCCCCGTCAG CCTCCGTAGCGGCGGAAGCGTCGAC 3′) coding for a 16-amino acid long peptide, SGSGGSIPVSLRSGGS containing the MMP2 recognition sequence IPVS↓LRSG [21]. The CFP-2RS (CFP+MMP2 recognition sequence encoding peptide) fusion was cloned in *Eco*RI and *Sal*I sites of pProEx-HTa bacterial expression vector. The chimeric protein was overexpressed in *E. coli* Origami™ strain and was purified from the soluble protein fraction using Ni^{+2}-NTA chromatography as a 33 kDa protein.

Acknowledgments

Financial assistance to DKS from Department of Science and Technology (EMR/2014/000997); Department of Biotechnology and Indian Institute of Science partnership program, and DST-FIST for equipment support and a Research Fellowship to RJ from University Grants Commission is acknowledged.

References

1. Massova I, Kotra LP, Fridman R, Mobashery S (1998) Matrix metalloproteinases: structures, evolution, and diversification. FASEB J 12(12):1075–1095
2. Baker AH, Edwards DR, Murphy G (2002) Metalloproteinase inhibitors: biological actions and therapeutic opportunities. J Cell Sci 115(Pt 19):3719–3727
3. Gialeli C, Theocharis AD, Karamanos NK (2011) Roles of matrix metalloproteinases in cancer progression and their pharmacological targeting. FEBS J 278(1):16–27
4. Schmalfeldt B, Prechtel D, Härting K, Späthe K, Rutke S, Konik E, Fridman R, Berger U, Schmitt M, Kuhn W, Lengyel E (2001) Increased expression of matrix metalloproteinases (MMP)-2, MMP-9, and the urokinase-type plasminogen activator is associated with progression from benign to advanced ovarian cancer. Clin Cancer Res 7(8):2396–2404

5. Vihinen P, Kähäri V-M (2002) Matrix metalloproteinases in cancer: prognostic markers and therapeutic targets. Int J Cancer 99(2):157–166
6. Roy R, Yang J, Moses MA (2009) Matrix metalloproteinases as novel biomarkers and potential therapeutic targets in human cancer. J Clin Oncol 27(31):5287–5297
7. Hadler-Olsen E, Winberg JO, Uhlin-Hansen L (2013) Matrix metalloproteinases in cancer: their value as diagnostic and prognostic markers and therapeutic targets. Tumour Biol 34(4):2041
8. Ra HJ, Parks WC (2007) Control of matrix metalloproteinase catalytic activity. Matrix Biol 26(8):587–596
9. Nagase H, Visse R, Murphy G (2006) Structure and function of matrix metalloproteinases and TIMPs. Cardiovasc Res 69(3):562–573
10. Steffensen B, Chen Z, Pal S, Mikhailova M, Su J, Wang Y, Xu X (2011) Fragmentation of fibronectin by inherent autolytic and matrix metalloproteinase activities. Matrix Biol 30(1):34–42
11. Wang X, Yi J, Lei J, Pei D (1999) Expression, purification and charaterization of recombinant mouse MT5-MMP protein products. FEBS Lett 462(3):261–266
12. Carrió MM, Villaverde A (2002) Construction and deconstruction of bacterial inclusion bodies. J Biotechnol 96(1):3–12
13. Tsumoto K, Ejima D, Kumagai I, Arakawa T (2003) Practical considerations in refolding proteins from inclusion bodies. Protein Expr Purif 28(1):1–8
14. Singh KK, Jain R, Ramanan H, Saini DK (2014) Matrix-assisted refolding, purification and activity assessment using a 'form invariant' assay for matrix metalloproteinase 2 (MMP2). Mol Biotechnol 56(12):1121–1132
15. Miroux B, Walker JE (1996) Over-production of proteins in *Escherichia coli*: mutant hosts that allow synthesis of some membrane proteins and globular proteins at high levels. J Mol Biol 260(3):289–298
16. Ferrer M, Chernikova TN, Yakimov MM, Golyshin PN, Timmis KN (2003) Chaperonins govern growth of *Escherichia coli* at low temperatures. Nat Biotechnol 21(11):1266–1267
17. Hawkes SP, Li H, Taniguchi GT (2010) Zymography and reverse zymography for detecting MMPs and TIMPs. Methods Mol Biol 622:257–269
18. Hu X, Beeton C (2010) Detection of functional matrix metalloproteinases by zymography. J Vis Exp 45:2445
19. Rossello A, Nuti E, Orlandini E, Carelli P, Rapposelli S, Macchia M, Minutolo F, Carbonaro L, Albini A, Benelli R, Cercignani G, Murphy G, Balsamo A (2004) New N-arylsulfonyl-N-alkoxyaminoacetohydroxamic acids as selective inhibitors of gelatinase A (MMP-2). Bioorg Med Chem 12(9):2441–2450
20. Saini DK, Pant N, Das TK, Tyagi JS (2002) Cloning, overexpression, purification, and matrix-assisted refolding of DevS (Rv 3132c) histidine protein kinase of Mycobacterium tuberculosis. Protein Expr Purif 25(1):203–208
21. Turk BE, Huang LL, Piro ET, Cantley LC (2001) Determination of protease cleavage site motifs using mixture-based oriented peptide libraries. Nat Biotechnol 19(7):661–667

Chapter 2

Expression and Purification of a Matrix Metalloprotease Transmembrane Domain in *Escherichia coli*

Charles A. Galea

Abstract

Membrane tethered matrix metalloproteases are bound to the plasma membrane by a glycosylphosphatidylinositol-anchor or a transmembrane domain. To date, most studies of membrane-bound matrix metalloprotease have focused on the globular catalytic and protein–protein interaction domains of these enzymes. However, the transmembrane domains have been poorly studied even though they are known to mediate intracellular signaling via interaction with various cellular proteins. The expression and purification of the transmembrane domain of these proteins can be challenging due to their hydrophobic nature. In this chapter we describe the purification of a transmembrane domain for a membrane-bound matrix metalloprotease expressed in *E. coli* and its initial characterization by NMR spectroscopy.

Key words Membrane-anchored MMP, Matrix metalloproteases, NMR spectroscopy, Transmembrane domain, Protein expression, Isotopically labeled

1 Introduction

Matrix metalloproteases comprise a large family of zinc-dependent endopeptidases that function at neutral pH to degrade extracellular matrix proteins, cleave cell surface receptors, release apoptotic ligands, and activate chemokines and cytokines [1–3]. Matrix metalloproteases (MMPs) participate in a wide range of cellular processes including tissue remodeling, cell proliferation, cell migration, differentiation, angiogenesis, apoptosis, and the immune response [4–9]. Aside from these normal physiological roles, MMPs have been implicated in a large number of pathological disorders such as arthritis, Alzheimer's disease, atherosclerosis, vascular disease, central nervous system disease, liver cirrhosis, and various cancers [10, 11].

MMP activity is regulated at multiple levels, including biosynthesis (transcription/translation) [12], zymogen activation [10, 12], compartmentalization [10, 13, 14], and inactivation [14–16]. Several MMPs are active during development and normal

Charles A. Galea (ed.), *Matrix Metalloproteases: Methods and Protocols*, Methods in Molecular Biology, vol. 1579,
DOI 10.1007/978-1-4939-6863-3_2, © Springer Science+Business Media LLC 2017

physiology and may play a role in homeostasis [7, 9, 17–20], while others play a role in tissue injury and infection [4, 7]. In chronically inflamed tissues and most cancers MMPs contribute to various pathological processes, including tissue degradation, tumor progression and invasion [21–27].

MMPs are expressed and transported through the cell as inactive zymogens and secreted or anchored to the plasma membrane, confining their activity to the extracellular environment or the cell surface [8]. Evidence suggests that secreted MMPs bind to specific cell-surface receptors, membrane-anchored proteins and cell-associated extracellular matrix (ECM) molecules, and function pericellularly at focused locations [14]. MMPs also function in the cell nucleus (MMP2, 3, 9, 13 and MT1-MMP), cytoplasm (MMP1, 2, 26, 23), and various organelles, where their localization is facilitated by interactions with other proteins, proteoglycan core proteins and/or their glycosaminoglycan chains, and other molecules [1, 10, 13, 28–31].

Unlike the majority of MMPs that function as soluble proteases, several are anchored to the cell membrane [32–34]. The membrane tethered MMPs include the membrane-type MMPs (MT-MMPs) and MMP23, a membrane anchored MMP containing a domain architecture that differs significantly from that of the MT-MMPs [35]. There are six MT-MMPs denoted MT1-MMP (MMP14), MT2-MMP (MMP15), MT3-MMP (MMP16), MT4-MMP (MMP17), MT5-MMP (MMP24), and MT6-MMP (MMP25). MT4-MMP and MT6-MMP are tethered to the plasma membrane via a glycosylphosphatidylinositol (GPI)-anchor while the remaining MT-MMPs are anchored via a transmembrane (TM) domain [32]. The MT-MMPs contain a common domain architecture consisting of an N-terminal signal peptide followed by a pro-domain, a catalytic domain, a hinge (linker-1) region, a hemopexin-like (Hpx) domain, and a stalk region (linker-2). Linker-2 is followed by a TM domain or GPI-anchor and a C-terminal cytoplasmic tail (Fig. 1) [32]. In contrast, the domain architecture of MMP23 comprises an N-terminal signal sequence followed by a cytoplasmic tail, a membrane-anchoring TM domain, a linker region, a pro-domain, a catalytic domain, a small cysteine-rich toxin-like domain (TxD), and a C-terminal immunoglobulin-like cell adhesion molecule (IgCAM) domain (Fig. 1) [35].

To date, studies of membrane-anchored MMPS have mostly focused on the characterization of the catalytic and globular protein–protein interaction domains. However, evidence indicates that the single helical TM domain and unstructured cytoplasmic domain of these enzymes can play a key role in the function of membrane-anchored MMPs. For example, the formation of MT1-MMP homo-dimers mediated by both the Hpx [36] and TM domains [37] is required for proMMP2 activation on the cell surface [36] and degradation of collagen [38].

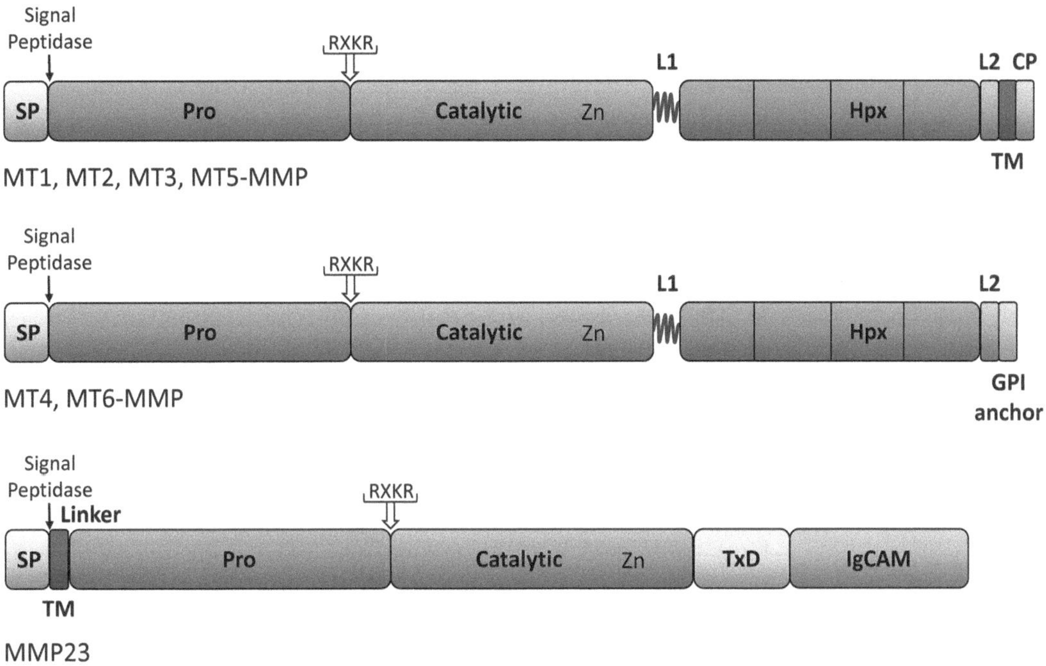

Fig. 1 Structural domains of the membrane-anchored MMPs. MT1-MMP, MT2-MMP, MT3-MM, MT4-MMP, MT5-MMP, and MT6-MMP share a common domain structure consisting of an N-terminal signal peptide (SP) that targets the protease to the cell surface, a pro-domain (Pro), a catalytic domain, a linker domain (L1), a hemopexin (Hpx) domain involved in protein–protein interactions, and a second linker region (L2). MT1-MMP, MT2-MMP, MT3-MMP, and MT5-MMP contain a C-terminal single helical transmembrane (TM) domain followed by a cytoplasmic (CP) domain while MT4-MMP and MT6-MMP possess a C-terminal GPI anchor. MMP23 differs in its domain structure, possessing an N-terminal signal peptide followed by a single type-II helical transmembrane domain, a pro-domain, a catalytic domain, a cysteine-rich toxin-like (TxD) domain, and a protein–protein interaction IgCAM domain. The pro-domain (Pro) in most membrane-anchored MMPs (except MMP23) contains an unpaired cysteine sulfhydryl group that interacts with the active site zinc atom and maintains the enzyme in a latent inactive form (the cysteine switch). The MT-MMPs are transported in their inactive latent form to the cell surface, where they are cleaved at the recognition motif (RXKR) within the pro-domain by a furin-like protease, resulting in activation of the enzyme

Previous studies have highlighted the importance of the cytoplasmic domain of membrane-anchored MMPs in mediating various intracellular interactions. For example, the 20 amino acid residue cytoplasmic tail of MT1-MMP has been reported to interact with a number of proteins and is also the target of posttranslational modifications. Interaction of the MT1-MMP cytoplasmic domain (CPD) with FIH-1 (Factor inhibiting HIF-1) triggers a cascade of protein–protein interactions resulting in activation of HIF-1α thereby regulating ATP production in macrophages during normoxia [39]. While interaction of phosphorylated Tyr573 in the MT1-MMP CPD with p130Cas regulates Rac1 signaling in osteoclast formation [40, 41]. The MT1-MMP CPD has also been shown to interact with p27RF-Rho (LAMTOR1) which regulates RhoA activation [42]. Other studies have shown that the MT1-MMP CPD is required for its localization to focal adhesions via interactions with the focal adhesion kinase (FAK)-p130Cas [43] and an eight amino acid residue loop structure, MT-Loop, located in the CPD is required to degrade underlying matrices in an efficient manner [44]. MT1-MMP has a half-life of less than 30 min on the cell surface and is endocytosed in a clathrin- and caveolae-dependent manner. The clathrin-dependent endocytosis of MT1-MMP is attributed to the interaction of residues LLY^{573} with adapter protein 2, a component of clathrin-coated pits [45]. Endocytosed MT1-MMP is also recycled to the cell surface and a sequence (DKV^{582}) within the CPD is required for recycling [46]. Other studies have shown that the last three residues (EWV^{645}) in the CPD of MT5-MMP play a similar role in its recycling to the cell surface [47]. These residues act as a PDZ binding motif and bind to Mint-3 which contains two PDZ domains. While our recent studies have shown that the pro-domain of MMP23, containing a TM and cytoplasmic domain, interacts with the voltage-gated potassium ion channel Kv1.3 and regulates the trafficking of this ion channel to the cell surface [35, 48].

Considering the important role played by the TM and cytoplasmic domains of membrane-anchored MMPs, very few studies have been undertaken to define their structure and dynamics. In this chapter we focus on the expression and purification of a truncated protein containing the transmembrane and cytoplasmic domains of a membrane-anchored MMP (MMP23) for the characterization of these domains by NMR spectroscopy.

2 Materials

2.1 Expression of Isotopically Labeled Peptides in E. coli in M9 Minimal Media

1. *E. coli* BL21(DE3) or *E. coli* BL21(DE3)pLysS chemically competent cells (Novagen).
2. pET32a expression vector (Novagen) containing cDNA for your peptide of interest covering the TM and cytoplasmic domains (TM/CPD) of your membrane-anchored MMP and an N-terminal 3C protease recognition site codon optimized for expression in *E. coli* (*see* **Note 1**).
3. Luria-Bertani (LB) broth: 10 g tryptone, 10 g NaCl, 5 g yeast extract per liter of deionized water. Adjust the pH to 7.0 with 0.5 N NaOH and autoclave. Add ampicillin to a final concentration of 100 μg/ml immediately prior to use.
4. Isopropyl β-D-1-thiogalactopyranoside (IPTG).
5. 50% (w/v) glucose: 125 g of D-glucose per 250 ml of deionized water. Sterilize by passing through a 0.2 μm filter and store at room temperature.

6. ^{13}C-labeled D-glucose (99% ^{13}C): This can be obtained from a number of suppliers including Cambridge Isotope Labs and Sigma-Aldrich.
7. 1 M $MgSO_4$: 24.7 g of $MgSO_4{\bullet}7H_2O$ per 100 ml of deionized water. Sterilize by autoclaving and store at room temperature.
8. ^{15}N–labeled ammonium chloride (NH_4Cl, 99% ^{15}N): This can be obtained from a number of suppliers including Cambridge Isotope Labs and Sigma-Aldrich.
9. Vitamin solution (5 mg/ml): 0.5 g thiamine hydrochloride and 0.5 g nicotinic acid per 100 ml of deionized water. Filter-sterilize and store at 4 °C.
10. Trace element solution (1,000×): 0.60 g of $CaCl_2{\bullet}2H_2O$, 0.60 g of $FeSO_4{\bullet}7H_2O$, 0.12 g of $MnCl_2{\bullet}4H_2O$, 0.08 g of $CoCl_2{\bullet}6H_2O$, 0.07 g of $ZnSO_4{\bullet}7H_2O$, 0.03 g of $CuCl_2{\bullet}2H_2O$, 2 mg of boric acid, 25 mg of $(NH_4)_6Mo_7O_{24}{\bullet}4H_2O$, and 0.50 g of EDTA per 100 ml of deionized water [49] (*see* **Note 2**).
11. M9 salts (5×): 2.5 g NaCl, 15 g anhydrous KH_2PO_4, 34 g anhydrous Na_2HPO_4 per liter of deionized water. Autoclave to sterilize (*see* **Note 3**).
12. Antibiotic for plasmid selection (1,000×): 100 mg/ml ampicillin in deionized water (*see* **Note 4**).
13. M9 minimal medium: 780 ml of deionized water, 200 ml of 5 × M9 salts, 6 ml of vitamins, 12 ml of D-glucose (50% w/v), 1 ml of antibiotic (1,000×), 1 ml of trace elements (1000×), 3 ml of 1 M $MgSO_4$ and 1 g of $^{15}NH_4Cl$ (*see* **Note 5**).
14. Filter-sterilize M9 minimal medium and add ampicillin to a final concentration of 100 μg/ml immediately prior to use.

2.2 Expression of Isotopically Labeled Peptides in E. coli in Autoinduction Media

This method was originally described by Studier [50].

1. 20 × N: 14.2 g of Na_2SO_4, 136 g of KH_2PO_4, and 142 g of Na_2HPO_4 per liter of deionized water. Add each in sequence and stir until dissolved. Sterilize by autoclaving and store at room temperature.
2. 50 × 5052: Dissolve 250 g of glycerol in 730 ml of deionized water then add 25 g of D-glucose and 100 g of α-lactose. Add each component in sequence and stir until dissolved. Lactose can take several hours to dissolve at room temperature. Sterilize by autoclaving and store at room temperature (*see* **Note 6**).
3. NG medium: 47 ml of sterile water, 300 μl of vitamins, 50 μl of 1 M $MgSO_4$, 50 μl of trace elements (1,000×), 500 μl of D-glucose (50% w/v), 2.5 ml of 20 × N, 50 μl of antibiotic, and 125 mg of $^{15}NH_4Cl$ (*see* **Note 7**).

4. N-5052 medium: 922 ml of sterile water, 6 ml of vitamins, 1 ml of $MgSO_4$, 1 ml of trace elements (1,000×), 20 ml of 50 × 5052, 50 ml of 20 × N, 1 ml of antibiotic (1,000×), and 2.5 g of $^{15}NH_4Cl$.
5. Additional components for preparing the autoinduction medium include 1 M $MgSO_4$, $^{15}NH_4Cl$, vitamins, trace element solution (1,000×), and antibiotics (1,000×) as described in Subheading 2.1.

2.3 Immobilized Metal Ion Affinity Chromatography (IMAC) Purification of Expressed Peptide

1. BugBuster protein extraction reagent (Merck Millipore).
2. cOmplete, EDTA-free protease inhibitor cocktail tablets (Sigma-Aldrich) (*see* **Note 8**).
3. Human Rhinovirus (HRV) 3C protease (Novagen) (*see* **Note 9**).
4. 3-((3-cholamidopropyl) dimethylammonio)-1-propanesulfonate (CHAPS) (Critical micelle concentration of 8–10 mM) (*see* **Note 10**).
5. Lysis buffer: BugBuster containing 10 mM CHAPS, 1.0 mg/ml lysozyme, 5 μg/ml DNaseI, and 1 × cOmplete EDTA-free protease inhibitor cocktail.
6. Ni^{2+} affinity column: 5 ml HiTRAP chelating HP column charged with nickel (GE Healthcare Life Sciences) (*see* **Note 11**).
7. NTA charging solution: 250 mM $NiSO_4$ in deionized water.
8. NTA buffer A: 20 mM Tris–HCl buffer, pH 8.0 containing 150 mM NaCl, 10 mM CHAPS and 5 mM imidazole.
9. NTA buffer B: 20 mM Tris–HCl buffer, pH 8.0 containing 150 mM NaCl, 10 mM CHAPS and 35 mM imidazole.
10. NTA buffer C: 20 mM Tris–HCl buffer, pH 8.0 containing 150 mM NaCl, 10 mM CHAPS and 350 mM imidazole.

2.4 HPLC Purification of Peptides

1. 0.2 μm PTFE syringe filters.
2. C8 reversed-phase semi-preparative HPLC column: 100 Å, C8, 5 μm (Phenomenex).
3. Trifluoroacetic acid (TFA).
4. Acetonitrile (HPLC grade).
5. Solvent A: 0.1% TFA (v/v). Add 1 ml of TFA to 1 L of deionized water.
6. Solvent B: acetonitrile–water (80:20) + 0.085% TFA. To 700 ml of acetonitrile add 300 ml of deionized water and 0.85 ml TFA.
7. Sample buffer: 20 mM sodium citrate, pH 5.0, containing 20 mM TCEP and 10 mM CHAPS.
8. Liquid nitrogen.

2.5 NMR Sample Preparation

1. 5 mm Shigemi tube (Shigemi Inc.).
2. Deuterated dodecylphosphocholine (d_{38}-DPC) (Sigma-Aldrich).
3. Tris(2-carboxyethyl)phosphine hydrochloride (TCEP).
4. Sodium azide (NaN_3).
5. Dueterium oxide (D_2O).
6. NMR buffer: 20 mM sodium citrate pH 5.0, 100 mM deuterated DPC, 20 mM TCEP, 0.02% NaN_3 in 90% H_2O/10% D_2O.
7. 1,4-Dioxane (NMR reference standard) (Sigma-Aldrich) (*see* **Note 12**).

3 Methods

The truncated membrane-anchored MMP containing the TM and CPD domains of the protein (the TM/CPD protein) can be overexpressed as a recombinant protein in *E. coli* BL21(DE3) following induction by the addition of IPTG. The production of uniformly isotopically labeled protein is achieved using minimal media containing ^{15}N–labeled NH_4Cl and ^{13}C–labeled glucose as the sole nitrogen and carbon sources, respectively.

3.1 Protein Overexpression of Isotopically Labeled Protein in M9 Minimal Media

1. Inoculate 2 ml of LB medium containing the appropriate antibiotic with a single colony of freshly transformed *E. coli* BL21(DE3)pLysS cells harboring the pET32a expression vector containing the TM/CPD cDNA (*see* **Notes 13** and **14**).
2. Grow overnight at 37 °C with shaking at approximately 200 rpm.
3. Inoculate 2 ml of M9 minimal medium with 20 μl of overnight culture. Grow at 37 °C for 6–8 h until the culture is visibly turbid (*see* **Note 15**).
4. Inoculate 48 ml of M9 minimal medium with the entire 2 ml of culture and grow overnight at 37 °C (*see* **Note 16**).
5. Inoculate 500 ml of M9 minimal medium in a 2 L Erlenmeyer flask (*see* **Note 17**) with 25 ml of overnight starter culture. Grow at 37 °C until the optical density at 600 nm (OD_{600}) is approximately 0.5–1.0 (about 2–3 h although it may take longer to reach this OD_{600}).
6. Add IPTG to a final concentration of 1 mM to induce the expression of your target protein.
7. Grow at 37 °C for 3–5 h or 20 °C overnight (*see* **Note 18**).
8. Centrifuge at 5,000 × *g* for 20 min at 4 °C to pellet the cells.
9. Decant the supernatant and store the pellet at −20 °C.

3.2 Protein Overexpression of Isotopically Labeled Peptides in Autoinduction Medium (*See* Notes 19 and 20)

Autoinduction typically produces significantly higher cell masses and yields of expressed protein compared to conventional induction protocols using IPTG (Fig. 2) [50].

1. Inoculate 2 ml of LB medium containing the appropriate antibiotic with a single colony from a fresh plate of *E. coli* BL21(DE3) cells harboring an expression plasmid containing your peptide (or protein) of interest.
2. Grow overnight at 37 °C with constant shaking at approximately 200 rpm.
3. Inoculate 2 ml of NG minimal medium with 20 μl of overnight culture.
4. Grow at 37 °C for 6–8 h until the culture is visibly turbid.
5. Inoculate 48 ml of NG minimal medium in a sterile 250 ml Erlenmeyer flask with the entire 2 ml of culture and grow overnight at 37 °C (*see* **Note 21**).
6. Inoculate 500 ml of N-5052 autoinduction medium in a sterile 2 L Erlenmeyer flask (*see* **Note 17**) with 25 ml of overnight starter culture and grow for 24–48 h at 30 °C with constant shaking at approximately 200 rpm (*see* **Note 22**).
7. Centrifuge for 20 min at 5,500 × *g* and 4 °C to pellet the cells.
8. Decant the supernatant and store the pellet at −20 °C.

3.3 Purification of Isotopically Labeled His$_6$-Tagged TM/CPD Peptide

1. Weigh the frozen pellet and add 10 ml of lysis buffer per 1 g of pellet.
2. Place on a rotary mixer and gently resuspend the pellet over 30–60 min.
3. Lyse the cells by using either sonication (*see* **Note 23**) or a French press.
4. Centrifuge the lysate at 16,000 × *g* for 20 min at 4 °C.

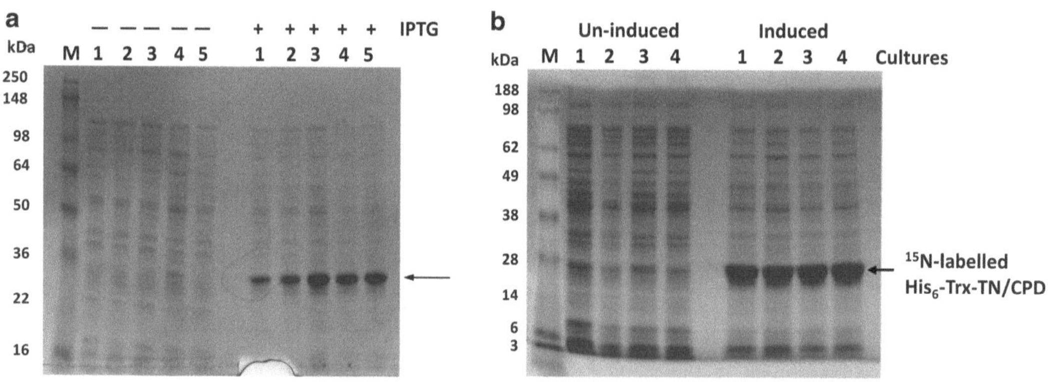

Fig. 2 Expression of ^{15}N-labeled His$_6$-tagged Trx-TM/CPD. Protein expression was induced by (**a**) the addition of 1 mM IPTG and the culture was grown for 3 h at 37 °C or (**b**) autoinduction at 28 °C over 48 h

5. Collect the supernatant and load onto a 5 ml nickel affinity column equilibrated with NTA buffer A (*see* **Note 24**). Collect the flow-through and take a 50 μl sample for analysis by SDS-PAGE.
6. Extensively wash the column with 50 ml of NTA buffer A followed by 50 ml of NTA buffer B (*see* **Note 25**). Collect the flow-through and take a 50 μl sample for analysis by SDS-PAGE.
7. Elute the His_6-tagged peptide using a linear gradient of 0–100% NTA buffer C over 30 min at a flow rate of 2 ml/min and collecting 5 ml fractions. Take a 50 μl sample of each fraction for analysis by SDS-PAGE.
8. Analyse samples taken during the purification of the TM/CPD protein by SDS-PAGE (Fig. 3).
9. Pool fractions containing the protein and store at 4 °C.
10. Add 3C protease to the pooled fractions and incubate with constant stirring at 4 °C to cleave the His_6-tagged protein (*see* **Note 26**).

3.4 Reverse-Phase HPLC Purification of TM/CPD Peptide

1. Add an equal volume of solvent A and then filter the cleaved peptide through a 0.22 μm membrane.
2. Load onto a semi-preparative C8 reverse-phase HPLC column at a flow-rate of 1 ml/min.
3. The column is developed at 0% solvent B for 5 min followed by a gradient of 0–40% solvent B over 10 min and 40–80% solvent B over 30 min. Monitor the absorbance at 214 nm and collect eluting peaks.
4. Freeze the peak fractions in liquid nitrogen and lyophilize.

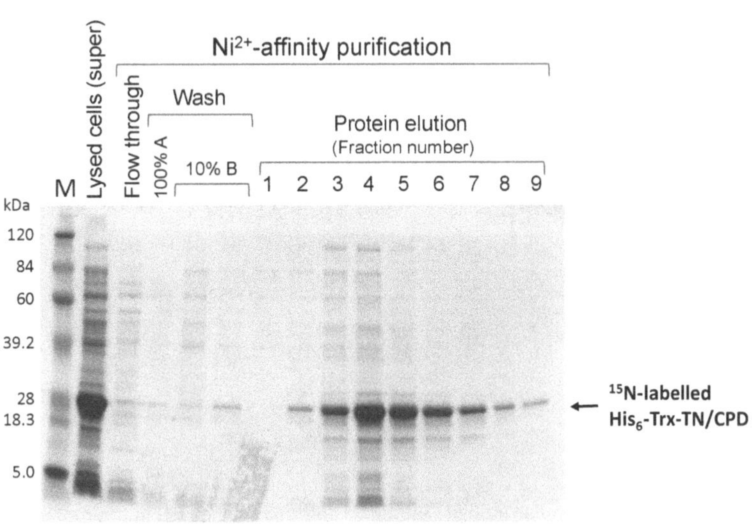

Fig. 3 Ni^{2+}-affinity (NTA) purification of the His_6-tagged TRX-TM/CPD protein. *A* and *B* refer to NTA buffers A and B, respectively

5. Dissolve the dried powder for each fraction in 0.5 ml of sample buffer and analyse by SDS-PAGE.
6. Pool peak fractions containing the TM/CPD peptide, add an equivalent volume of solvent A and then load onto the same semi-preparative C8 reverse-phase HPLC column.
7. Develop the column with a gradient of 0–50% solvent B over 5 min followed by 50–70% solvent B over 40 min. Monitor the absorbance at 214 nm and collect eluting peaks.
8. Freeze the peak fractions in liquid nitrogen, lyophilize, and store dried powder at −20 °C (*see* **Note 27**).
9. Dissolve a small amount of dried powder for each peak fraction in sample buffer and analyse by SDS-PAGE (Fig. 4a) (*see* **Note 28**).
10. Determine the purity and mass of the purified TM/CPD peptide by analytical C8 reverse-phase HPLC and MALDI-TOF mass spectrometry.

3.5 NMR Sample Preparation

1. Dissolve the dried TM/CPD peptide in 550 μl of NMR buffer (*see* **Notes 29** and **30**).
2. Vortex briefly for about 30 s to promote the incorporation of the TM/CPD peptide into the DPC micelles.
3. Take a 50 μl sample to determine the concentration of the protein and for analysis by circular dichroism (CD) spectroscopy (Fig. 4b) (*see* **Note 31**).
4. Transfer the remaining 500 μl of material to a 5 mm NMR tube (*see* **Note 32**). Store at 4 °C prior to NMR data acquisition.

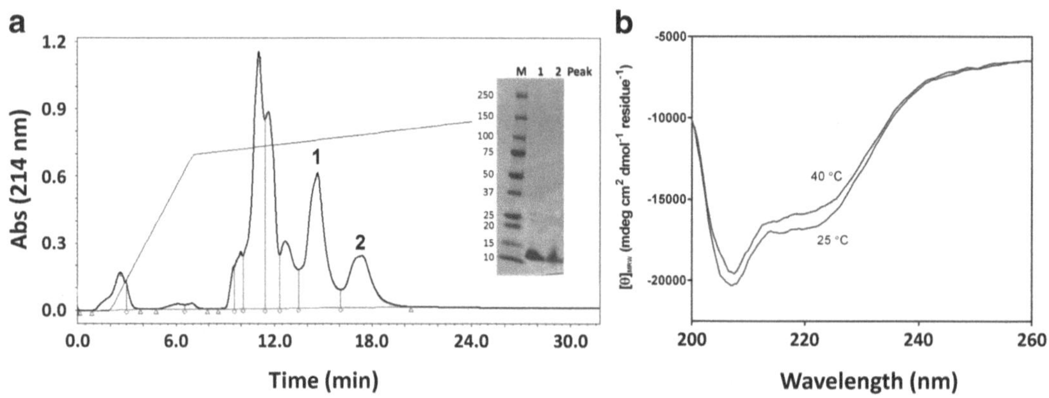

Fig. 4 The purified TM/CPD protein. (**a**) C8 reverse-phase HPLC chromatogram for the purification of the TM/CPD protein. The *blue line* indicates the percentage of solvent B in the mobile phase. *Inset*: SDS-PAGE analysis of HPLC peaks 1 and 2 showing the presence of the TM/CPD protein in both fractions. Native PAGE gel analysis shows peak 2 contains the dimeric form of the TM/CPD protein (data not shown). (**b**) CD spectrum recorded at 25 and 40 °C for the TM/CPD protein prepared in the NMR buffer which contains DPC micelles. The spectrum indicates that part of the protein adopts a helical structure (minima at 208 and 222 nm)

3.6 NMR Screening

1. Record a 1D ^1H spectrum and a 2D ^1H-^{15}N-TROSY spectrum of the TM/CPD protein at different temperatures to determine the optimal temperature to acquire NMR spectra (*see* **Notes 33** and **34**).
2. Process the NMR data using TopSpin (Bruker), NMRView (One Moon Scientific Inc.) [51], NMRPipe/NMRDraw [52], or another suitable software package.
3. Examine the spectra (i.e., spectral dispersion, line widths, peak intensities and number of resonance peaks compared to number expected based on the amino acid sequence) to determine whether they are of sufficient quality for structure determination (Fig. 5).

4 Notes

1. The pET32a expression vector contains an ampicillin resistance gene and N-terminal His_6-tag and thioredoxin fusion tag. There are a wide range of commercially available expression vectors that contain other selectable markers for expression in the presence of an antibiotic (e.g., kanamycin, chloramphenicol), protease recognition sites (e.g., tobacco etch virus (TEV) protease, thrombin, enterokinase, and factor Xa), and one or more N- or C-terminal fusion tags. Fusion tags can facilitate detection on Western blots (e.g., FLAG, S-tag, c-myc, HSV, glutathione S-transferase (GST), His_6), solubility (e.g., thioredoxin (Trx), SUMO, maltose binding protein (MBP), NusA), and purification (e.g., His_6, GST, MBP) of the target protein.
2. Trace metals are required for maximal growth in minimal media. Each salt should be added in the order outlined in **item 9** in Subheading 2.1 after the previous salt has fully dissolved into the solution. Filter-sterilize and store in the dark at room temperature.
3. M9 salts can be stored at room temperature for up to 6 months after autoclaving.
4. The majority of T7-based expression vectors use either ampicillin or kanamycin (35 mg/ml in deionized water) for plasmid selection.
5. $^{15}NH_4CL$ and $U^{13}C$ D-glucose can be substituted as the nitrogen and carbon source instead of unlabeled NH_4Cl and D-glucose for the expression of uniformly $^{13}C/^{15}N$-labeled peptides (or proteins).
6. Uniformly labeled ^{13}C D-glucose can be added instead of D-glucose for the expression of ^{13}C-labeled peptides.
7. $^{13}C/^{15}N$-labeled peptides (or proteins) can be produced if ^{13}C-glycerol is used as the carbon source instead of unlabeled glycerol.

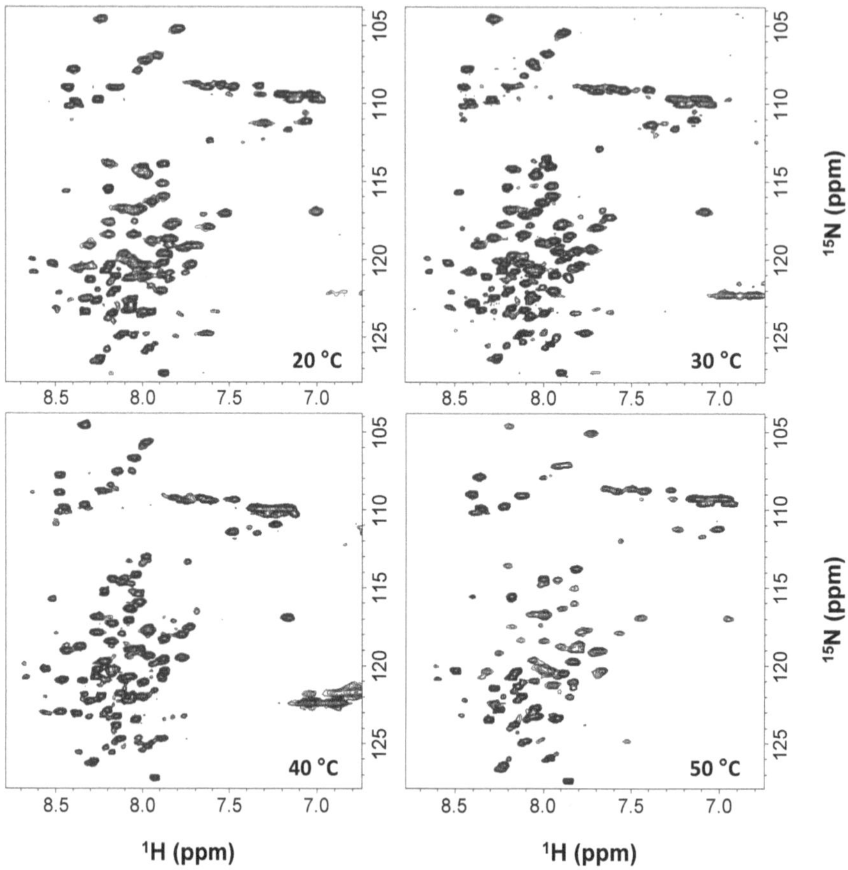

Fig. 5 NMR spectra for the TM/CPD protein in DPC micelles. 2D ^{1}H/^{15}N TROSY spectra were acquired at 20, 30, 40, and 50 °C on a Varian Inova 600 MHz NMR spectrometer equipped with a cryogenically cooled triple-resonance probe. Optimal spectral resolution and peak intensities was obtained at 40 °C

8. The cOmplete EDTA-free protease inhibitor cocktail comprises tablets that contain inhibitors for serine and cysteine proteases. The cOmplete EDTA-free cocktail is recommended since EDTA interferes with the IMAC (immobilized metal affinity chromatography) purification of His_6-tagged fusion proteins.
9. HRV 3C protease is a recombinant cysteine protease used to remove fusion tags from proteins containing the HRV 3C protease cleavage site (Leu-Glu-Val-Phe-Gln*Gly-Pro). This protease exhibits a high degree of specificity and has optimal activity at 4 °C.
10. CHAPS is a zwitterionic detergent that is used for the solubilization of membrane proteins over a wide range of pH (2–12). It has a high critical micelle concentration (CMC) and for this reason can be easily removed by dialysis or reverse-phase HPLC.

11. The HiTrap chelating HP column is prepacked with chelating sepharose high performance agarose. The medium is charged with nickel for Immobilized Metal Ion Affinity Chromatography (IMAC) and subsequent purification of polyhistidine tagged proteins.
12. Chemical shifts are measured relative to the reference compound 1,4-dioxane (3.69 ppm). Other reference compounds that are commonly used include: 2,2-dimethyl-2-silapentane-5-sulfonate (DSS, 0.00 ppm) and tetramethylsilane (TMS, 0.00 ppm). D_2O is included as a deuterium lock to correct for drift in the NMR spectrometers magnetic field with time. NaN_3 (sodium azide) is an antimicrobial that is included to prevent microbial growth.
13. It is often convenient to prepare a 50% glycerol stock of *E. coli* cells transformed with the expression vector containing your gene of interest and store frozen at −80 °C. To do this inoculate 5 ml of LB broth containing an appropriate antibiotic with a single colony from an LB agar plate and grow overnight at 37 °C. Take 0.5 ml of overnight culture and add 0.5 ml of sterile glycerol. Freeze on liquid nitrogen and then store at −80 °C. Fresh cultures can be prepared by scraping a small amount of material from the surface of the frozen glycerol stock and using this to inoculate the LB broth.
14. *E. coli* BL21(DE3)pLysS contains the pLysS plasmid which expresses T7 lysozyme that binds to the T7 promoter and represses expression of the target protein (or peptide) in the absence of IPTG. Inhibition of basal expression of proteins (or peptides) prior to induction with IPTG is especially important if the expressed protein (or peptide) is toxic to the cells.
15. This step allows the cells to adapt to M9 minimal medium before inoculating into a large volume of minimal media.
16. This provides a starter culture for inoculating a much larger volume of M9 minimal medium the following day. The volume of this initial starter culture should be approximately 5% of the final larger volume.
17. Good aeration is particularly important during induction. The use of baffled flasks and culture volumes no greater than 25% of the flask volume will help to improve yields.
18. Prior to isotopic labeling, optimal protein expression conditions (i.e., temperature and concentration of IPTG) should be determined using small-scale cultures. Parameters such as temperature, IPTG concentration, OD_{600} at induction, expression strain, and induction time can critically influence the yield and solubility of the expressed protein or peptide.
19. The autoinduction medium contains glucose, glycerol, and lactose as carbon sources. Glucose is the initial primary carbon and energy source promoting culture growth to high cell den-

sities while inhibiting protein expression. Glucose inhibits recombinant protein expression by preventing induction of operons responsible for metabolizing lactose to allolactose, the actual inducer. Once the glucose has been consumed lactose is imported into the host cells and induces the expression of the recombinant protein. This allows the cell culture to reach very high OD_{600} values of ~20 prior to the expression of the target protein. Glycerol is provided in the autoinducing media as a good carbon and energy source that does not prevent glucose depletion during growth, glucose inhibition of protein expression, or the uptake of lactose upon glucose depletion.

20. For autoinduction, do not use an *E. coli* strain harboring a pLysS plasmid if your expression vector contains a T7*lac* promoter. The presence of the *lac* repressor in combination with T7 lysozyme expressed by the pLysS plasmid (which binds to and inactivates the T7 promoter) leads to significantly lower levels of protein expression in the autoinduction medium.
21. This starter culture is used to inoculate 1 L of medium the next day. Increase the size of the starter culture proportionally (i.e., 50 ml of starter culture per 1 L of minimal medium) if you plan to inoculate a larger batch of medium.
22. Grow the cells to saturation when using autoinduction medium. The cultures will grow to a high OD_{600} before induction starts. It may take at least 24 h or more for the cultures to reach saturation when grown at lower temperatures.
23. Sonication should be performed on ice with five rounds of short bursts of 1.0 min at 2.0 min intervals. It is important to avoid frothing or overheating the sample as these may cause protein denaturation.
24. Load the lysate onto the column at a flow-rate of about 2–3 ml/min. The clear lysate can be kept at room temperature at this stage as it contains protease inhibitors and becomes cloudy at lower temperatures due to the presence of the CHAPS detergent.
25. The same flow-rate should be used as for loading the lysate onto the column. Monitor the absorbance at 280 nm and keep washing the column until OD_{280} no longer decreases. It should be noted that imidazole exhibits a significant absorbance at 280 nm.
26. Perform a small-scale pilot trial in order to determine ratio of protease to purified protein to use and the time of incubation. We prefer to use 3C protease since it is a highly specific protease and has an optimal activity to 4 °C. However, expression vectors containing a variety of protease recognition sites (e.g., TEV, thrombin, enterokinase, and Factor Xa) are also available

but it is important that you check to ensure that the protein that you wish to express does not also contain an identical protease cleavage site.

27. Most proteins and peptides are very stable when store for long periods as a lyophilized powder.
28. Place the sealed container containing the lyophilize powder on the bench for 10–20 min to equilibrate to room temperature before opening to avoid the powder absorbing moisture from the atmosphere.
29. The final concentration of the TM/CPD peptide should be at least 1 mM. Concentrations lower than 1 mM are acceptable but will result in weaker NMR signals and an increase in the amount of time required to acquire NMR spectra. If possible concentrate the protein to about 300 μl using a centrifugal concentrator such as an Amicon Ultra-0.5 Centrifugal Unit (Merck Millipore) with an appropriate molecular weight cut-off. The high concentration of DPC included in the NMR buffer ensures that only one molecule of peptide will be incorporated into each DPC micelle.
30. The buffer and pH used for the NMR buffer will depend on the pI of the TM/CDP peptide, as proteins tend to aggregate close to their pI value where they have neutral charge. If sufficient protein is available screen a range of buffer conditions.
31. Determine the concentration of the TM/CPD protein by measuring the absorbance at 280 nm and using the equation $A = \varepsilon bc$ (where A is the absorbance, ε is the molar extinction coefficient, b is the path length, and c is the concentration).
32. If the sample volume is less than 500 μl transfer to a 5 mm Shigemi NMR tube. These NMR tubes can take a minimum volume of 300 μl.
33. We usually record spectra at 10, 20, 30, 40, and 50 °C (depending on the stability of the protein). Single helical transmembrane proteins often exhibit higher quality spectra (improved spectral resolution and peak intensity) at higher temperatures. However, transient structure within the extra-membrane regions (e.g., the cytoplasmic domain) may only be observed at lower temperatures.
34. 2D–TROSY spectra exhibit improved spectral resolution compared to 2D-HSQC (heteronuclear single quantum coherence) spectra due to the large size of the TM/CPD protein-DPC micelle complex (typically ≥ 30 kDa). Typical 2D datasets are usually recorded with 1024 × 256 complex points with 8 to 32 scans per increment (depending on the sample concentration) on a 600 MHz NMR spectrometer. Data acquisition times are typically 30–60 min.

References

1. Butler GS, Overall CM (2009) Updated biological roles for matrix metalloproteinases and new "intracellular" substrates revealed by degradomics. Biochemistry 48(46):10830–10845
2. Morrison CJ, Butler GS, Rodríguez D, Overall CM (2009) Matrix metalloproteinase proteomics: substrates, targets, and therapy. Curr Opin Cell Biol 21(5):645–653
3. Rodríguez D, Morrison CJ, Overall CM (2010) Matrix metalloproteinases: what do they not do? New substrates and biological roles identified by murine models and proteomics. Biochim Biophys Acta 1803(1):39–54
4. Gill SE, Parks WC (2008) Metalloproteinases and their inhibitors: regulators of wound healing. Int J Biochem Cell Biol 40(6–7):1334–1347
5. Mott JD, Werb Z (2004) Regulation of matrix biology by matrix metalloproteinases. Curr Opin Cell Biol 16(5):558–564
6. Nagase H, Visse R, Murphy G (2006) Structure and function of matrix metalloproteinases and TIMPs. Cardiovasc Res 69(3):562–573
7. Page-McCaw A, Ewald AJ, Werb Z (2007) Matrix metalloproteinases and the regulation of tissue remodelling. Nat Rev Mol Cell Biol 8(3):221–233
8. Sternlicht MD, Werb Z (2001) How matrix metalloproteinases regulate cell behavior. Annu Rev Cell Dev Biol 17:463–516
9. Vu TH, Werb Z (2000) Matrix metalloproteinases: effectors of development and normal physiology. Genes Dev 14(17):2123–2133
10. Hadler-Olsen E, Fadnes B, Sylte I, Uhlin-Hansen L, Winberg J-O (2011) Regulation of matrix metalloproteinase activity in health and disease. FEBS J 278(1):28–45
11. Sbardella D, Fasciglione GF, Gioia M, Ciaccio C, Tundo GR, Marini S, Coletta M (2012) Human matrix metalloproteinases: an ubiquitarian class of enzymes involved in several pathological processes. Mol Aspects Med 33(2):119–208
12. Piperi C, Papavassiliou AG (2012) Molecular mechanisms regulating matrix metalloproteinases. Curr Top Med Chem 12(10):1095–1112
13. Mannello F, Medda V (2012) Nuclear localization of matrix metalloproteinases. Prog Histochem Cytochem 47(1):27–58
14. Murphy G, Nagase H (2011) Localizing matrix metalloproteinase activities in the pericellular environment. FEBS J 278(1):2–15
15. Bourboulia D, Stetler-Stevenson WG (2010) Matrix metalloproteinases (MMPs) and tissue inhibitors of metalloproteinases (TIMPs): Positive and negative regulators in tumor cell adhesion. Semin Cancer Biol 20(3):161–168
16. Brew K, Nagase H (2010) The tissue inhibitors of metalloproteinases (TIMPs): an ancient family with structural and functional diversity. Biochim Biophys Acta 1803(1):55–71
17. Gustafsson T (2011) Vascular remodelling in human skeletal muscle. Biochem Soc Trans 39(6):1628–1632
18. Kraiem Z, Korem S (2000) Matrix metalloproteinases and the thyroid. Thyroid 10(12):1061–1069
19. Ortega N, Behonick D, Stickens D, Werb Z (2003) How proteases regulate bone morphogenesis. Ann N Y Acad Sci 995:109–116
20. Parks WC, Shapiro SD (2001) Matrix metalloproteinases in lung biology. Respir Res 2(1):10–19
21. Apte SS, Parks WC (2015) Metalloproteinases: a parade of functions in matrix biology and an outlook for the future. Matrix Biol 44-46:1–6
22. Rohani MG, Parks WC (2015) Matrix remodeling by MMPs during wound repair. Matrix Biol 44-46:113–121
23. Hu Y-FF, Chen Y-JJ, Lin Y-JJ, Chen S-AA (2015) Inflammation and the pathogenesis of atrial fibrillation. Nat Rev Cardiol 12(4):230–243
24. Nissinen L, Kähäri V-MM (2014) Matrix metalloproteinases in inflammation. Biochim Biophys Acta 1840(8):2571–2580
25. Houghton AM (2015) Matrix metalloproteinases in destructive lung disease. Matrix Biol 44-46:167–174
26. Duarte S, Baber J, Fujii T, Coito AJ (2015) Matrix metalloproteinases in liver injury, repair and fibrosis. Matrix Biol 44-46:147–156
27. Mukherjee A, Swarnakar S (2015) Implication of matrix metalloproteinases in regulating neuronal disorder. Mol Biol Rep 42(1):1–11
28. Ip YC, Cheung ST, Fan ST (2007) Atypical localization of membrane type 1-matrix metalloproteinase in the nucleus is associated with aggressive features of hepatocellular carcinoma. Mol Carcinog 46(3):225–230
29. Kwan JA, Schulze CJ, Wang W, Leon H, Sariahmetoglu M, Sung M, Sawicka J, Sims DE, Sawicki G, Schulz R (2004) Matrix metalloproteinase-2 (MMP-2) is present in the nucleus of cardiac myocytes and is capable of cleaving poly (ADP-ribose) polymerase (PARP) in vitro. FASEB J 18(6):690–692
30. Limb GA, Matter K, Murphy G, Cambrey AD, Bishop PN, Morris GE, Khaw PT (2005) Matrix metalloproteinase-1 associates with

intracellular organelles and confers resistance to lamin A/C degradation during apoptosis. Am J Pathol 166(5):1555–1563

31. Si-Tayeb K, Monvoisin A, Mazzocco C, Lepreux S, Decossas M, Cubel G, Taras D, Blanc J-FF, Robinson DR, Rosenbaum J (2006) Matrix metalloproteinase 3 is present in the cell nucleus and is involved in apoptosis. Am J Pathol 169(4):1390–1401
32. Itoh Y (2015) Membrane-type matrix metalloproteinases: Their functions and regulations. Matrix Biol 44-46:207–223
33. Zucker S, Pei D, Cao J, Lopez-Otin C (2003) Membrane type-matrix metalloproteinases (MT-MMP). Curr Top Dev Biol 54:1–74
34. Hernandez-Barrantes S, Bernardo M, Toth M, Fridman R (2002) Regulation of membrane type-matrix metalloproteinases. Semin Cancer Biol 12(2):131–138
35. Galea C, Nguyen H, Chandy K, Smith B, Norton R (2014) Domain structure and function of matrix metalloprotease 23 (MMP23): role in potassium channel trafficking. Cell Mol Life Sci 71(7):1191–1210
36. Itoh Y, Takamura A, Ito N, Maru Y, Sato H (2001) Homophilic complex formation of MT1 MMP facilitates proMMP-2 activation on the cell surface and promotes tumor cell invasion. EMBO J 20(17):4782–4793
37. Itoh Y, Ito N, Nagase H, Seiki M (2008) The second dimer interface of MT1-MMP, the transmembrane domain, is essential for ProMMP-2 activation on the cell surface. J Biol Chem 283(19):13053–13062
38. Itoh Y, Ito N, Nagase H, Evans RD (2006) Cell surface collagenolysis requires homodimerization of the membrane-bound collagenase MT1-MMP. Mol Biol Cell 17(12):5390–5399
39. Sakamoto T, Seiki M (2010) A membrane protease regulates energy production in macrophages by activating hypoxia-inducible factor-1 via a non-proteolytic mechanism. J Biol Chem 285(39):29951–29964
40. Gonzalo P, Guadamillas MC, Hernández-Riquer MV (2010) MT1-MMP is required for myeloid cell fusion via regulation of Rac1 signaling. Dev Cell 18(1):77–89
41. Gingras D, Michaud M, Tomasso DG, Béliveau E (2008) Sphingosine-1-phosphate induces the association of membrane-type 1 matrix metalloproteinase with p130Cas in endothelial cells. FEBS Lett 582(3):399–404
42. Hoshino D, Tomari T, Nagano M, Koshikawa N (2009) A novel protein associated with membrane-type 1 matrix metalloproteinase binds p27kip1 and regulates RhoA activation, actin remodeling, and matrigel invasion. J Biol Chem 284(40):27315–27326
43. Wang Y, McNiven MA (2012) Invasive matrix degradation at focal adhesions occurs via protease recruitment by a FAK-p130Cas complex. J Cell Biol 196(3):375–385
44. Woskowicz AM, Weaver SA, Shitomi Y, Ito N (2013) MT-LOOP-dependent localization of membrane type I matrix metalloproteinase (MT1-MMP) to the cell adhesion complexes promotes cancer cell invasion. J Biol Chem 288(49):35126–35137
45. Uekita T, Itoh Y, Yana I, Ohno H, Seiki M (2002) Cytoplasmic tail-dependent internalization of membrane-type 1 matrix metalloproteinase is important for its invasion-promoting activity. J Cell Biol 155(7):1345–1356
46. Wang X, Ma D, Keski-Oja J, Pei D (2004) Co-recycling of MT1-MMP and MT3-MMP through the trans-Golgi network identification of DKV582 as a recycling signal. J Biol Chem 279(10):9331–9336
47. Wang P, Wang X, Pei D (2004) Mint-3 regulates the retrieval of the internalized membrane-type matrix metalloproteinase, MT5-MMP, to the plasma membrane by binding to its carboxyl end motif EWV. J Biol Chem 279(19):20461–20470
48. Nguyen H, Galea C, Schmunk G, Smith B, Edwards R, Norton R, Chandy K (2013) Intracellular trafficking of the K(V)1.3 potassium channel is regulated by the prodomain of a matrix metalloprotease. J Biol Chem 288(9):6451–6464
49. Cai M, Huang Y, Sakaguchi K, Clore MG, Gronenborn AM, Craigie R (1998) An efficient and cost-effective isotope labeling protocol for proteins expressed in shape *Escherichia coli*. J Biomol NMR 11(1):97–102
50. Studier FW (2005) Protein production by auto-induction in high-density shaking cultures. Protein Expr Purif 41(1):207–234
51. Johnson BA (2004) Using NMRView to visualize and analyze the NMR spectra of macromolecules. Methods Mol Biol 278:313–352
52. Delaglio F, Grzesiek S, Vuister GW, Zhu G, Pfeifer J, Bax A (1995) NMRPipe: a multidimensional spectral processing system based on UNIX pipes. J Biomol NMR 6(3):277–293

Chapter 3

Heterologous Expression of the Astacin Protease Meprin β in *Pichia pastoris*

Dagmar Schlenzig and Stephan Schilling

Abstract

Meprins are zinc-dependent proteases of the metzincin superfamily of metalloproteases. The enzymes are extracellular multi-domain proteins which are stabilized by disulfide bridges, dimerization, and glycosylation. Due to their complex structure, recombinant expression was first established in mammalian and insect cells. However, these methods have several disadvantages such as high costs and the low yields. For this reason, yeast is often considered a preferable expression system. Here, we describe the manipulation and secretory expression of human meprin β in the methylotrophic yeast *P. pastoris*. We show that the position of the affinity tag strongly influences the yield of expression, favoring fusion of the affinity tag at the C-terminus.

Key words Metzincin, Endoprotease, Meprin, yeast, *Pichia pastoris*

1 Introduction

Meprins (meprin α and β) are endoproteases of the astacin family. Together with matrix metalloproteases (MMPs), astacins constitute the superfamily of metzincins. Meprin α and meprin β are two evolutionary related enzymes only found in vertebrata. Both are formed as zymogenes containing a transmembrane sequence (type 1 transmembrane proteins) and a distinct multi-domain structure. The signal sequence is followed by an inhibitory pro-sequence and an astacin-like catalytic domain bearing the typical zinc binding motif HExxHxxGFxHExxRxDRD [1]. C-Terminal to the catalytic domain are a *MAM* (meprin/A5 protein/protein tyrosine phosphatase μ) domain and a *TRAF* (tumor necrosis factor receptor-associated factor) domain which are thought to play a role in protein–protein interactions, an *EGF* (epidermal growth factor)-like domain, the transmembrane domain and a cytosolic region. Additionally, meprin α contains a so-called "inserted" domain N-terminal to the *EGF*-like domain. Proteolytic processing occurs

Charles A. Galea (ed.), *Matrix Metalloproteases: Methods and Protocols*, Methods in Molecular Biology, vol. 1579,
DOI 10.1007/978-1-4939-6863-3_3, © Springer Science+Business Media LLC 2017

at this site during passage through the secretory pathway, prior to secretion of soluble meprin α into the extracellular matrix. Meprin α and β are disulfide-linked homodimers which in the case of meprin α tends to form large oligomers complexes up to 6 MDa in size [2]. Both enzymes are heavily glycosylated and the protein–protein interaction domains as well as glycosylation are necessary for correct folding, stability, and localization.

In men and mouse, meprins are highly expressed in epithelial cells of the kidney and the intestine, and they have been localized in intestinal leukocytes, skin, and certain cancer cells [2]. Both enzymes show a unique substrate specificity with a preference for acidic amino acids in the P1′-position [3]. A number of in vitro substrates of meprins have been identified including extracellular matrix proteins (procollagen I + III, occludin, E-cadherin) and biologically active peptides (interleukin 1β, interleukin 18, transforming growth factor-α, vascular endothelial growth factor-A and fibroblast growth factor 19) [4–6]. Due to the nature of these substrates, there is compelling evidence linking meprins with collagen assembly, inflammation, intestinal immune response, and neurodegeneration [7].

Unlike MMPs, meprins are not inhibited by tissue inhibitors of metalloproteinases (TIMPs). However, since the catalytic mechanism and active site architecture of MMPs, ADAMs and meprins are similar many small molecule inhibitors exhibit limited selectivity between these enzymes. Considering the compelling evidence for a role of meprins in pathophysiological processes like fibrosis, nephropathy, and inflammatory bowel disease [7, 8] the lack of effective and selective small molecule inhibitors is an obstacle. To meet the requirements of large amounts of protein for the development of selective, potent, and drug-like inhibitors, we report here the utilization of yeast for the cost-effective production of recombinant meprin β [9].

2 Materials

If not otherwise stated, all chemicals and solvents used are of gradient grade. All materials are sterilized prior to use. Procedures are carried out under semi-sterile conditions.

2.1 Plasmid Linearization and Transformation of Pichia pastoris X-33 Cells

1. Yeast expression vector pPICZα (Thermo Fischer Scientific, Waltham, MA USA).
2. PmeI and 10× CutSmart buffer (NEB BioLabs, Ipswich, USA).
3. Isopropyl alcohol.
4. Electro competent X33 cells (Life Technologies GmbH, Darmstadt, Germany).
5. 0.2 cm electroporation cuvettes.

6. YPDS medium: 10 g yeast extract, 20 g peptone, 182 g sorbitol, 900 ml mono-distilled water, supplemented with 100 ml 20% Dextrose (filter sterilized) after autoclaving.
7. YPDS-agar: add 20 g agar to the YPDS medium before autoclaving.
8. YPDS agar plates containing increasing concentrations of Zeocin: 100, 200, and 500 μg/ml.

2.2 Test Expression

1. BMGY medium: 10 g yeast extract, 20 g peptone, 700 ml mono-distilled water supplemented with 100 ml 1 M K_2SO_4 pH 6.0, 100 ml YNB (34 mg yeast nitrogen base w/o amino acids, 100 mg $(NH_4)_2SO_4$, and 1 l of water), 2 ml 0.02% biotin, and 100 ml 10% glycerol (filter sterilized) after autoclaving.
2. BMMY medium: the same as BMGY but replace the glycerol with 100 ml of 5% methanol.
3. Methanol (99% HPLC grade).

2.3 Activity Assay.

1. Assay buffer 1: 50 mM Tris–HCl, pH 7.5, 20 mM $CaCl_2$.
2. Assay buffer 2: 40 mM Tris, pH 8.0.
3. Trypsin: 400 μg/ml trypsin in assay buffer 2 (*see* **Note 1**).
4. Substrate: 50 μM solution of Abz-YVADAP(Dnp)G-OH (or Mca-YVADAPK(Dnp)-OH) in assay buffer 2 (*see* **Note 2**).

2.4 Purification

1. Equilibration buffer 3: 3 M NaCl, 300 mM Tris, pH 7.6.
2. Equilibration buffer 4: 300 mM NaCl, 30 mM Tris, pH 7.6.
3. Wash buffer 5: 20 mM imidazole, 300 mM NaCl, 30 mM Tris, pH 7.6.
4. Elution buffer 6: 200 mM imidazole, 100 mM NaCl, 30 mM Tris, pH 7.6.
5. 1 mM $CaCl_2$ in water.
6. Trypsin: 2 mg/ml trypsin in assay buffer 2.
7. Trypsin inhibition: 1 M 4-(2-aminoethyl)benzenesulfonylfluoride (AEBSF) in water.
8. Equilibration buffer 7: 30 mM Tris, 1.5 M $(NH_4)_2SO_4$.
9. Elution buffer 8: 30 mM Tris, 100 mM NaCl.
10. Equilibration buffer 9: 40 mM Tris, 100 mM NaCl, pH 7.6.

3 Methods

The open reading frame of meprin α or β are integrated into pPICZαC using the *Cla*I and *Not*I restriction sites. The start codon and signal sequences are replaced by the plasmid-encoded

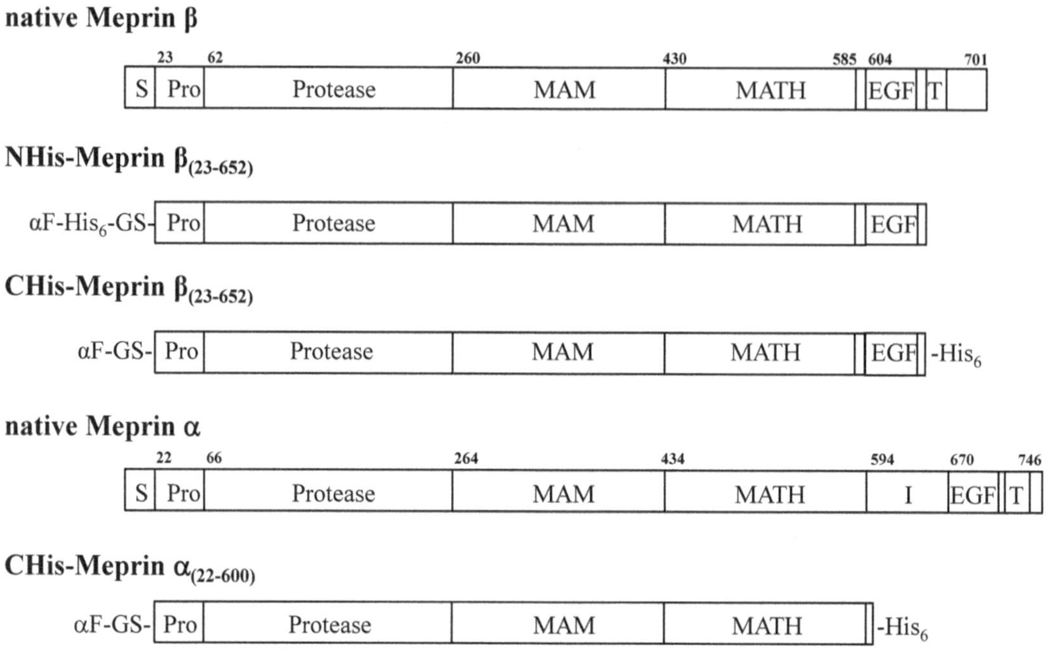

Scheme 1 Schematic representation of constructs used for expression of human meprins α and β in *P. pastoris*. The native N-terminal signal sequence was replaced by the vector-encoded α-factor pre pro-sequence of *Saccharomyces* to enable efficient secretion of the protein. The C-terminal transmembrane and cytosolic regions were omitted. A His_6-tag was introduced either N-terminally between the α-factor and pro-sequence or C-terminally to accelerate the purification process. Domain structure of meprins according to UniProt Knowledgebase: *S* signal sequence, *Pro* pro-sequence, *MAM* meprin, 5A protein, tyrosine phosphatase μ domain, *MATH* meprin and TRAF homology domain, *EGF* epidermal growth factor-like domain, *T* transmembrane region, *I* insertion in meprin α which contains a cleavage site. Modified from [9] with permission of Elsevier

α-factor prepro-sequence of *S. cerevisiae*. The transmembrane and cytosolic regions are removed to yield secreted proteins. The resulting sequences correspond to meprin α (22–600) and meprin β (23–652). A His_6-tag is fused to the N- or C-terminus by means of appropriate primers in order to facilitate the purification (*see* Scheme 1). For termination of translation, a stop codon is introduced at the end of the sequence. All cloning procedures are performed according to standard methods described in the technical guidelines obtained from New England Biolabs. The ligation efficiency could be enhanced by a dephosphorylation step utilizing shrimp alkaline phosphatase after restriction digestion of the vector.

3.1 Plasmid Linearization and Transformation of P. pastoris

1. Incubate expression vector (2 μg) with 20 U of *Pme*I and 5 μl 10× CutSmart buffer in a total volume of 50 μl at 37 °C for 1 h. Check for complete linearization using agarose gel electrophoresis (0.8% gel).

2. Add 500 μl of isopropanol to the reaction for precipitation of DNA and incubate for at least 2 h at −20 °C. Centrifuge the mixture at 4 °C and 14400 × *g* for 30 min, decant the supernatant, air-dry the pellet for about 2 h and dissolve it in 30 μl of water. Estimate the DNA concentration.
3. Thaw 50 μl electro competent X33 cells on ice and add 5 μl of linearized plasmid. Place the electroporation cuvette on ice and transfer the cells. Place the dried cuvettes in the electroporator and apply a pulse of 2 kV (*see* **Note 3**).
4. Immediately add 1 ml of ice-cold YPDS medium to the cells and incubate for 2 h at 30 °C while gently shaking from time to time.
5. Spread 300 μl of cells per YPDS-agar plate with increasing Zeocin concentrations (100, 200, 400, and 600 μg/ml) and incubate for 3 days at 30 °C (*see* **Note 4**).
6. Select in total 70–100 colonies from the plates and transfer them onto labeled master plates containing 100 μg/ml Zeocin (*see* **Note 5**). Incubate at 30 °C for 2 days.

3.2 Test Expression

1. Prepare 70–100 cultivation tubes containing 2 ml of BMGY. Inoculate each with a single colony from the master plates and let them grow in a shaking incubator (200 rpm) at 30 °C for 24 h (*see* **Notes 6** and **7**).
2. Centrifuge the tubes at 2,000 × *g* for 5 min and decant the supernatant. Add 2 ml of BMMY to the cells, resuspend the cells and incubate with shaking (200 rpm or higher) at 30 °C for 24 h.
3. Maintain the methanol concentration at about 0.5% (v/v) for the next 2 days by daily addition of 10 μl of 100% methanol.
4. Centrifuge the samples at 2,000 × *g* for 10 min and collect the supernatant. Keep it on ice until performing the activity assay.

3.3 Activity Assay

1. Pipet 50 μl of media supernatant to a 96-well plate, add 100 μl of assay buffer 1, 50 μl trypsin and incubate for 30 min at 37 °C.
2. Transfer 50 μl of each well to a black clear bottom 96-well plate, add 150 μl of assay buffer 2 and incubate for 10 min at 30 °C. Start kinetic fluorescence measurement immediately after addition of 50 μl of substrate solution and measure time-dependent conversion of substrate at excitation and emission wavelengths of 340 and 420 nm, respectively (Fig. 1, *see* **Note 8**).
3. Select samples with the highest RFU/min and save the corresponding colonies as glycerol stocks (Figs. 1 and 2, *see* **Note 9**).

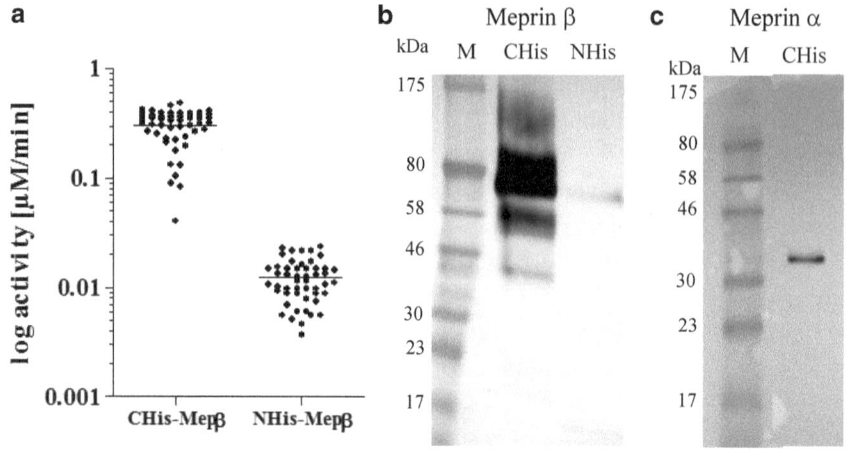

Fig. 1 (**a**) Meprin β activity determined in the medium after small-scale expression of C-terminally or N-terminally His_6-tagged meprin β in *P. pastoris*. Meprin β activity in the media of at least 50 clones after an activation step with trypsin depicted as logarithmic scatter plot. On average, the activity for NHis-meprin β is more than 20-fold lower compared with CHis-meprin β. (**b**) The difference in expression efficiency is also reflected by Western blot analysis. A total amount of 100 µg protein from the fermentation medium was applied to each lane. Meprin β was detected using a polyclonal antibody. (**c**) No enzymatic activity could be detected after test expression of CHis-meprin *α* in *P. pastoris*. On the protein level, a signal of reduced size was detected with a polyclonal antibody. ServaGelTM TG PrimeTM 4–20%, Modified from [9] with permission of Elsevier

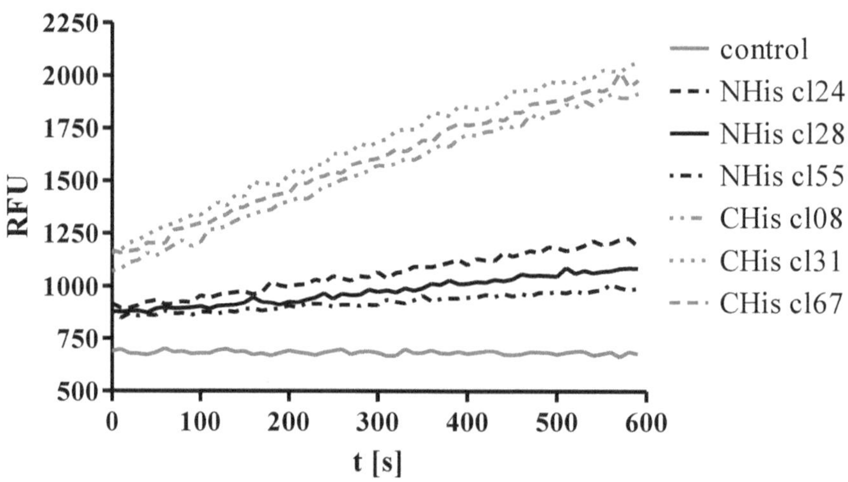

Fig. 2 Activity measurement of three selected colonies expressing CHis-meprin *β* or NHis-meprin *β* (Fig. 1), respectively. Prior to measurement, samples were activated by incubation with trypsin. Assays were started by addition of substrate. Samples of CHis-meprin *β* were diluted 1:10

3.4 Scale Up of Expression (See **Note 10***)*

1. Expression was scaled up by fermentation in a bioreactor, basically as described in the Pichia Fermentation Process Guidelines (Thermo Fischer Scientific, Waltham, MA USA). Briefly, fermentation was carried out in a 5 l reactor. The cells were grown in fermentation basal salt medium (pH 5.5, $T = 30$ °C) supplemented with trace salts and glycerol. The process was started by inoculation with 200 ml pre-culture with an OD_{600} of about 2–3. A 24 h initial batch phase followed by a 6 h glycerol feed batch phase results in an accumulation of cell mass (OD_{600} of about 400–450). Application of the three-step methanol feeding protocol outlined in the Process Guideline for 45–50 h is accompanied by a steady increase of meprin activity [9] (*see* **Note 11**).
2. Separate the cells from the supernatant by centrifugation at 6,000 × *g* for 15 min at 4 °C. Store the supernatant at −80 °C. Long-term storage for months does not negatively affect the yield of meprin.

3.5 Purification (Fig. 3, See **Note 12***)*

1. Thaw the supernatant and dilute it with 1:10 of the final total volume with equilibration buffer 3. Centrifuge at 30,000 × *g* for 20 min at 4 °C and apply the supernatant to a Streamline chelating expanded bed column (25 × 280 mm in non-expanded state; GE Healthcare Life Science, Piscataway, USA, *see* **Note 13**) previously equilibrated with equilibration buffer 4 in reverse flow at a flow rate of 12 ml/min. After a washing step with wash buffer 5 the enzyme is eluted in forward flow with elution buffer 6. The eluted protein is stored at 4 °C overnight.
2. The eluted enzyme is supplemented with $CaCl_2$ (final concentration of 10 mM). One milligram of trypsin is added per 15 mg of protein and incubated with gentle shaking at room temperature for 15 min. Afterwards, add AEBSF (final concentration of 5 mM) for inhibition of trypsin.
3. The volume of the sample is doubled by the addition of 3 M $(NH_4)_2SO_4$ in the cold (*see* **Note 14**). Syringe filter through a 0.46 μM filter membrane to remove turbidity. Load the sample onto a pre-equilibrated Butyl Sepharose column (25 × 100 mm, GE Healthcare Life Science, Piscataway, USA, equilibrated with buffer 7) at a flow rate of 3 ml/min. Elute with an ammonium sulfate gradient (1.5–0 M ammonium sulfate acheived by mixing buffers 7 and 8) for 10 column volumes at the same flow rate.
4. Eluted protein is concentrated by ultrafiltration (MWCO 30,000 Da) and applied onto a Superdex 200 column (26 × 850 mm, material GE Healthcare Life Science, Piscataway, USA) equilibrated with buffer 9 (*see* **Notes 15** and **16**). Eluted

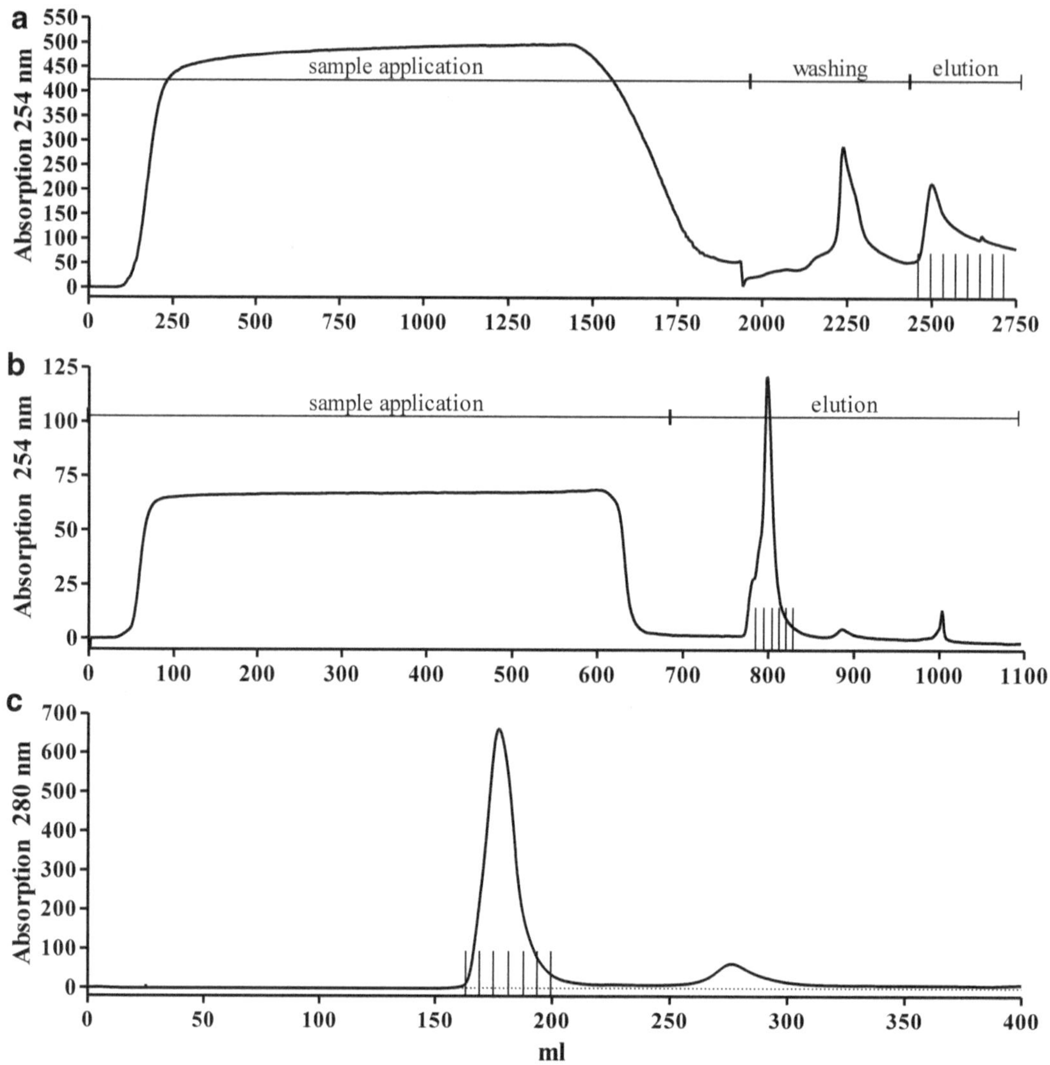

Fig. 3 Purification of CHis-meprin *β* expressed in *P. pastoris*. (**a**) Chromatogram of purification of the fermentation broth by immobilized metal chelating chromatography (IMAC). (**b**) Chromatogram of purification of the eluate from IMAC using hydrophobic interaction chromatography. (**c**) Final purification performing size exclusion chromatography. *Bars* indicate meprin containing fractions

protein can be concentrated again, aliquoted and stored at −80 °C (*see* **Note 17**). Analyze the eluted protein by SDS-PAGE to determine its purity (Fig. 4, *see* **Note 18**).

4 Notes

1. Aliquots of 20 mg/ml trypsin stock solution can be prepared in 0.1 M HCl for long term storage at −20 °C. Prepare fresh trypsin working solution from these aliquots prior to use by the addition of assay buffer 1.

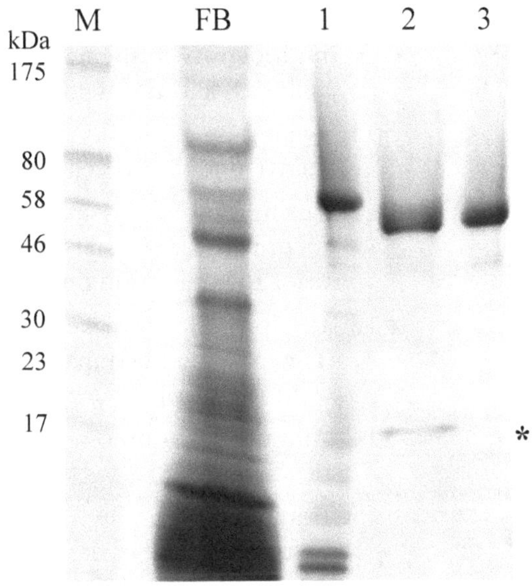

Fig. 4 SDS-PAGE analysis illustrating the purification of CHis-meprin β from the fermentation supernatant of *P. pastoris*. The combined fractions after each step have been applied. *Lanes*: *M* molecular mass standards, *FB* fermentation broth (50 μg), *1* after IMAC (5 μg), *2* HIC following zymogen activation by trypsin (5 μg), *3* after final SEC (5 μg). *Asterisk* trypsin, ServaGel™ TG Prime™ 4–20%. Reproduced from [9] with permission of Elsevier

2. Aliquots of 1 mM substrate stock solution can be prepared in DMSO for long term storage at −20 °C. Prepare fresh substrate working solution from these aliquots prior to use by the addition of assay buffer 2.
3. A pulse time of about 5 ms is optimal for electroporation.
4. Increasing the Zeocin concentration for selection of transformants to more than 500 μg/ml leads to fewer colonies which did not show higher expression rates.
5. For longer storage keep master plates at 4 °C. After 1 month, transfer the colonies on new plates if further use is intended.
6. Addition of Zeocin to test expression medium is not essential.
7. Include untransformed X33 cells in the test expression to perform negative control assays.
8. Subtract the slope obtained for a negative control test expression from that of the samples.
9. Further evaluation of selected colonies before scale-up is advised. The protein concentration in the supernatant after test expression is very low. Since the target protein can hardly be detected on Coomassie-stained SDS gels a Western blot should be used to verify the molecular mass and integrity of the expressed protein (i.e., presence of His-tag protein).

10. Due to the low amount of expressed protein, process-controlled high density fermentation is appropriate for fast expansion of cell mass and production of larger amounts of enzyme.
11. Scaling up expression to 2 l of media in shaking flasks is ineffective due to low cell mass accumulation.
12. Purification steps are performed at room temperature as meprin β is very stable. However, samples were kept on ice prior to application onto columns and after elution.
13. It is possible to make use of other standard IMAC columns. The expanded bed material has the advantage of application of crude material at very high flow rates.
14. Add the 3 M ammonium sulfate dropwise and slowly with gentle stirring.
15. Equilibrate the column with 4–5 column volumes (CV) equilibration buffer. If the column is stored in buffer containing ethanol, start with 2 CV of water prior to buffer.
16. Apply not more than 4.5 ml of sample (1% of total CV) to the gel filtration column for efficient separation.
17. The protein is highly glycosylated. About 15% of the protein is glycosylated with N-linked carbohydrates. Therefore, the protein can be concentrated to 30 mg/ml without precipitation. Due to the high stability of human meprin β, freezing aliquots at −80 °C does not lead to decreased enzyme activity.
18. Human meprin β has been successfully expressed in *P. pastoris.* The location of the His_6-tag dramatically influenced the overall yield. Thirty milligram of functional enzyme purified to homogeneity could be produced from 2 l of process-controlled high density fermentation of C-terminally tagged protein. For N-terminally His_6-taged enzyme the yield was tenfold lower. Due to protein degradation, human meprin α could not be isolated, despite a high sequence homology of about 45%. The utilization of protease A deficient SMD5896 cells and the variation of culture conditions like pH, temperature, and supplementation with casein hydrolysate did not show any positive effect on expression of meprin α.

References

1. Dumermuth E, Sterchi EE, Jiang WP et al (1991) The astacin family of metalloendopeptidases. J Biol Chem 266(32):21381–21385
2. Sterchi EE, Stocker W, Bond JS (2008) Meprins, membrane-bound and secreted astacin metalloproteinases. Mol Aspects Med 29(5):309–328
3. Becker-Pauly C, Barre O, Schilling O et al (2011) Proteomic analyses reveal an acidic prime side specificity for the astacin metalloprotease family reflected by physiological substrates. Mol Cell Proteomics 10(9):M111.009233
4. Kronenberg D, Bruns BC, Moali C et al (2010) Processing of procollagen III by meprins: new players in extracellular matrix assembly? J Invest Dermatol 130(12):2727–2735
5. Bao J, Yura RE, Matters GL et al (2013) Meprin A impairs epithelial barrier function, enhances

monocyte migration, and cleaves the tight junction protein occludin. Am J Physiol Renal Physiol 305(5):F714–F726
6. Huguenin M, Muller EJ, Trachsel-Rosmann S et al (2008) The metalloprotease meprinbeta processes E-cadherin and weakens intercellular adhesion. PLoS ONE 3(5):e2153
7. Broder C, Becker-Pauly C (2013) The metalloproteases meprin α and meprin β: unique enzymes in inflammation, neurodegeneration, cancer and fibrosis. Biochem J 450(2): 253–264
8. Biasin V, Marsh LM, Egemnazarov B et al (2014) Meprin β, a novel mediator of vascular remodelling underlying pulmonary hypertension. J Pathol 233(1):7–17
9. Schlenzig D, Wermann M, Ramsbeck D et al (2015) Expression, purification and initial characterization of human meprin β from *Pichia pastoris*. Protein Expr Purif 116:75–81

Part II

Structural Characterization of Matrix Metalloproteases

Chapter 4

Structural Studies of Matrix Metalloproteinase by X-Ray Diffraction

Elena Decaneto, Wolfgang Lubitz, and Hideaki Ogata

Abstract

Matrix Metalloproteinases (MMPs) are a family of proteolytic enzymes whose endopeptidase activity is dependent on the presence of specific metal ions. MT1-MMP (or MMP-14), which has been implicated in tumor progression and cellular invasion, contains a membrane-spanning region located C-terminal to a hemopexin-like domain and an N-terminal catalytic domain. We recombinantly expressed the catalytic domain of human MT1-MMP in *E. coli* and purified it from inclusion bodies using a refolding protocol that yielded significant quantities of active protein. Crystals of MT1-MMP were obtained using the vapour diffusion method. Here, we describe the protocols used for crystallization and the data analysis together with the resulting diffraction pattern.

Key words Metalloproteinase, MMP-14, MT1-MMP, Crystallization, Diffraction

1 Introduction

Matrix metalloproteinases (MMPs) are a family of proteolytic enzymes whose endopeptidase activity is dependent on the presence of specific metal ions [1]. These enzymes are conserved across the animal kingdom and are mostly involved in various physiological processes such as embryonic development, morphogenesis, tissue remodeling and wound healing. Abnormalities in the mechanisms of regulation that control the multiple functions of MMPs may lead to severe inflammatory and degenerative conditions such as arthritis, cancer, and cardiovascular diseases [2, 3]. MMPs share a common domain structure characterized by (1) an N-terminal pro-peptide which inhibits protein activity prior to zymogenic activation, (2) a catalytic domain containing two zinc binding sites and at least two calcium binding sites, and (3) an hemopexin-like domain connected to the catalytic domain through a flexible "hinge region." While MT1-MMP (or MMP-14), an important MMP involved in tumor progression and cellular invasion, contains an additional membrane-spanning domain located

Charles A. Galea (ed.), *Matrix Metalloproteases: Methods and Protocols*, Methods in Molecular Biology, vol. 1579, DOI 10.1007/978-1-4939-6863-3_4,

C-terminal to the hemopexin-like domain [4, 5]. Of the 176 X-ray structures of human MMP catalytic domains, only MMP-1 [6], MMP-3 [7], MMP-8, and MMP-12 [8] were crystallized in the active form. In addition, no catalytic domain structure for a membrane-associated MMP has been solved except for MT1-MMP and MT3-MMP complexed with natural and synthetic inhibitors, respectively [8–10]. Detailed structural knowledge of the catalytic domain of MMPs in all stages of enzymatic activity is indispensable for understanding the catalytic process and thus for controlling enzymatic activity, improving the design criteria for specific inhibitors, and enhancing the sensitivity of ligand screening [7, 11].

Herein, we describe the recombinant expression of the catalytic domain of human MT1-MMP in *E. coli* and its purification from inclusion bodies utilizing a refolding protocol that yields significant quantities of active protein. The purified protein was used to obtain crystals of MT1-MMP [12]. We also outline protocols used for the crystallization of the protein and analysis of data to generate the diffraction pattern.

2 Materials

2.1 Chemicals and Solutions for Purification

1. pET-3 expression system (Novagen; Madison, WI).
2. Human MT1-MMP cDNA.
3. *E. coli* strains BL21 (DE3).
4. SOC (Super optimal broth) sterile media: 20 g tryptone, 5 g yeast extract, 2 mL of 5 M NaCl, 2.5 mL of 1 M $MgCl_2$, 10 mL of 1 M $MgSO_4$, and 20 mL of 1 M glucose for 1 L.
5. 2YT agar plates: 2.5 g NaCl, 8 g tryptone, 5 g yeast extract, and 7.5 g agar for 0.5 L.
6. LB (Lysogeny broth) sterile media: 10 g NaCl, 10 g tryptone, and 5 g yeast extract for 1 L.
7. Potter-Elvehjem homogenizer (Heidolph; Elektro GmbH, Kelheim, Germany).
8. Sonifier cell disrupter (Branson, Danbury, CT).
9. 0.9% NaCl.
10. Wash buffer I: 30 mM Tris–HCl pH 7.4, 30 mM NaCl, 1 mM EDTA, 2% (v/v) Tween 20, and 5 mM β-mercaptoethanol.
11. Wash buffer II: 30 mM Tris–HCl pH 7.4, 500 mM NaCl, 1 mM EDTA, and 5 mM β-mercaptoethanol.
12. Wash buffer III: 30 mM Tris–HCl pH 7.4, 30 mM NaCl, 1 mM EDTA, and 5 mM β-mercaptoethanol.
13. Refolding buffer I: 50 mM Tris–HCl pH 8.5 and 6 M urea.

14. Refolding buffer II: 50 mM Tris–HCl pH 8.5, 100 mM NaCl, 10 mM $CaCl_2$, 10 µM $Zn(OAc)_2$, and 2 M urea.
15. Refolding buffer III: 50 µM Tris–HCl pH 7.5, 100 mM NaCl, 10 mM $CaCl_2$, 10 µM $Zn(OAc)_2$, and 1 M urea.
16. Refolding buffer IV: 50 mM Tris–HCl pH 7.5, 100 mM NaCl, 5 mM $CaCl_2$ and 1 µM $Zn(OAc)_2$.
17. Refolding buffer V: 50 µM Tris–HCl pH 7.5, 100 mM NaCl, and 5 mM $CaCl_2$.
18. 10 kDa Amicon Ultra-0.5 centrifugal filters (EMD Millipore, Darmstadt, Germany).
19. SDS-PAGE (sodium dodecyl sulfate–polyacrylamide gel electrophoresis) equipment.
20. Running buffer: 25 mM Tris–HCl pH 6.8, 1 mM Na_2EDTA, 10% (v/v) glycerol, 5% (v/v) β-mercaptoethanol, and 0.025% (w/v) bromophenol blue.
21. Staining solution: 0.1% (w/v) Coomassie Brilliant Blue G-250 in 10% (v/v) acetic acid and 40% (v/v) methanol.
22. Destaining solution: 10% (v/v) acetic acid and 20% (v/v) methanol.
23. Protein Test Mixture 6 (Serva Electrophoresis, Heidelberg, Germany).
24. Fluorogenic substrate Mca-Lys-Pro-Leu-Gly~Leu-Lys(Dnp)-Ala-Arg-NH_2.

2.2 Chemicals and Instrumentations for Crystallization

1. Cryoprotectant buffer (0.3 M ammonium nitrate, 25% (w/v) PEG 3350 (pH 6.2), and 20% glycerol).
2. Thermal Seal RT film, 2mil (Sigma-Aldrich).
3. Clear tape, HD clear (ShurTech Brands, LLC).
4. Microscope, Leica M125 (Leica Microsystems).
5. 96-well "Intelli-Plate," LBR, Low profile, three wells (Dunn Labortechnik GmbH, Germany).
6. Liquid dispensing robot, PHOENIX (Art Robbins Instruments, CA, USA).
7. Crystallization buffer kits (PEG ION/PEG ION2) (Hampton research, CA, USA).
8. Compressor-cooled incubator, IPP500 (Memmert, Schwabach, Germany).
9. Izit Crystal Dye (Hampton research, CA, USA).

2.3 Instrumentation for X-Ray Data Collection

1. Magnetic cryovials and cryocaps (Molecular Dimensions, UK).
2. Foam dewar (Molecular Dimensions, UK).
3. Vial clamp, curved (Hampton Research, USA).

4. Crystalwand Magnetic (Hampton Research, USA).
5. Cryo tong, 18 mm (Hampton Research, USA).
6. Cryoloop 10 μm (0.1–1.0 mm) (Hampton Research, USA).
7. Cryocane (Molecular Dimensions, UK).
8. Dryshipper, CX100B-11 M (Taylor Wharton).

2.4 Computer Programs

The computer programs for data analysis can be found on the Internet websites.

1. Linux operation system (Ubuntu12.04).
2. XDS (X-ray Detector Software), XSCALE and XDSCONV (http://xds.mpimf-heidelberg.mpg.de/).
3. CCP4i (http://www.ccp4.ac.uk/).
4. PHENIX (http://www.phenix-online.org/).
5. COOT (http://www2.mrc-lmb.cam.ac.uk/personal/pemsley/coot/).
6. PYMOL (https://www.pymol.org/).

3 Methods

3.1 Purification

3.1.1 Induction and Harvest

1. Transform *E. coli* BL21 (DE3) chemically competent cells with the pET-3a plasmid containing the catalytic domain of human MT1-MMP together with the hinge-linker to the hemopexin-like domain (residues 112–292) [5]. Transfer 1 μL of DNA to 100 μL of *E. coli* strain BL21 (DE3) competent cells in a sterile tube and kept on ice for 30 min.
2. Incubate the cells at 40 °C for 90 s and then place on ice for 2 min.
3. Add 60 μL of cells to 1 mL of SOC medium in a 15 mL sterile tube (*see* **Note 1**).
4. Incubate the cells at 37 °C for 1 h at 180 rpm.
5. Streak cells on 2YT agar plates containing ampicillin (100 μg/mL) and incubate overnight at 37 °C.
6. Inoculate a single colony into LB media containing 100 μg/mL of ampicillin and grow overnight at 37 °C (*see* **Note 2**).
7. Inoculate 1 L of LB media containing ampicillin with an aliquot (50–100 mL) of the overnight culture and grow at 37 °C to an OD_{600} of 0.8.
8. Induce cells with IPTG (1 mM) for a further 5 h at 37 °C.
9. Following induction, harvest the cells by centrifugation (6,000 × *g*, 4 °C, 15 min) and suspend the cell pellet in 0.9% NaCl (*see* **Note 3**).

3.1.2 Refolding and Purification

1. Homogenize the cells using a Potter–Elvehjem homogenizer.
2. Sonicate the cells on ice for five cycles of 1 min with a sonifier with a 1 s pulse duration at 70% output power (*see* **Note 4**).
3. Centrifuge (6,000 × *g*, 4 °C, 15 min) to harvest the inclusion bodies in the pellet.
4. Homogenize the inclusion bodies and wash them twice with Wash buffer I.
5. Then, wash the inclusion bodies again twice with Wash buffer II.
6. Finally wash them twice with Wash buffer III.
7. Homogenize the inclusion bodies in Refolding buffer I and keep at 4 °C for 2 h.
8. Ultracentrifuge the inclusion bodies (100,000 × *g*, 4 °C, 30 min).
9. Dilute the supernatant in the same buffer to a protein concentration of less than 1 mg/mL (*see* **Note 5**).
10. Dialysis (12–14 kDa molecular weight cutoff membrane) against a ten-fold volume of Refolding buffer II for 12 h at room temperature.
11. Change the solution to Refolding buffer III for 24 h.
12. Dialyze the protein against Refolding buffer IV for 24 h.
13. Concentrate the solution to approximately 10 mg/mL using an Amicon Ultra-0.5 centrifugal filter device (10 kDa molecular weight cutoff) and then exchange the solution for zinc-free Refolding buffer V.
14. Incubate the protein solution for 24 h at room temperature until the C-terminal hinge region is fully digested (Fig. 1). The band at ~25 kDa in Fig. 1 corresponded to the full-length construct while the lower band at 19–20 kDa correlated with the size of the domain without the C-terminal hinge region. After approximately 1 day, only the lower band corresponding to the MT1-MMP catalytic domain was observed and this was stable over the next 3 days. The freshly purified MT1-MMP catalytic domain lacking the hinge region was used for crystallization and structure determination (*see* **Note 6**).
15. Measure the activity of the mature MT1-MMP using a fluorogenic assay (λ_{ex} = 340 nm, λ_{em} = 400 nm). We typically use an assay which involves monitoring the change in fluorescence upon cleavage of the peptide Mca-Lys-Pro-Leu-Gly~Leu-Lys(Dnp)-Ala-Arg-NH_2 (*see* **Note 7**).

3.2 Crystallization

Screening of crystallization conditions was carried out by the sitting vapour-diffusion method with a liquid dispensing robot, PHOENIX (Art Robbins Instruments, CA, USA).

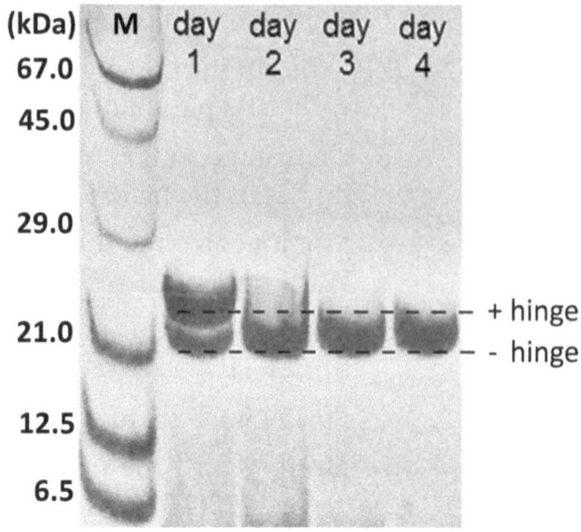

Fig. 1 Autoproteolytic activity of MT1-MMP. Day 1: recombinantly expressed human MT1-MMP catalytic domain together with the hinge region following dialysis in refolding buffer. Day 2, 3, and 4: the sample shown in lane 1 was kept at 20 °C for 1, 2, and 3 days, respectively prior to analysis. *Lane M* contains the molecular-mass marker (labeled in kDa). Samples were run on a 12% SDS–PAGE gel

1. Precool the crystallization screening buffer kits (PEG ION and PEG ION2) at 4 °C.
2. Spin the concentrated MT1-MMP samples (10 mg/mL) at 10,000 rpm (~9,000 g) for 5 min at 4 °C.
3. Wash the nano-needle and the 96-syringes of the dispensing robot PHOENIX with distilled water.
4. Aspirate each 100 μL reservoir solution of the 96 different buffers (PEG ION and PEG ION2) with the 96-syringes. Aspirate the 50 μL protein solution with the nano-needle.
5. Dispense 80 μL of the 96 different buffers into the reservoir well of the "Intelli-plate" with a dispensing robot, PHOENIX (*see* **Note 8**).
6. Dispense 0.3 μL of the protein solution into the middle of the three wells of the "Intelli-plate" (*see* **Note 9**).
7. Dispense 0.3 μL of the screening buffer into the middle of the three wells of the "Intelli-plate" (*see* **Note 9**).
8. Seal the plate with "ThermalSeal" and store at 4 °C.
9. Recover the remaining crystallization buffer in the 96-syringes and the remaining protein solution in the nano-needle.
10. Wash the 96-syringes and the nano-needle with distilled water.
11. Check the crystallization plates with a microscope (×50–×100 magnitude). Crystals of MT1-MMP were obtained in a solu-

tion containing 0.2 M ammonium nitrate and 20% (w/v) PEG (polyethylene glycol) 3350 (pH 6.2) (Fig. 2a).

12. To stain the protein crystals add Izit Crystal Dye to one of the crystallization wells and incubate for a few hours (Fig. 2b).

3.3 Crystal Handling for the Cryo Experiments

1. Prepare the cryoprotectant buffer (*see* **Note 10**).
2. Open the seal of the crystallization plate.
3. Pick the crystals with a cryo-loop under the microscope.
4. Soak the crystals in the cryoprotectant buffer and wait for several minutes (1–5 min).
5. Pick a single crystal and freeze in liquid nitrogen (*see* **Note 11**).
6. Put the cryocap on the cryo-loop vial in liquid nitrogen and position it on the cryo-cane (*see* **Note 12**).
7. Store the cryo-cane in the precooled dryshipper for future analysis (*see* **Note 13**).

3.4 X-Ray Diffraction Data Collection

Diffraction data sets were collected on beamline BL14.2 at BESSYII (Hermholtz-Zentrum Berlin, Germany).

1. The CCD detector is a Rayonix MX-225 (Rayonix, USA). Set the detector-to-crystal distance to 300 mm.
2. Set the cryo nitrogen gas-stream temperature at 100 K.
3. Mount the crystal with the cryo-loop on the goniometer. Then align the crystal to the center of the X-ray beam position.
4. Set the X-ray wavelength of 0.91841 Å (maximum X-ray beam intensity).
5. Collect a diffraction image and check the highest resolution diffraction spot. If the resolution is higher, set the detector-to-crystal distance shorter.
6. Repeat **step 5** until the maximum resolution of the diffraction spots is covered.

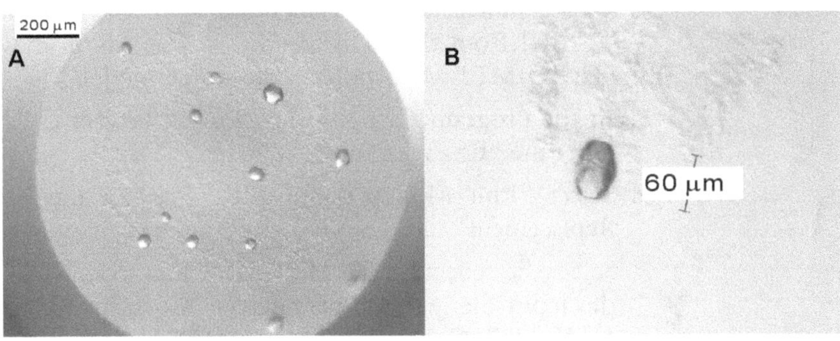

Fig. 2 MT1-MMP crystals. (**a**) The original MT1-MMP crystals and (**b**) stained with the blue colored Izit dye

7. Collect a complete data set. The data collection conditions of the MT1-MMP crystal are: X-ray exposure time of 3 s, oscillation angle of 1.0°, and total number of diffraction images of 100. The detector-to-crystal distance was 250 mm.
8. Copy the image files to a USB hard drive.

3.5 Data Analysis

The computer programs described below were used for data analysis on a Linux operating system (Ubuntu12.04).

1. The diffraction images are indexed and integrated using the program XDS [13]. The input file template to use is supplied by the beamline facility. The input parameters are as follows; JOB = XYCORR INIT COLSPOT IDXREF DEFPIX XPLAN INTEGRATE CORRECT, DETECTOR_DISTANCE = 250.0, OSCILLATION_RANGE = 1.0, X-RAY_WAVELENGTH = 0.91841, NAME_TEMPLATE_OF_DATA_FRAMES = mmp_1_???.img, DATA_RANGE = 1100 (XDS will substitute the appropriate image number for the question marks), SPOT_RANGE = 1 11, SPOT_RANGE = 42 52, SPOT_RANGE = 73 78, SPACE_GROUP_NUMBERS = 96, UNIT_CELL_PARAMETERS = 63 63 123 90 90 90, INCLUDE_RESOLUTION_RANGE = 50.0 2.24, FRIEDEL'S_LAW = TRUE (*see* **Note 14**).
2. Scaling is carried out using XSCALE. In the input file, select the integrated data (XDS_ASCII.HKL) as the input file. The output file is named as MT1-MMP.xscale. Set the parameters, INCLUDE_RESOLUTION_RANGE = 50 2.24, OUTPUT_FILE = MT1-MMP.xscale.
3. The scaled file (MT1-MMP.xscale) is converted to the MTZ file format (CCP4) by XDSCONV. Set the parameters: GENERATE_FRACTION_OF_TEST_REFLECTIONS = 0.05, FRIEDEL'S_LAW = TRUE.

   (a) For molecular replacement, an initial model, the MT1-MMP/TIMP-2 complex (PDB entry 1BQQ) [10], was chosen from the PDB database (http://www.rcsb.org/).
4. Before molecular replacement, remove the coordinates for TIMP-2 from the PDB file with a text editor and save this (1BQQ-MT1-MMP.pdb) as the search model.
5. In the program CCP4i [14], create a project (MT1-MMP) using the "Change Project" tab.
6. Select "Run Molrep –auto MR" [15] in the "Molecular Replacement" tab.
7. Select the reflection file MT1-MMP.mtz from XDSCONV as the input file. Set the parameters; Model in "1BQQ-MT1-MMP.pdb", Coords out "MT1-MMP-MR.pdb", Use data to maximum resolution "2.24" in the Experimental Data tab, Search for "1" monomers in the asymmetric unit in the Search Parameters, then run the program.

8. In the output file from MOLREP, check the molecular replacement solution.
9. In the refinement program PHENIX [16], create the new project "MT1-MMP" and select "refinement" in the menu tab.
10. For the first refinement, select the files (MT1-MMP.mtz, MT1-MMP-MR.pdb, ZN.cif and CA.cif) as the input data (*see* **Note 15**). Select the parameter, "Rigid-body" in the Refinement strategy. Run the refinement setting the number of cycles to 3 and the output file name as MT1-MMP_refine_1.pdb.
11. For the second refinement, select the files MT1-MMP.mtz, MT1-MMP_refine_1.pdb, ZN.cif and CA.cif as the input data. Set the parameters; "xyz coordinates", "Real-space", "Individual B-factors", "Occupancies" in the Refinement strategy. Set the other parameters: "Update waters" and "Automatically correct N/Q/H error". Run the refinement with three cycles.
12. For the model building, run COOT [17]. Open the coordinate file and the reflection file, MT1-MMP_refine_2.pdb and MT1-MMP_refine_2.mtz (including the phases from the PHENIX output), respectively. Set the contour levels (e.g., for the 2Fo-Fc map: 1 rmsd, for the Fo-Fc map: 3 rmsd).
13. Check the electron density maps and the refined model (Fig. 3). Modify the model to fit into the electron density. Save the coordinates as MT1-MMP_refine_2-coot1.pdb.
14. Repeat **steps 11–13** until the R-factor is converged.
15. Check the refinement statistics using POLYGON and the stereochemical geometry (Ramachandran plot) using MOLPROBITY in the output tab of the refinement in PHENIX.

4 Notes

1. The nutrient-rich microbial broth SOC instead of LB is recommended for transformation because of its refined glucose and salts balance which maximizes the transformation efficiency.
2. Care should be taken during the selection of a single colony from the plate to avoid cross contamination from the other colonies.
3. This step was repeated three times. Usually about 3–4 g of pellet was obtained per liter of media. Note that after resuspension of the pellet, the cells can either be directly extracted or frozen at −80 °C for up to a month before extraction without a significant degradation of the protein.
4. Alternatively, cells were also lysed using a high pressure Emulsiflex C5 homogenizer (Avestin, Mannheim, Germany).

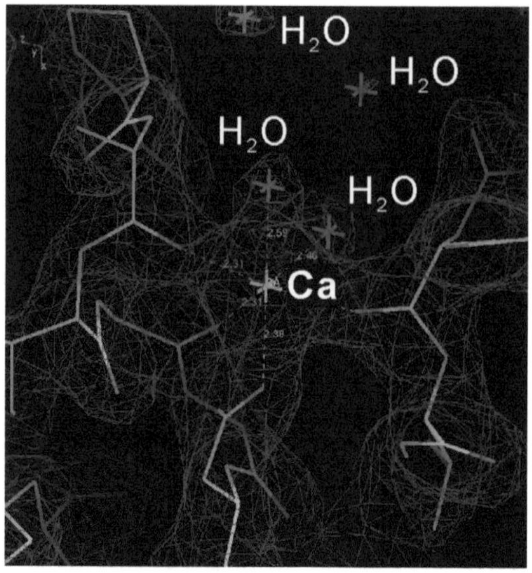

Fig. 3 MT1-MMP catalytic domain model generated using the program COOT. Snapshot showing the stick model for the MT1-MMP catalytic domain and the corresponding electron density map (2Fo-Fc map in blue mesh). The coordination of a calcium ion to various protein residues and water molecules is also illustrated

The cells were passed twice through a French press at 10,000–15,000 psi to ensure complete cell lysis.

5. Dilution is important to reduce steric hindrance by providing space for refolding to occur, thus allowing for more efficient protein refolding and decreased protein precipitation. The protein concentration was determined by measuring the absorbance at 280 nm using the calculated molar extinction coefficient of 35,410 M^{-1} cm^{-1} at 280 nm in water [18].
6. Following digestion of the hinge region, the protein can be frozen in liquid nitrogen after addition of glycerol (10% final concentration) and stored at −80 °C until use. Repeated freezing and thawing cycles should be avoided.
7. Mca is (7-methoxycoumarin-4-yl)-acetyl and Dnp is *N*-3-(2,4-dinitrophenyl)-l-2,3-diaminopropionyl. This is a FRET-based activity assay for MMPs in which peptide hydrolysis can be monitored by fluorescence spectroscopy [19, 20].
8. If a dispensing robot is not available, dispense each 80 μL aliquot manually.
9. The crystallization drop size can be increased up to 1.0 μL in the 96-well Intelli-plate. If a dispensing robot is not available, dispense each 0.5 μL aliquot manually. A 24-well "Cryschem Plate" (Hampton research) can be used to obtain larger crystals.

10. Optimize the concentration of the cryoprotectant. It is better to use a slightly higher concentration of salts and precipitants.
11. Wear gloves and goggles for protection when you use liquid nitrogen.
12. The cryo cap should be precooled in liquid nitrogen and the frozen crystal should be kept at liquid nitrogen temperature.
13. The dryshipper should be precooled with liquid nitrogen before use (at least 1 day before).
14. A graphical user interface, XDSAPP [21], can be used. It automates the data processing (XDS, SCALE, and XDSCONV) and generates the suitable reflection files (https://www.helmholtz-berlin.de/forschung/oe/em/soft-matter/forschung/bessy-mx/xdsapp/index_en.html).
15. The cif files can be found under the monomer library directory of PHENIX (e.g., /usr/local/phenix/chem_data/mon_lib/z/ZN.cif).

Acknowledgment

We thank Dr. Moran Grossman and Prof. Irit Sagi for providing the plasmid containing human MT1-MMP and Yvonne Brandenburger and Ingeborg Heise for technical assistance. This work was supported by the Cluster of Excellence RESOLV (grant No. EXC 1069) funded by the Deutsche Forschungsgemeinschaft and the Max Planck Society.

References

1. Tallant C, Marrero A, Gomis-Ruth FX (2010) Matrix metalloproteinases: Fold and function of their catalytic domains. Biochim Biophys Acta Molecular Cell Research 1803:20–28
2. Vandenbroucke RE, Libert C (2014) Is there new hope for therapeutic matrix metalloproteinase inhibition? Nat Rev Drug Discov 13:904–927
3. Gialeli C, Theocharis AD, Karamanos NK (2011) Roles of matrix metalloproteinases in cancer progression and their pharmacological targeting. FEBS J 278:16–27
4. Decaneto E et al (2015) Pressure and temperature effects on the activity and structure of the catalytic domain of human MT1-MMP. Biophys J 109:2371–2381
5. Grossman M et al (2011) Correlated structural kinetics and retarded solvent dynamics at the metalloprotease active site. Nat Struct Mol Biol 18:1102–1108
6. Bertini I et al (2012) The catalytic domain of MMP-1 studied through tagged lanthanides. Febs Lett 586:557–567
7. Pavlovsky AG et al (1999) X-ray structure of human stromelysin catalytic domain complexed with nonpeptide inhibitors: implications for inhibitor selectivity. Protein Science 8:1455–1462
8. Bertini I et al (2006) Snapshots of the reaction mechanism of matrix metalloproteinases. Angew Chem Int Ed 45:7952–7955
9. Grossman M et al (2010) The intrinsic protein flexibility of endogenous protease inhibitor TIMP-1 controls its binding interface and affects its function. Biochemistry 49:6184–6192

10. Fernandez-Catalan C (1998) Crystal structure of the complex formed by the membrane type 1-matrix metalloproteinase with the tissue inhibitor of metalloproteinases-2, the soluble progelatinase A receptor. EMBO J 18:5238–5248
11. Bertini I et al (2003) X-ray structures of binary and ternary enzyme-product-inhibitor complexes of matrix metalloproteinases. Angew Chem Int Ed 42:2673–2676
12. Ogata H et al (2014) Crystalization and preliminary X-ray crystallographic analysis of the catalytic domain of membrane type 1 matrix metalloproteinase. Acta Cryst F 70:232–235
13. Kabsch W (2010) XDS. Acta Cryst D66:125–132
14. Winn MD et al (2011) Overview of the CCP4 sute and current developments. Acta Cryst Sec D 67:235–242
15. Vagin AA, Teplyakov A (2010) An approach to multi-copy search in molecular replacement. Acta Cryst D 56:1622–1624
16. Adams PD, Afonine PV, Bunkoczi G et al (2010) PHENIX: a comprehensive Python-based system for macromolecular structure solution. Acta Cryst Sec D 66:213–221
17. Emsley P, Lohkamp B, Scott WG et al (2010) Features and development of coot. Acta Cryst Sec D 66:486–501
18. Wilkins MR et al (1999) Protein identification and analysis tools in the ExPASy server. Methods Mol Biol 112:531–552
19. Decaneto E et al (2015) A caged substrate peptide for matrix metalloproteinases. Photochem Photobiol Sci 14:300–307
20. Neumann U et al (2004) Characterization of Mca-Lys-Pro-Leu-Gly-Leu-Dpa-Ala-Arg-NH2, a fluorogenic substrate with increased specificity constants for collagenases and tumor necrosis factor converting enzyme. Analytical Biochemistry 328:166–173
21. Krug M, Weiss MS, Heinemann U, Mueller U (2012) XDSAPP: a graphical user interface for the convenient processing of diffraction data using XDS. J Appl Cryst 45:568–572

Chapter 5

Mapping Lipid Bilayer Recognition Sites of Metalloproteinases and Other Prospective Peripheral Membrane Proteins

Tara C. Marcink*, Rama K. Koppisetti*, Yan G. Fulcher*, and Steven R. Van Doren*

Abstract

Peripheral binding of proteins to lipid bilayers is critical not only in intracellular signaling but also in metalloproteinase shedding of signaling proteins from cell surfaces. Assessment of how proteins recognize fluid bilayers peripherally using crystallography or structure-based predictions has been important but incomplete. Assay of dynamic protein–bilayer interactions in solution has become feasible and reliable using paramagnetic NMR and site-directed fluor labeling. Details of preparations and assay protocols for these spectroscopic measurements of bilayer proximity or contact, respectively, are described.

Key words Protein–lipid interactions, Peripheral membrane protein, Metalloproteinase, NMR, Paramagnetic relaxation enhancement, Environment-sensitive fluorophore, Structure-based prediction

1 Introduction

Protein-mediated membrane-proximal events are critical in cell biology. The best characterized peripheral membrane interactions involve intracellular signaling proteins, and include many studies of proteins that bind phosphoinositides or head groups of phospholipids [1]. Fewer studies have examined proteins that bind peripherally to the cell surface and the extracellular leaflet of plasma membranes. The matrix metalloproteinases (MMPs) have important proteolytic targets on the cell surface such as receptors, growth factors, growth factor binding proteins, and cell adhesion molecules [2] while the ectodomains of some of these proteins may be shed from the cell surface by MMPs [2, 3]. Related ADAM

*These authors contributed equally to this work.

Charles A. Galea (ed.), *Matrix Metalloproteases: Methods and Protocols*, Methods in Molecular Biology, vol. 1579, DOI 10.1007/978-1-4939-6863-3_5, © Springer Science+Business Media LLC 2017

proteases (a disintegrin and metalloproteinase) are prominent as sheddases [3–5] while soluble MMP-2, -7, -8, -9, and -12 have been observed in compartments at or near cell surfaces [6–12] and may also be involved in cleavage of these cell surface proteins.

Paramagnetic NMR and fluorescence studies have revealed the dual modes of peripheral binding of MMP-12 and -7 to model membranes and cellular membranes [13, 14]. Comparison of four methods for mapping the interfaces of these proteins with membranes (and interfaces of other proteins under study), suggests the following relative reliability and priority of the mapping methods: paramagnetic NMR (Subheading 3.6) > site-directed fluor labeling (Subheading 3.8) > bilayer-induced shifts or broadening of NMR peaks (Subheading 3.6) ≥ predictions from structural coordinates (Subheading 3.1). The approaches developed for studying bilayer interactions for soluble MMP-7 and -12 [13, 14] should be applicable for other soluble MMPs and potential peripheral membrane-binding proteins. The methodology we propose for studying peripheral membrane interactions involves the following steps.

1. Anticipate potential sites for peripheral binding to membranes using recently reported predictive methods that use high-resolution structural coordinates for the protein of interest.
2. Isotopically label the protein and assign NMR spectral peaks.
3. Prepare spin-labeled membrane bilayer mimics.
4. Localize and quantify paramagnetic NMR line broadening resulting from proximity to the spin-labeled lipid.
5. Use site-directed fluor labeling (SDFL) to verify the NMR-mapped binding sites for recognition of (A) liposomes and (B) cells, as well as (C) effects of lipid composition.

The cornerstone of this strategy for accurately determining membrane binding sites and orientation is to measure strongly distance-dependent NMR line broadening emanating from mobile electron spin-labeled phospholipids placed in the membrane [13–16]. Our experience in mapping protein–protein and other protein–macromolecular interfaces is that paramagnetic NMR line broadening is superior in accuracy and interpretability to more conventional mapping of shifted NMR peaks onto the structure of the protein [17–23]. Paramagnetic relaxation enhancements (PREs) provide insightful assessment of dynamic, transient and light occupation of binding modes [24–28], including with bilayer-like partners [13, 14]. SDFL can be used to confirm modes of binding determined by NMR methods [13, 14]. Alternatively, other investigators have utilized site-directed mutagenesis to confirm these NMR-derived phospholipid binding interfaces [29].

Two types of disk-like mimics of lipid bilayers are promising for NMR studies in solution: (a) a class of small bicelles and (b) nanodiscs. Bilayered micelles or bicelles form disks containing an interior long-chain phospholipid bilayer circumscribed by an annulus of short-chain phospholipids [30]. Bicelles are well-suited to solution NMR studies of membrane-associated proteins [31–34]. The bicelles are discoidal, tumble rapidly and isotropically, and have a low critical micelle concentration (CMC) [33].

Nanodiscs are disks of lipid bilayers encircled by two belts of molecular scaffolding proteins, which define their size and planarity [35]. There are several reasons why nanodiscs are well suited for biophysical studies of membrane proteins. These include: (1) a diameter near 10 nm that is large enough to accommodate two interacting proteins, (2) homogeneity, (3) long-term stability, and (4) high fluidity at the high concentrations needed for NMR experiments [36]. A marked disadvantage for solution NMR of proteins embedded in nanodiscs is their high molecular weight, resulting in slow tumbling, rapid decay of NMR signals, and spectra with broad peaks [37–40]. This led to efforts to decrease their size by engineering shorter molecular scaffolding proteins which has in turn resulted in enhanced NMR spectra [40]. The peripheral binding of proteins to nanodiscs is transient or dynamic so their NMR spectra are usually less affected by slow tumbling associated with embedded membrane proteins. The solubility limit of nanodiscs is ~500 μM which constrains mixtures and titrations with proteins binding them peripherally to concentrations that are comparatively low for NMR studies. To summarize, nanodiscs provide a stabilizing environment for membrane proteins but this comes at the cost of weaker NMR signal intensity due to the slow tumbling of the protein/nanodisc complex.

2 Materials

2.1 Protein Expression and Purification

1. Beckman ultracentrifuge, and an ultracentrifuge rotor such as a Ti 70.
2. Bath sonicator (Laboratory Supplies Co., Hicksville, NY).
3. Sorvall or other centrifuge.
4. Cell homogenizer such as a motor-driven Potter-Elvehjem tissue grinder.
5. Bacterial cell disruptor such as a French press.
6. Low to medium-pressure chromatography system.
7. Column for gel permeation chromatography such as Superdex 200 10/300 GL.
8. Ni-NTA column.
9. Trace minerals [41] (5,000× stock).
10. MEM vitamin solution (100× stock; HyClone).

11. Minimal growth medium: 50 mM Na_2HPO_4, 50 mM KH_2PO_4, 5 mM Na_2SO_4, 2 mM $MgSO_4$, 50 mM NH_4HCl, 0.5% (w/v) glucose, and 0.2× trace metals (*see* **Note 1**).
12. Isopropyl β-D-1-thiogalactopyranoside (IPTG, Sigma-Aldrich).
13. Lysis buffer: 20 mM Tris–HCl, 10 mM EDTA, pH 7.5.
14. 1,2-dimyristoyl-*sn*-glycero-3-phosphocholine (DMPC) (Avanti Polar Lipids).
15. 1,2-diheptanoyl-*sn*-glycero-3-phosphocholine (D7PC) (Avanti Polar Lipids).
16. Dipalmitoylphosphatidylcholine (DPPC) variants with doxyl spin-label at 5, 10, or 14 position in one acyl chain (i.e., 1-pal mitoyl-2-stearoyl-(*X*-doxyl)-*sn*-glycero-3-phosphocholine, where *X* is 5, 10, or 14) (Sigma-Aldrich). Store lipids at 20 °C in a tightly sealed container.
17. Chloroform (ACS-grade) (Sigma-Aldrich).
18. Sodium cholate (Sigma-Aldrich).
19. Bio-Beads SM2 (BioRad).
20. NMR buffer for MMPs: 20 mM imidazole (pH 6.6), 10 mM $CaCl_2$, 20 μM $ZnCl_2$, 0.02% NaN_3, and 5% ^2H_2O.
21. Molecular scaffolding protein MSP1D1 (Sigma-Aldrich). *See* ref. 42 and Subheading 3.5, **step 1**.
22. MSP buffer: 20 mM Tris–HCl (pH 7.4), 0.1 M NaCl, 0.5 mM EDTA, 0.01% NaN_3.
23. TEV protease (Sigma-Aldrich).
24. Tris buffer: 20 mM Tris–HCl (pH 7.2).
25. TNC buffer: 20 mM Tris–HCl (pH 7.4), 150 mM NaCl, 10 mM $CaCl_2$.

2.2 Fluorescence Spectroscopy

1. Freshly prepared 10 mM *N*,*N*′-dimethyl-*N*-(iodoacetyl)-*N*′-(7-nitrobenz-2-oxa-1,3-diazol-4-yl) ethylenediamine (IANBD) (ThermoFisher).
2. Dimethylsulfoxide (DMSO; Sigma-Aldrich).
3. G-25 desalting resin (Sigma-Aldrich).
4. Dulbecco's modified Eagle's medium (DMEM) with 10% fetal bovine serum, supplemented with L-glutamine and nonessential amino acids (NEAA).
5. Cell lines (ATCC).
6. Biotek Synergy MX plate reader (*see* **Note 2**).
7. Zeiss LSM 510 Meta confocal microscope for imaging live cells [13, 14].

2.3 NMR Spectroscopy

1. NMR spectrometer operating at a ^{1}H frequency of 600 MHz or higher (*see* **Note 3**).
2. 5 mm NMR tubes (Norell) (*see* **Note 4**).

3 Methods

3.1 Navigating Servers for Predicting Protein Binding Peripherally to Membranes

Two excellent servers are available to predict the atomic structural coordinates of proteins that peripherally bind to membranes: Membrane Optimal Docking Area (MODA) [43] and Positioning of Proteins in Membranes (PPM) [44]. MODA is designed for predicting membrane binding sites of peripheral membrane proteins [43]. PPM is designed for predicting the positioning of both peripheral and integral membrane proteins with respect to bilayers [44]. PPM was preceded by the well-developed OPM database of predictions of positions and orientations of proteins in and on membranes [44–47]. The predictions may be used to determine whether it is worthwhile undertaking experimental testing of membrane interactions. For example, a comparison of a PPM prediction for the catalytic domain of MMP-12 with results of experimental docking studies using paramagnetic NMR [13] is shown in Fig. 1. The PPM prediction failed to predict the α-interface (Fig. 1a). However, it did predict one of the three experimentally determined loops (the II–III loop) of the β-interface [13]. This prediction bears resemblance to the measured orientation, but approaches the bilayer at a different angle (Fig. 1b). While MODA predicted residues for the other two experimentally defined loops (III–IV and IV–V loops) within the β-interface [13].

3.1.1 MODA and PPM Predictions

1. First obtain the PDB file containing the structural coordinates for the protein of interest from the Protein Data Bank (www.rcsb.org).
2. MODA predictions (http://molsoft.com/~eugene/moda/modamain.cgi).
   (a) Enter the PDB accession code or upload the PDB coordinate file. If the PDB file contains multiple chains, indicate which chain should be used for the prediction, e.g., A, B, C. Select "Predict" and wait for the results (*see* **Note 5**).
   (b) After the server identifies potential sites of membrane binding, it outputs a tabulated list of MODA scores for each of the amino acid residues in the sequence. The plain MODA score is scaled down by curvIndex to the more conservative curvMODA score.
   (c) If the curvMODA score for a residue is above 40 [48] and it has one or more neighbors (in sequence or space) with a high curvMODA score, that patch of residues is likely to

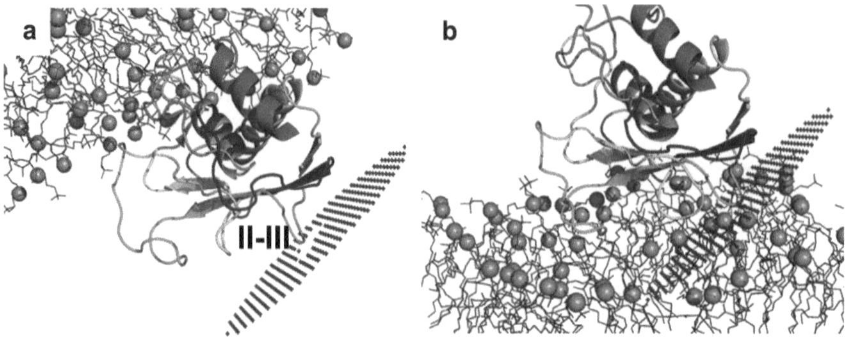

Fig. 1 Comparison of the PPM predicted and experimentally determined mode of bilayer binding by MMP-12. The protein backbone shown is used both to illustrate the experimental dockings and the predicted docking. The lipid chains plotted are from the experimental docking. The ***blue grid*** indicates the predicted location of bilayer head groups where the bilayer contacts the II–III loop of MMP-12. (**a**) The experimental structural model bound to the membrane via the α-interface (PDB: 2MLR, with lipid chains) is superposed with predicted interfaced marked by the *blue grid*. Molecules comprising the bilayer are shown and phosphorous atoms of the DMPC head groups are illustrated as *orange spheres*. The protein chain is *rainbow colored* from *blue* at the N-terminus to *red* at the C-terminus. (**b**) The experimental structural model bound to the membrane via the β-interface (PDB: 2MLS, with lipid chains) which includes not only the II–III loop but also the III–IV and IV–V loops, resulting in a different tilt angle than predicted by PPM (*blue grid*)

interact with membrane bilayers. Residues with curv-MODA score between 20 and 40 that cluster with other high-scoring residues possibly also interact with lipid bilayers.

(d) Download the spreadsheet readable text file in .CSV format. One should either manually mark high scoring residues on the protein structure with a molecular graphics software such as Pymol. Alternately, residues predicted to interact with membranes can be semi-automatically denoted by opening the ICB file in the program MolSoft.

3. PPM predictions (http://opm.phar.umich.edu/server.php).

(e) Upload the PDB file of interest.

(f) Specify the topology of the protein (N-terminus being in or out of the membrane) and whether you wish to include nonstandard residues or atoms in the prediction. Submit and wait for the results to appear on a fresh webpage.

(g) The calculated predictions include the depth that the protein penetrates into the membrane, the water-to-membrane transfer free energy (ΔGtrans in kcal/mol), and the tilt angle (°) of the protein in the membrane.

(h) To display the predicted model of the protein–membrane interaction select the "jmol" link under the heading "Image of the protein in membrane". This model is best viewed using a web browser that supports Java applets, such as Firefox.

(i) Download the PDB file generated for the predicted model to examine the hypothetical protein–membrane interface in a molecular graphics software package like PyMol or Chimera. The projected location of the membrane bilayer in the predicted protein–membrane model is defined by a grid of dummy atoms.

3.2 Preparation of E. coli Inclusion Bodies Harboring Recombinant Isotopically Labeled MMP (or Other Eukaryotic Protein) for NMR Studies

1. Prepare at least 100 ml of unlabeled PG medium [41] and at least 250 ml of PG medium using 99% D_2O in lieu of H_2O, as well as $^{15}NH_4Cl$ (Sigma-Isotec or Cambridge Isotope Laboratories) as the sole nitrogen source (*see* **Note 6**).
2. Transform competent *E. coli* BL21(DE3) Gold cells with the expression plasmid and plates 20–50 μl of the mixture onto a PG medium agar plate containing an appropriate antibiotic.
3. After incubating at 37 °C overnight, pick a few colonies from the plate and add to 1 ml of PG medium containing the antibiotic and incubate for 8 h at 37 °C on a shaker operating at 250 rpm.
4. Transfer the entire 1 ml of culture into 50 ml of PG medium containing antibiotics in a 250 ml flask. Incubate overnight at 37 °C in the shaker at 250 rpm.
5. Subculture 1–2% of the overnight culture into 250 ml of labeled PG medium containing antibiotic in a 2-l shake flask.
6. When A_{600} reaches 0.4–0.8, add 0.5 mM IPTG to induce protein expression and incubated overnight at 37 °C in a shaking incubator at 250 rpm.
7. Collect the bacterial cell pellet by centrifugation and store at −20 °C until needed.
8. Resuspend the cell pellet with lysis buffer and homogenize the suspension with a motor-driven Potter-Elvehjem tissue grinder.
9. Rupture the cell suspension, preferably by two slow passes through a French pressure cell, until the milky suspension darkens due to breakage of the cells.
10. Centrifuge at 20,000 × *g* for 40 min at 4 °C. Discard the supernatant as the MMP construct should be in the pellet in inclusion bodies. This pellet may optionally be washed by resuspension in lysis buffer containing 0.01% Triton X-100 and centrifuging again.
11. The inclusion bodies are dissolved in concentrated urea or guanidinium•HCl, centrifuged to remove insoluble debris, refolded, chromatographically purified, and concentrated in a centrifugal filter unit (Millipore). Optimal procedures for the

purification and refolding of the protein are specific to the construct and have been described for MMP-1 [49, 50], MMP-3 [51], MMP-7 [52], MMP-12 [21, 53, 54], MMP-13 [55], and MMP-14 [56, 57].

3.3 Assignment of NMR Spectral Peaks

1. Initially check to determine whether the assignments for the backbone amide NMR peaks are not available in the literature or in the BioMagResBank database (http://www.bmrb.wisc.edu/).
2. If de novo NMR peak assignments are required, prepare the $^{15}N/^{13}C$ labeled protein at a concentration of at least 150 μM using the procedure outlined in Subheading 3.2 (*see* **Notes 7** and **8**).
3. Acquire standard NMR spectra for spectral peak assignments (including HNCA, HN(CO)CA, CBCA(CO)NH and HNCACB triple resonance spectra) (*see* **Note 9**).
4. Semi-automated assignments for a majority of the backbone resonance peaks can be readily determined by non-specialists using the program PINE, a downloadable, integrated software environment which comes with a set of explanatory tutorials [58]. The software relies heavily upon NMR spectral analysis software NMRFAM-Sparky for visualizing the spectra [59] and PONDEROSA for automation of peak assignments [60, 61].
5. Assignments of methyl peaks for hydrophobic side chains are optional, but are desirable in order to provide additional distance constraints between protein residues and atoms within the lipid bilayer. This typically involves selective "ILV" labeling of isoleucine, leucine, and valine residues in a perdeuterated medium supplemented with α-ketoacids containing $^{1}H/^{13}C$–labeled methyl group(s) and optional deuteration [62] (*see* **Note 10**).
6. Methyl peaks can be assigned for smaller proteins using HMCM[CG]CBCA (HMCACB) [63] and CC(CO)NH or HNCACB triple resonance spectra, supplemented with NOESY-derived NOEs to the backbone [64] and methyl-amides [56]. This strategy works well for single domain proteins. To extend assignments of methyl peaks to multiple domain protein constructs such as MMPs, the assignments of methyl peaks from the single domains can simply be combined with NOEs from methyl groups in the larger construct [64, 65]. Software designed to aid in the determination of methyl peak assignments that utilizes additional experimental data such as structural coordinates, methyl-methyl NOEs, and PREs is also available [66].

3.4 Preparation of Small Bicelles for NMR Studies

Bilayered micelles or bicelles that are small, discoidal, and isotropic in tumbling are preferred for solution NMR [30, 32, 33, 67]. These are composed of 2–3 equivalents of detergent-like short chain lipids per equivalent of long chain phospholipid, which are designated $q = 0.5$ to $q = 0.33$ (ratio of long to short chains), respectively. Short chain lipids are amendable to forming curved surfaces, in this case, the rims of the bicelle disks [30] (Fig. 2b). Long chain phospholipids, such as DMPC, form the planar bilayers in the interior of the disks (Fig. 2b). Dihexanoylphosphatidylcholine (D6PC) with its 6-carbon acyl chains has dominated biophysical studies [30] but suffers the limitation of not aggregating into micelles or bicelles until reaching a critical micelle concentration (CMC) as high 14 mM in monomers [33]. D7PC with its 7-carbon acyl chains has the much lower and more favorable CMC of 1.2 mM allowing D7PC-DMPC bicelles ($q \leq 0.5$) to form at much lower lipid concentrations with only 1.2 mM monomers in solution [33]. Preparation of bicelles incorporating the latter and using our recommended formulation is outlined below.

1. Weigh 20 mg of the powder form of D7PC into a clean 5 ml glass tube and dissolve in 1.5 ml of chloroform.
2. Blow a gentle stream of argon or nitrogen gas into the glass tube for 30–45 min to evaporate the chloroform. Disturbance

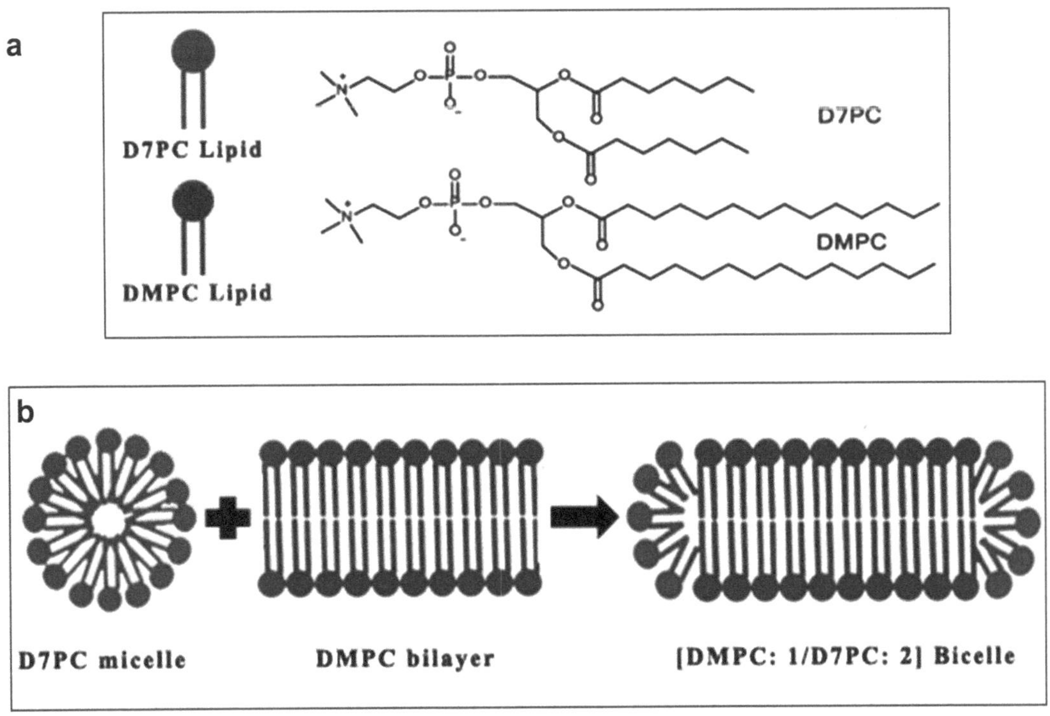

Fig. 2 Phosphatidylcholine molecules with 7- and 14-carbon acyl chains (**a**) form the annulus and center, respectively, of disk-like bicelles (**b**)

of the liquid by the nitrogen (or argon) stream should be slight and barely visible, with no splashing of liquid. Upon drying, a solid lipid film will form at the bottom of the glass tube. Dried lipids should be white in appearance. If the film appears glassy, it means that some residual chloroform remains in the lipid film.

3. Cover the glass tube with a piece of parafilm and pierce this several times. Place the covered tube in a jar containing desiccant or under vacuum for drying overnight.
4. To the dried D7PC, add 35 μl of NMR buffer. Cover with parafilm and hydrate at 42 °C for 1 h. Frequently and gently centrifuge using either a hand-cranked centrifuge or a microfuge on its slowest setting in order to collect the micelles from the sides of the tube.
5. Dissolve 20 mg of DMPC in 2 ml of Tris buffer. Hydrate at 42 °C for 2 h. Vortex occasionally to ensure an even suspension.
6. Place the glass tube in a bath sonicator. Adjust the glass tube so that the water barely covers the liposome dispersion.
7. Sonicate for 10 min (*see* **Note 11**). The solution will turn from milky white to translucent (*see* **Note 12**). Incubate at 42 °C for 1 h.
8. Ultracentrifuge the DMPC liposomes in a Ti 70 rotor for 1 h at ~64,000 × *g* and 20 °C. Decant the supernatant which contains the small unilamellar vesicles (SUVs), freeze and store at −80 °C for later use in fluorescence assays. The nearly white flocculent material at the bottom of the centrifuge tube contains large unilamellar vesicles (LUVs) and medium-sized unilamellar vesicles (MUVs) which will be used for preparing bicelles in the next step.
9. Resuspend the DMPC liposomes (LUVs and MUVs) with the solution containing the D7PC micelles. Incubate at 42 °C until an even suspension is achieved. Be careful to minimize frothing.
10. Plunge the suspension into liquid nitrogen and freeze. Thaw at 42 °C.
11. Transfer to a 500 μl microfuge tube and bring the volume up to 100 μl with NMR buffer. Freeze in liquid nitrogen again and thaw at 42 °C.
12. Transfer 25 μl aliquots to a fresh tube and freeze in liquid nitrogen. Store at −80 °C until needed.

3.5 Preparation of Nanodiscs for NMR or Fluorescence Assays

This is a synopsis of the previously reported protocol for preparing nanodiscs [42, 68, 69]. This protocol uses the MSP1D1 variant of the molecular scaffolding protein to encircle a DMPC phospholipid disc. Other lipids and MSP variants may also be assembled into nanodiscs.

1. MSP1D1 [42] may be purchased from Sigma-Aldrich or the His-tagged protein can be expressed from a plasmid (Addgene) and purified by affinity chromatography on Ni-NTA agarose in a buffer containing sodium cholate [68].
2. Pool fractions containing the His-tagged MSP1D1 and dialysis against MSP buffer.
3. Treat with TEV protease to release the His-tag. Remove the peptide containing the His tag by capturing it on the Ni-NTA column. Measure the concentration of the cleaved protein using a molar extinction coefficient (ε_{280}) of 18,200 M^{-1} cm^{-1} [69].
4. Dissolve sodium cholate in MSP1D1 or NMR buffer to make a solution that is 20–40 mM in sodium cholate monomers (*see* **Note 13**). Add this solution to 40 mg of dry DMPC at a molar ratio of two detergent molecules per phospholipid molecule. Vortex and heat in a water bath at 60 °C for 1 h. Then sonicate in a bath sonicator until clear.
5. Add the MSP1D1 to the cholate-solublized DMPC solution to achieve a final ratio of 1:80, while keeping the concentration of sodium cholate above 20 mM. Then mix (*see* **Note 14**).
6. Incubate with 0.5–0.8 g of BioBeads SM2 per ml of sample for 2–3 h to remove the detergent micelles.
7. Wash the BioBeads with an excess of MSP or NMR buffer in order to release more nanodiscs.
8. If turbidity or precipitation is present, pellet the particles by briefly spinning in a microcentrifuge. Inject the clarified solution of nanodiscs onto a gel filtration column such as Superdex 200 10/300 GL and run with a flow rate of ~0.5 ml/min. Collect 1 ml fractions. The elution profile for a preparation of nanodiscs composed of DMPC and MSP1D1 superimposed with that for several molecular weight standards is shown in Fig. 3.

3.6 NMR to Measure Proximity to Mimics of Membrane Bilayers

The main idea here is to estimate distances from protein amide and methyl groups to doxyl spin labeled DPPC, with around one or two such DPPC added per leaflet of the bilayer mimetic. PRE (paramagnetic relaxation enhancement) measurements from the unpaired electron of the doxyl group to the amide or methyl proton of the protein depends strongly (α r^{-6}) on the distance r between them and less strongly on the time constant τ_c of their rotational diffusion. The PREs are measured from the increased paramagnetic relaxation rate enhancement Γ_2 resulting from the addition of an unpaired electron.

1. Prepare a ^{15}N-^{2}H labeled protein sample according to Subheading 3.1.1 and concentrate to between 200 and 600 μM in NMR buffer in a volume of at least 400 μl, typically using a centrifugal membrane concentrator (Millipore or Vivaspin).

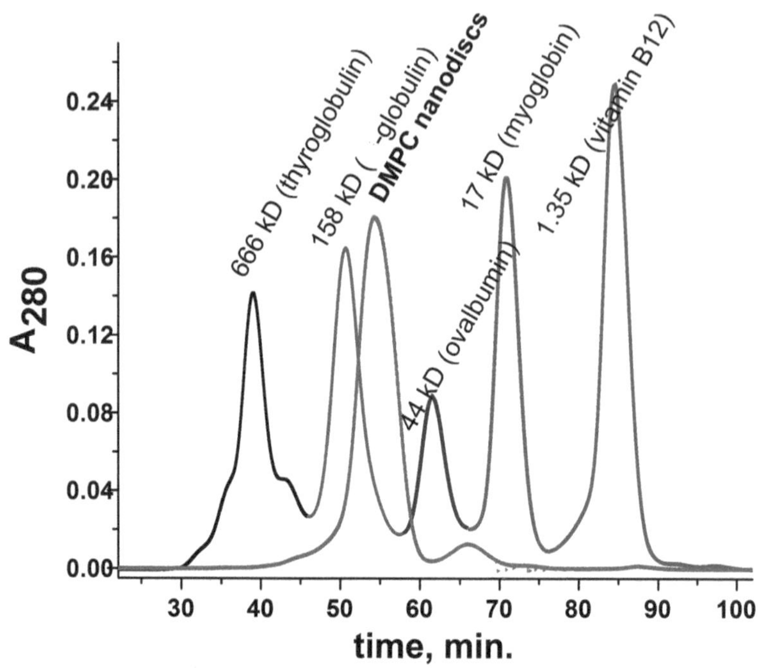

Fig. 3 Gel permeation chromatography of the assembled nanodiscs comprised of DMPC and MSP1D1. The elution profiles for several hydrodynamic standards are superimposed onto the chromatogram for the nanodiscs. The nanodiscs migrated with an apparent molecular weight that was less than that of γ-globulin (158 kDa) on a Superdex 200 10/300 GL column. The major peaks comprised uniformly assembled nanodiscs while a significantly smaller peak corresponding to incomplete assemblies was also observed

2. Titrate the protein sample with the bicelles, nanodiscs, or liposomes. Record ^{1}H-^{15} N TROSY NMR spectra of the protein amide spectral region at each lipid concentration in the titration. If protein methyl groups are specifically labeled, also collect ^{13}C HMQC spectra of the methyl spectral region at each concentration. Titrate the membrane mimic to at least half of the molar equivalents of the discoidal bilayer assembly per equivalent of protein, in order to provide at least one discoidal leaflet as a potential binding site for every protein molecule (expecting instances of two protein molecules per disc).
3. Continue the titration to a ratio of 1:1 and beyond, in order to maximize the number of assemblies harboring one protein bound per disc-like bilayer (*see* **Notes 15** and **16**).
4. Estimate the rotational correlation time τc to confirm bilayer-protein association and calibrate distance estimates. ^{15}N NMR relaxation indicative of τ_c is measured, preferably at each concentration point in the titration of Subheading 3.4, **step 1**

(*see* **Note 17**) Acquire a 1D relaxation series with at least five increments of constant-time relaxation using an appropriate pulse sequence for measuring ^{15}N NMR transverse dipole–dipole/CSA cross-correlated relaxation [70]. Set the longest relaxation period to attenuate peak heights to around 35–50% of the initial intensity.

5. Integrate the 1D spectral envelope at each relaxation time, while omitting the central region from 7.6 to 8.4 ppm which is enriched in signal from unstructured loops. An exponential relaxation rate constant is fitted to these integrals. This constant is known as the transverse cross-correlated relaxation rate, or ηxy.

6. Use ηxy to estimate τ_c. The relationship between ηxy and τ_c is given by the following equations of ref. 71:

$$\eta_{xy} = p\delta_N(4J(0) + 3J(\omega_N))(3\cos^2\theta - 1) \quad (1)$$

where p is the dipole–dipole coupling between the ^1H and ^{15}N of the amide group.

$$p = \mu_0\gamma_H\gamma_N h / (16\pi^2\sqrt{2}r_{HN}^3) \quad (2)$$

δ_N is the chemical shift anisotropy of the ^{15}N nucleus defined as

$$\delta_N = \gamma_N B_0 \Delta\delta_N / (3\sqrt{2}) \quad (3)$$

where γ_H are γ_N are the respective ^1H and ^{15}N gyromagnetic ratios, h is Planck's constant, r_{HN} is the distance between amide ^1H and ^{15}N nuclei, $\Delta\delta_N$ is the difference between the two main components of the ^{15}N chemical shift tensor, and $J(\omega)$ is the spectral density function that depends on frequency ω.

$$J(\omega) = 0.4\tau_C / [1 + (\omega\tau_C)^2] \quad (4)$$

These equations, the constants, and the ^{15}N resonance frequency of an 800 MHz spectrometer define the relationship of ηxy and τ_c as:

$$\eta_{xy} = 9.075e8 \times 0.4 \times \tau_c \times (4 + 3/(1 + \tau_c \times 5.0948e8)^2) \quad (5)$$

Using Eq. (5), ηxy is plotted versus τ_c and is almost linear. At the measured ηxy rate, the τ_c value corresponding to this point on the line is read from this plot. This estimate of τc represents the hydrodynamics. It can be monitored throughout the titration. An increase in τc during the titration indicates the binding of protein to the membrane and slower tumbling of the protein due to its association with the membrane.

7. Measure the ^{1}H NMR transverse (R_2) relaxation rate constants of the diamagnetic state of the lipid–protein mixture. (These exponential rate constants will be used as the reference values to which the distance-dependent Γ_2 values increase relaxation.) Use an NMR pulse sequence modified to include a PROJECT-CPMG train that suppresses proton-proton J-couplings [72]. For amide ^{1}H relaxation, either a ^{15}N TROSY sequence with a prepended PROJECT-CPMG train or an ^{15}N HSQC with the PROJECT-CPMG train appended are recommended [13, 14, 56]. For methyl ^{1}H relaxation, use a ^{13}C HMQC incorporating PROJECT-CPMG [13, 56]. Set the ^{1}H transmitter frequency offset to the water resonance peak near 4.7 ppm. Determine and set the ^{1}H 90° pulse length at high power. Attenuate the ^{1}H power level for the PROJECT-CPMG train by 3 dB and set its constituent 90° pulse widths $\sqrt{2}$-fold longer than the high power pulse width. Acquire trial 1D spectra as a series of 4 ms steps for the length of the CPMG train in order to monitor the degree of exponential reduction in overall peak height. Use this spectral series to determine the CPMG periods (multiples of 4 ms) to use for acquiring a series of 2D spectra. For relaxation of amide peaks, try CPMG periods from 0 to 32 ms. For methyl relaxation, try CPMG periods out to 40 or 48 ms. Choose a maximum CPMG period in which the spectrum has decayed to 35–50% of the intensity with a CPMG period of 0 ms.
8. Copy the optimized parameters into a number of 2D experiments with increasing the value of the CPMG period by a multiple of 4 ms for each successive experiment. Increase the number of transients acquired to achieve a $S/N \geq 20$ for the spectrum with the longest CPMG period. Measure the ^{1}H NMR transverse relaxation with a series of five or six relaxation delays [13, 14, 56].
9. To prepare the paramagnetic lipid DPPC (modified with the doxyl spin label at the 5, 10, or 14 position) dissolve 1 mg of doxyl-substituted DPPC in 250 μl of methanol to make a stock solution of 4.6 mM. Store at 4 °C.
10. Estimate the volume of this solution needed to incorporate an average of one (possibly two) doxyl-DPPC molecules per leaflet of the bicelles or nanodiscs. Transfer this volume of solution to a fresh tube and evaporate using a gentle stream of nitrogen or argon gas. Add the sample of labeled protein with bicelles or nanodiscs to the dried doxyl-DPPC and incubate for 10–15 min to incorporate the spin label in the bilayers (*see* **Note 18**). The incorporation of the doxyl-DPPC introduces its unpaired electron to the membrane bilayer. This broadens the ^{1}H NMR peaks of nearby hydrogen atoms. The breadth of the NMR peak is measured in the time domain as relaxation

rate constant R_2, which decays exponentially with the length of the CPMG time period used in the NMR pulse sequence. The increase in ^{1}H R_2 relaxation due to the presence of the unpaired electron is Γ_2 and results in the distance-dependent PRE. Γ_2 values may be measureable on protons on the protein as far as 25 Å from the doxyl group.

11. Repeat the series of CPMG experiments for the paramagnetic doxyl-DPPC incorporated membrane. Use the same buffer, temperature and membrane mimic to protein ratio as used for the diamagnetic sample (*see* **Notes 19** and **20**). Fit an exponential decay function and rate constant of decay to each resolved and confidently assigned peak. The exponential decay rate in the paramagnetic sample minus the exponential decay rate of the diamagnetic sample is equal to $\Gamma 2$, the rate constant of the paramagnetic relaxation for a given NMR resonance peak.
12. These experiments should be repeated using other DPPC analogues, where the doxyl label is incorporated at different positions along the acyl chain, in order to sample varying depths and magnitudes of PREs (Fig. 4). This provides additional information about the depth (or lack of) penetration of the protein into the membrane. These experiments provide additional distance constraints that can be used for structural modeling of the peripheral membrane binding.
13. Estimate the distance r from the mobile doxyl spin label to each amide or methyl resonance peak using the expression:

$$r = 4K\tau_c / \Gamma_2 \quad (6)$$

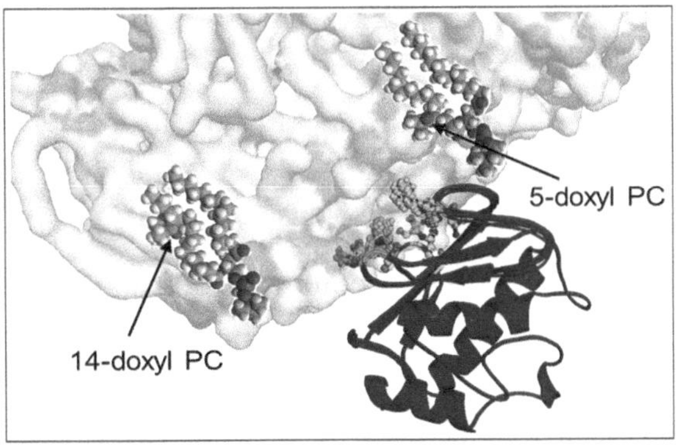

Fig. 4 Options for the location of doxyl substitution on phosphatidylcholine (PC) inserted into the membrane bilayer. Doxyl substitution at the 5-position is the closest choice in the acyl chain to the head group. It introduces more numerous PREs to the protein interacting peripherally. Doxyl substitution at the 14-position deep in the bilayer results in fewer PREs in the protein, but which are more likely to be central to the interface. The blue ribbon depicts MMP-12. Interfacial residues are plotted by ball-and-stick

Here $K = 1.23 \times 10^{-44}$ m^6 s^{-2} and τ_c is the rotational correlation time. These estimates can then be used as distance constraints in rigid body structural calculations using a docking program, such as HADDOCK [13, 29], to define areas of interaction between the protein and the membrane mimic, and in subsequent restrained molecular dynamics calculations. Computational details of the docking calculations are complex and beyond this chapter's scope, but have been introduced previously [13, 14].

3.7 Preparation of Liposomes for Fluorescence Assays

Fluorescence assays are recommended to confirm modes of bilayer binding determined by the paramagnetic NMR measurements described in Subheading 3.6 [13, 14]. These fluorescence assays can be undertaken using classic liposomes. Small unilamellar vesicles (SUVs) are recommended over larger vesicles in order to decrease light scattering that adds unwanted background to the detected fluorescence signal. Overall, the protocol outlined below involves preparing a stock suspension of SUVs of DMPC that is 10 mM in lipid monomers. SUVs may be also prepared from other lipids.

1. Dissolve 3.4 mg of DMPC (MW 677.94) in 500 μl of assay buffer (*see* **Note 21**).
2. Hydrate the lipids by placing the solution in a 42 °C water bath for 2 h. Vortex every 30 min in order to suspend the lipids.
3. Sonicate the suspension until it is clear. This should take 10–15 min. If it does not become clear or if the concentration of lipid is low, use freeze–thaw cycles in a plastic tube with liquid nitrogen as described for preparing bicelles in Subheading 3.4. That is, incubate the plastic tube for 3 min in liquid nitrogen followed by 6 min in a 42 °C water bath. Remove MUVs by spinning for 1–2 min in a microcentrifuge at 9300 × *g*. The SUVs can be stored in the 42 °C water bath. It is best to use them within one to 2 days of preparation, but they may be useable for up to a week. SUVs stored at 4 °C will form MUVs or aggregates (*see* **Notes 22** and **23**).

3.8 Site-Directed Fluor Labeling (SDFL) to Interrogate Binding to Liposomes

3.8.1 Preparation of Fluorescently Labeled Protein

A fluorophore that exhibits enhanced fluorescence upon insertion into lipid bilayers has proven useful to experimentally determine the site and orientation of interaction of peripheral proteins with membrane bilayers of both liposomes and cells [13, 14, 73, 74] (Fig. 5). We recommend using site-directed fluor labeling (SDFL) to confirm peripheral membrane binding sites determined by paramagnetic NMR. Alternately, SDFL can be utilized as the primary method for mapping the membrane–protein binding interface when it is not feasible to use paramagnetic NMR. The general approach for using SDFL to define the interaction of peripheral binding protein with membrane bilayers is outlined below:

1. To confirm membrane-binding sites identified by paramagnetic NMR, a number of single cysteine residue protein mutants need to be prepared containing cysteine residues located at positions corresponding to the interfacial binding site and one or more distant locations to serve as negative control(s) (Fig. 5) (*see* **Notes 24** and **25**). The substituted cysteine residues should be placed at a surface-accessible site, typically in a loop, inside or outside the proposed interface (*see* **Note 26**).
2. Reduce the single reactive Cys residue of the mutated protein (typically ≥1 μM) with at least a tenfold molar excess of thiol reductant, e.g., 1 mM β-mercaptoethanol (BME) (*see* **Note 27**).
3. Dialyze against TNC buffer to remove excess reductant (*see* **Note 28**).
4. To a solution containing 1 μM protein add a tenfold molar excess of IANBD (*see* **Note 29** and Fig. 6).
5. Let the conjugation reaction proceed for 2 h at room temperature or overnight at 4 °C. Perform this in an anaerobic chamber under argon.
6. Add excess BME or glutathione to consume excess thiol-reactive IANBD.

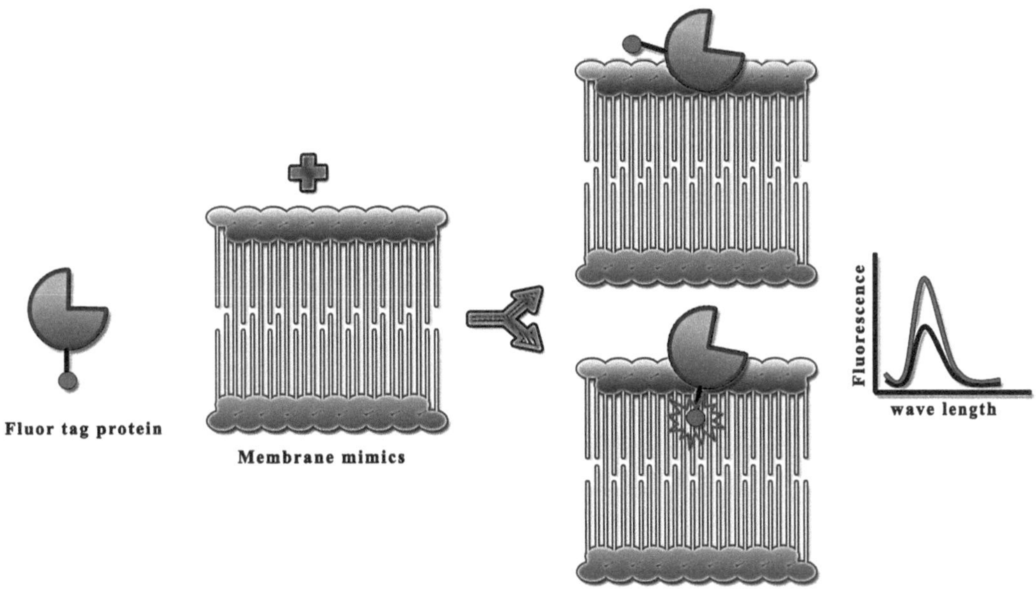

Fig. 5 The site-directed fluor labeling (SDFL) approach for defining peripheral membrane-binding sites for a protein. When the IANBD (*red*) conjugated to the protein (*blue*) inserts in a hydrophobic compartment such as the bilayer, its fluorescence emission increases

7. Concentrate the reaction mixture before loading onto a G-25 desalting column equilibrated in 10 column volumes of Tris buffer containing between 5 and 10 mM $CaCl_2$.
8. Collect the eluted conjugated protein and determine the extent of the conjugation reaction as follows. Determine the protein concentration using a protein assay such as a Bradford assay or absorbance at 280 nm. Determine the concentration of IANBD groups by measuring the maximum absorbance between 472 and 480 nm and using an extinction coefficient of 23,700 M^{-1} cm^{-1} [75]. The ratio of the concentration of protein to that of the fluorphore (IANBD) equals the apparent percentage labeling, which is typically ~85–105%. The excess over 100% could suggest incomplete removal of free IANBD.

3.8.2 Fluorescent Binding Assay

1. In a 96-well plate prepare a series of 200 μl solutions (in quadruplicate) containing 10–100 nM fluorophore-conjugated protein. Measure fluorescence at 541 nm using an excitation wavelength of 478 nm to establish sensitivity and fluorescence linearity (*see* **Note 30**).
2. Each 200 μl solution should contain 250–300 μM liposome monomers, typically prepared by diluting 5–6 μl from a 10 mM stock solution. Measure fluorescence intensity every 10–30 min for 2–3 h in order to allow the protein time to equilibrate (i.e., insert into the liposomes), before photobleaching takes a toll on emission from the IANBD. Normalize the fluorescence intensity in the presence of the liposomes (*FSUV*) by dividing it by the fluorescence in the absence of membranes (*F0*), (i.e., *F0/FSUV*). Bilayer insertion of the NBD moiety is accompanied by an increase in fluorescence intensity with $FSUV/F0 \geq 1.5$ [13, 73] (*see* **Note 31**).

3.8.3 Confocal Imaging of Binding to Live Cells

1. The IANBD-tagged protein may also be used to investigate interactions with cell membranes and their compartments. We have used human HeLa cells or Raw264.7 murine macrophages grown in suspension [13, 14], (Macrophages secrete

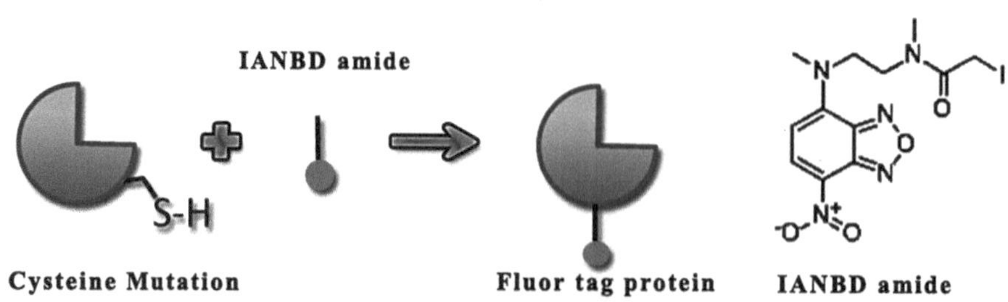

Fig. 6 Conjugation of the iodoacetamide-containing, environment-sensitive NBD fluorochrome (IANBD) to a single surface-exposed, reactive cysteine thiol

several MMPs). Like the assays with liposomes, the degree of association with cell membranes is quantified using the plate reader measurements of the fluorescence emission with cells present (*Fcells*) divided by the emission in their absence (i.e., *Fcells/F0*). Use of three to five million cells per well has been effective in our experiments [13].

2. The mammalian cell line is seeded on sterile MatTek glass bottom culture dishes and grown overnight at 37 °C in a 5% CO_2 incubator until reaching about 60–80% confluency.
3. The cells are washed three times with DMEM without phenol red and incubated with media containing a fluorescent staining reagents specific for the membrane compartment of interest and nuclei for 10 min at 37 °C in a 5% CO_2 incubator.
4. The cells are then washed three times with 20 mM Tris, pH 7.2, 5 mM $CaCl_2$, 0.1 mM ZnCl.
5. The culture dish is then placed on the stage of a confocal microscope (in our case a Leica TCP SP8 MP equipped with 405 nm and tunable white light lasers). The excitation/emission bandpass wavelengths used to detect IANBD, Hoechst 33,342, and Alexa Fluor 594 WGA are set to 472/485–525, 405/415–470, and 594/610–680 nm, respectively.
6. Confocal fluorescence images are taken immediately (control images). IANBD-labeled protein is then added to the dish and images are collected every 90 s for 15–30 min.

4 Notes

1. This minimal growth medium for high level expression in *E. coli* is Studier's non-inducing PG medium (P-0.5G) [41], with the addition of vitamins and use of an alternative choice of ammonium salts.
2. The authors used a Biotek Synergy MX plate reader, an instrument with the advantage of monochromator selection of excitation and detections wavelengths 478 and 541 nm, respectively. With a number of other plate readers, an appropriate set of filters are used instead on the excitation and emission channels.
3. A cryogenic probe is highly recommended for the sensitivity needed given the dilution and line broadening accompanying association with a slowly tumbling bilayer disc, and the quantitative nature of measurements of paramagnetic NMR relaxation. The authors have used a Bruker Avance III 800 MHz NMR spectrometer with TCI cryoprobe for proteins of 20

kDa and greater, and a Bruker Avance HD 600 MHz system with TCI/F cryoprobe for proteins less than 20 kDa.

4. Five millimeter NMR tubes are recommended for sufficient sample volume to accommodate the dilution of the protein sample upon mixing with a discoidal membrane mimic.
5. The option to incorporate electrostatics appears to be inconsequential. The option of clustering spreads the scores across more residues, without filling in the omissions of binding modes that we have observed anecdotally; we do not recommend clustering.
6. To prepare double-labeled protein for assignment of NMR chemical shift peaks also substitute 0.3% (w/v) $^{13}C_6$-glucose as the sole carbon source. In some circumstances (e.g., severe spectral overlap) it is possible to selectively label specific amino acids using α-keto acids or other precursors [62].
7. Some assignment projects may be especially challenging due to line broadening caused by the slow tumbling of the membrane-associated protein, low protein concentration, or overlapped peaks. In these cases, combinatorial or selective labeling strategies for partial assignment of specific amino acid residues or protein segments can be worthwhile [76–78]. If insect cells are required for high-level expression and proper folding, media for labelling in insect cells can now be obtained affordably by supplementing with $^{15}N/^{13}C/^{2}H$–labeled yeast extracts [79].
8. Proteins exceeding 20 kDa in mass should be fractionally deuterated to at least 60% (preferably 99%) by inclusion of D_2O in the growth medium [79] in order to achieve narrower line-widths and increased sensitivity [80].
9. For proteins greater than 20 kDa, TROSY versions of these and other triple resonance spectra are recommended [81, 82]. For proteins less than 20 kDa with sharper line widths, rapid acquisition BEST versions of triple resonance spectra are recommended [83, 84].
10. Selective labeling of the methyl groups of alanine, methionine, and threonine is also feasible [62, 85].
11. While using the bath sonicator, monitor the temperature of the water to ensure that it does not rise above about 40 °C. Turning the sonicator off for a while will allow it to cool toward room temperature.
12. A translucent, blue tinted solution is a clear sign of the formation of liposomes. If the suspension has not clarified, then liposomes have not yet formed adequately. Continue sonicating for a further 5–10 min.
13. This forms detergent micelles since the CMC of sodium cholate is 3 mM [86].
14. The optimal molar ratio of MSP1D1 to DMPC is 1:80.

15. The 1:1 assemblies should tumble faster and have sharper NMR peaks than 2:1 assemblies.
16. Note residues with NMR peaks broadened or shifted by addition of the membrane mimics. Such spectral perturbations suggest that the protein is binding to the membrane mimic and the general location of binding or conformational perturbations linked to binding.
17. This is performed before doping the bilayers with doxyl-substituted DPPC, which is paramagnetic.
18. We have had success with D7PC/DMPC ($q = 0.5$) bicelles at 300 μM discoidal aggregates, corresponding to 100 mM in total lipid monomers. These conditions require addition of the doxyl-DPPC to 600 μM for an average of one per leaflet.
19. The R_2 ^{1}H NMR relaxation rate constants fitted to the series should be insensitive to experimental choices of length of signal averaging, spectral window, and exact concentrations.
20. The R_2 values do depend on the hydrodynamics, which are sensitive to temperature and degree of molecular association, which should be kept uniform for making comparisons. Using similar protein and lipid concentrations for the diamagnetic (control) and paramagnetic experiments ensures similar degrees of molecular association are obtained in both experiments.
21. Minimize exposure to air while weighing the dry lipids. Weigh out a 10% excess of DMPC in order to obtain the desired concentration of SUVs since ~10% of the DMPC will form MUVs which will subsequently be removed by microcentrifugation.
22. Lipids dissolved in chloroform can be used to prepare the SUVs, provided an additional day is allowed to evaporate the chloroform and dry the lipids overnight under vacuum.
23. The aggregates could be sonicated to recover SUVs, but it is not recommended.
24. If one or more such cysteine already exists in the protein at an experimentally undesirable location, it should first be removed by site-directed mutagenesis.
25. If mapping by NMR (preferably paramagnetic NMR) is not available, the interfaces predicted by MODA and PPM should be regarded as the hypotheses to test, realizing that these algorithms have each missed one of the dual membrane-binding sites that we located experimentally [13, 14] (Fig. 1a).
26. Since basic and hydrophobic residues are frequently found within peripheral membrane interfaces, care should be taken not to substitute Arg, Lys, Phe, Tyr, Leu, and Ile residues within this region as this could disrupt binding at the protein–membrane interface.

27. We recommend performing these steps in an anaerobic or semi-anaerobic environment in order to prevent oxidation of thiols to disulfides. All buffers should be deoxygenated, e.g., by bubbling through argon and reactions should be carried out under an inert atmosphere of argon.
28. Constructs containing an MMP catalytic domain require 5–10 mM $CaCl_2$ to stabilize the catalytic domain.
29. Use a fresh 10 mM stock solution of IANBD.
30. Linearity of fluorescence deteriorates as the sum of the absorbance at the excitation and emission wavelengths exceeds 0.08 [87].
31. Examination of other lipids that are found in cell membranes is also recommended in order to explore biological specificity. These may include anionic lipids like phosphatidylserine [88], signaling lipids potentially relevant to your target protein (e.g., phosphoinositides) [1], unsaturated lipids, sterols, sphingholipids, glycolipids, etc.).
32. Our studies have been utilyzing the Image-It™ kit (Life Technologies) containing the Alexa Fluor 594 conjugate of wheat germ agglutinin (WGA) at 5 μg/ml to dye the glycosylation red on the outer face of the plasma membranes and 2 μM of Hoechst 33342 to label chromatin and nuclei blue.

Acknowledgment

This work was supported primarily by NIH grant GM057289, but also by NIH grant CA098799.

References

1. Moravcevic K, Oxley CL, Lemmon MA (2012) Conditional peripheral membrane proteins: facing up to limited specificity. Structure 20:15–27
2. Sternlicht MD, Werb Z (2001) How matrix metalloproteinases regulate cell behavior. Annu Rev Cell Dev Biol 17:463–516
3. Clark P (2014) Protease-mediated ectodomain shedding. Thorax 69:682–684
4. Rose-John S (2013) ADAM17, shedding, TACE as therapeutic targets. Pharmacol Res 71:19–22
5. Edwards DR, Handsley MM, Pennington CJ (2008) The ADAM metalloproteinases. Mol Aspects Med 29:258–289
6. Sato H, Takino T, Okada Y, Cao J, Shinagawa A, Yamamoto E, Seiki M (1994) A matrix metalloproteinase expressed on the surface of invasive tumour cells. Nature 370:61–65
7. Strongin AY, Collier I, Bannikov G, Marmer BL, Grant GA, Goldberg GI (1995) Mechanism of cell surface activation of 72-kDa type IV collagenase. Isolation of the activated form of the membrane metalloprotease. J Biol Chem 270:5331–5338
8. Yu WH, Woessner JF Jr, McNeish JD, Stamenkovic I (2002) CD44 anchors the assembly of matrilysin/MMP-7 with heparin-binding epidermal growth factor precursor and ErbB4 and regulates female reproductive organ remodeling. Genes Dev 16:307–323
9. Berton A, Selvais C, Lemoine P, Henriet P, Courtoy PJ, Marbaix E, Emonard H (2007) Binding of matrilysin-1 to human epithelial

cells promotes its activity. Cell Mol Life Sci 64:610–620

10. Owen CA, Hu Z, Lopez-Otin C, Shapiro SD (2004) Membrane-bound matrix metalloproteinase-8 on activated polymorphonuclear cells is a potent, tissue inhibitor of metalloproteinase-resistant collagenase and serpinase. J Immunol 172:7791–7803
11. Owen CA, Hu Z, Barrick B, Shapiro SD (2003) Inducible expression of tissue inhibitor of metalloproteinases–resistant matrix metalloproteinase-9 on the cell surface of neutrophils. Am J Respir Cell Mol Biol 29:283–294
12. Cobos-Correa A, Trojanek JB, Diemer S, Mall MA, Schultz C (2009) Membrane-bound FRET probe visualizes MMP12 activity in pulmonary inflammation. Nat Chem Biol 5:628–630
13. Koppisetti RK, Fulcher YG, Jurkevich A, Prior SH, Xu J, Lenoir M, Overduin M, Van Doren SR (2014) Ambidextrous binding of cell and membrane bilayers by soluble matrix metalloproteinase-12. Nat Commun 5:5552
14. Prior SH, Fulcher YG, Koppisetti RK, Jurkevich A, Van Doren SR (2015) Charge-triggered membrane insertion of matrix metalloproteinase-7, supporter of innate immunity and tumors. Structure 23:2099–2110
15. Kutateladze TG, Capelluto DG, Ferguson CG, Cheever ML, Kutateladze AG, Prestwich GD, Overduin M (2004) Multivalent mechanism of membrane insertion by the FYVE domain. J Biol Chem 279:3050–3057
16. Hilty C, Wider G, Fernandez C, Wuthrich K (2004) Membrane protein-lipid interactions in mixed micelles studied by NMR spectroscopy with the use of paramagnetic reagents. Chembiochem 5:467–473
17. Arumugam S, Van Doren SR (2003) Global orientation of bound MMP-3 and N-TIMP-1 in solution via residual dipolar couplings. Biochemistry 42:7950–7958
18. Arumugam S, Hemme CL, Yoshida N, Suzuki K, Nagase H, Berjanskii M, Wu B, Van Doren SR (1998) TIMP-1 contact sites and perturbations of stromelysin 1 mapped by NMR and a paramagnetic surface probe. Biochemistry 37:9650–9657
19. Takeda M, Terasawa H, Sakakura M, Yamaguchi Y, Kajiwara M, Kawashima H, Miyasaka M, Shimada I (2003) Hyaluronan recognition mode of CD44 revealed by cross-saturation and chemical shift perturbation experiments. J Biol Chem 278:43550–43555
20. Garimella R, Liu X, Qiao W, Liang X, Zuiderweg ER, Riley MI, Van Doren SR (2006) Hsc70 contacts helix III of the J domain from polyomavirus T antigens: addressing a dilemma in the chaperone hypothesis of how they release E2F from pRb. Biochemistry 45:6917–6929
21. Palmier MO, Fulcher YG, Bhaskaran R, Duong VQ, Fields GB, Van Doren SR (2010) NMR and bioinformatics discovery of exosites that tune metalloelastase specificity for solubilized elastin and collagen triple helices. J Biol Chem 285:30918–30930
22. Gobl C, Madl T, Simon B, Sattler M (2014) NMR approaches for structural analysis of multidomain proteins and complexes in solution. Prog Nucl Magn Reson Spectrosc 80:26–63
23. Hennig J, Warner LR, Simon B, Geerlof A, Mackereth CD, Sattler M (2015) Structural analysis of protein–RNA complexes in solution using NMR paramagnetic relaxation enhancements. In: Sarah AW, Frédéric HTA (eds) Methods in Enzymology. Academic, New York, NY, pp 333–362
24. Prior, S. H., Byrne, T. S., Tokmina-Roszyk, D., Fields, G. B., and Van Doren, S. R. (2016) Path to collagenolysis: collagen V triple-helix model bound productively and in encounters by matrix metalloproteinase-12. J Biol Chem 291:7888
25. Iwahara J, Clore GM (2006) Detecting transient intermediates in macromolecular binding by paramagnetic NMR. Nature 440:1227–1230
26. Fawzi NL, Doucleff M, Suh JY, Clore GM (2010) Mechanistic details of a protein-protein association pathway revealed by paramagnetic relaxation enhancement titration measurements. Proc Natl Acad Sci U S A 107:1379–1384
27. Schilder J, Ubbink M (2013) Formation of transient protein complexes. Curr Opin Struct Biol 23:911–918
28. Bashir Q, Volkov AN, Ullmann GM, Ubbink M (2010) Visualization of the encounter ensemble of the transient electron transfer complex of cytochrome c and cytochrome c peroxidase. J Am Chem Soc 132:241–247
29. Dancea F, Kami K, Overduin M (2008) Lipid interaction networks of peripheral membrane proteins revealed by data-driven micelle docking. Biophys J 94:515–524
30. Glover KJ, Whiles JA, Wu G, Yu N, Deems R, Struppe JO, Stark RE, Komives EA, Vold RR (2001) Structural evaluation of phospholipid bicelles for solution-state studies of membrane-associated biomolecules. Biophys J 81:2163–2171

31. Morrison EA, DeKoster GT, Dutta S, Vafabakhsh R, Clarkson MW, Bahl A, Kern D, Ha T, Henzler-Wildman KA (2012) Antiparallel EmrE exports drugs by exchanging between asymmetric structures. Nature 481:45–50
32. Liu Y, Kahn RA, Prestegard JH (2010) Dynamic structure of membrane-anchored Arf*GTP. Nat Struct Mol Biol 17:876–881
33. Lu Z, Van Horn WD, Chen J, Mathew S, Zent R, Sanders CR (2012) Bicelles at low concentrations. Mol Pharm 9:752–761
34. Song Y, Mittendorf KF, Lu Z, Sanders CR (2014) Impact of bilayer lipid composition on the structure and topology of the transmembrane amyloid precursor C99 protein. J Am Chem Soc 136:4093–4096
35. Schuler MA, Denisov IG, Sligar SG (2013) Nanodiscs as a new tool to examine lipid-protein interactions. Methods Mol Biol 974:415–433
36. Denisov IG, Sligar SG (2016) Nanodiscs for structural and functional studies of membrane proteins. Nat Struct Mol Biol 23:481–486
37. Raschle T, Hiller S, Yu TY, Rice AJ, Walz T, Wagner G (2009) Structural and functional characterization of the integral membrane protein VDAC-1 in lipid bilayer nanodiscs. J Am Chem Soc 131:17777–17779
38. Gluck JM, Wittlich M, Feuerstein S, Hoffmann S, Willbold D, Koenig BW (2009) Integral membrane proteins in nanodiscs can be studied by solution NMR spectroscopy. J Am Chem Soc 131:12060–12061
39. Yu TY, Raschle T, Hiller S, Wagner G (2012) Solution NMR spectroscopic characterization of human VDAC-2 in detergent micelles and lipid bilayer nanodiscs. Biochim Biophys Acta 1818:1562–1569
40. Hagn F, Etzkorn M, Raschle T, Wagner G (2013) Optimized phospholipid bilayer nanodiscs facilitate high-resolution structure determination of membrane proteins. J Am Chem Soc 135:1919–1925
41. Studier FW (2005) Protein production by auto-induction in high-density shaking cultures. Protein Expr Purif 41:207–234
42. Denisov IG, Grinkova YV, Lazarides AA, Sligar SG (2004) Directed self-assembly of monodisperse phospholipid bilayer Nanodiscs with controlled size. J Am Chem Soc 126:3477–3487
43. Kufareva I, Lenoir M, Dancea F, Sridhar P, Raush E, Bissig C, Gruenberg J, Abagyan R, Overduin M (2014) Discovery of novel membrane binding structures and functions. Biochem Cell Biol 92:555–563
44. Lomize MA, Pogozheva ID, Joo H, Mosberg HI, Lomize AL (2012) OPM database and PPM web server: resources for positioning of proteins in membranes. Nucleic Acids Res 40:D370–D376
45. Lomize MA, Lomize AL, Pogozheva ID, Mosberg HI (2006) OPM: orientations of proteins in membranes database. Bioinformatics 22:623–625
46. Lomize AL, Pogozheva ID, Lomize MA, Mosberg HI (2007) The role of hydrophobic interactions in positioning of peripheral proteins in membranes. BMC Struct Biol 7:44
47. Lomize AL, Pogozheva ID, Lomize MA, Mosberg HI (2006) Positioning of proteins in membranes: a computational approach. Protein Sci 15:1318–1333
48. Kufareva, I., and Overduin, M. (2015) http://www.slideshare.net/overduin/moda-slideshare.
49. Spurlino JC, Smallwood AM, Carlton DD, Banks TM, Vavra KJ, Johnson JS, Cook ER, Falvo J, Wahl RC, Pulvino TA et al (1994) 1.56 A structure of mature truncated human fibroblast collagenase. Proteins 19:98–109
50. Bertini I, Fragai M, Luchinat C, Melikian M, Mylonas E, Sarti N, Svergun DI (2009) Interdomain flexibility in full-length matrix metalloproteinase-1 (MMP-1). J Biol Chem 284:12821–12828
51. Ye Q-Z, Johnson LJ, Baragi V (1992) Gene synthesis and expression in *E. coli* for PUMP, a human matrix metalloproteinase. Biochem Biophys Res Commun 186:143–149
52. Fulcher YG, Sanganna Gari RR, Frey NC, Zhang F, Linhardt RJ, King GM, Van Doren SR (2014) Heparinoids activate a protease, secreted by mucosa and tumors, via tethering supplemented by allostery. ACS Chem Biol 9:957–966
53. Zheng X, Ou L, Tong X, Zhu J, Wu H (2007) Over-expression and refolding of isotopically labeled recombinant catalytic domain of human macrophage elastase (MMP-12) for NMR studies. Protein Expr Purif 56:160–166
54. Bertini I, Calderone V, Fragai M, Jaiswal R, Luchinat C, Melikian M, Mylonas E, Svergun DI (2008) Evidence of reciprocal reorientation of the catalytic and hemopexin-like domains of full-length MMP-12. J Am Chem Soc 130:7011–7021
55. Lovejoy B, Welch AR, Carr S, Luong C, Broka C, Hendricks RT, Campbell JA, Walker KA, Martin R, Van Wart H, Browner MF (1999) Crystal structures of MMP-1 and -13 reveal the structural basis for selectivity of collagenase inhibitors. Nat Struct Biol 6:217–221
56. Zhao Y, Marcink TC, Sanganna Gari RR, Marsh BP, King GM, Stawikowska R, Fields GB, Van Doren SR (2015) Transient collagen

triple helix binding to a key metalloproteinase in invasion and development. Structure 23:257–269
57. Udi Y, Fragai M, Grossman M, Mitternacht S, Arad-Yellin R, Calderone V, Melikian M, Toccafondi M, Berezovsky IN, Luchinat C, Sagi I (2013) Unraveling hidden regulatory sites in structurally homologous metalloproteases. J Mol Biol 425:2330–2346
58. Lee W, Cornilescu G, Dashti H, Eghbalnia HR, Tonelli M, Westler WM, Butcher SE, Henzler-Wildman KA, Markley JL (2016) Integrative NMR for biomolecular research. J Biomol NMR 64:307–332
59. Lee W, Tonelli M, Markley JL (2015) NMRFAM-SPARKY: enhanced software for biomolecular NMR spectroscopy. Bioinformatics 31:1325–1327
60. Lee W, Kim JH, Westler WM, Markley JL (2011) PONDEROSA, an automated 3D-NOESY peak picking program, enables automated protein structure determination. Bioinformatics 27:1727–1728
61. Lee W, Stark JL, Markley JL (2014) PONDEROSA-C/S: client-server based software package for automated protein 3D structure determination. J Biomol NMR 60:73–75
62. Ruschak AM, Kay LE (2010) Methyl groups as probes of supra-molecular structure, dynamics and function. J Biomol NMR 46:75–87
63. Tugarinov V, Kay LE (2003) Ile, Leu, and Val methyl assignments of the 723-residue malate synthase G using a new labeling strategy and novel NMR methods. J Am Chem Soc 125:13868–13878
64. Sprangers R, Kay LE (2007) Quantitative dynamics and binding studies of the 20S proteasome by NMR. Nature 445:618–622
65. Sinha K, Jen-Jacobson L, Rule GS (2013) Divide and conquer is always best: sensitivity of methyl correlation experiments. J Biomol NMR 56:331–335
66. Chao FA, Kim J, Xia Y, Milligan M, Rowe N, Veglia G (2014) FLAMEnGO 2.0: an enhanced fuzzy logic algorithm for structure-based assignment of methyl group resonances. J Magn Reson 245:17–23
67. Poget SF, Cahill SM, Girvin ME (2007) Isotropic bicelles stabilize the functional form of a small multidrug-resistance pump for NMR structural studies. J Am Chem Soc 129:2432–2433
68. Bayburt TH, Grinkova YV, Sligar SG (2002) Self-assembly of discoidal phospholipid bilayer nanoparticles with membrane scaffold proteins. Nano Lett 2:853–856
69. Bayburt, T. H., Dennisov, I. G., Grinkova, Y. V., and Sligar, S. G. Nanodisc technology: protocols for preparation of nanodiscs. University of Illinois, Urbana-Champaign, IL. http://sligarlab.life.uiuc.edu/nanodisc/protocols.html
70. Liu Y, Prestegard JH (2008) Direct measurement of dipole-dipole/CSA cross-correlated relaxation by a constant-time experiment. J Magn Reson 193:23–31
71. Lee D, Hilty C, Wider G, Wuthrich K (2006) Effective rotational correlation times of proteins from NMR relaxation interference. J Magn Reson 178:72–76
72. Aguilar JA, Nilsson M, Bodenhausen G, Morris GA (2012) Spin echo NMR spectra without J modulation. Chem Commun 48:811–813
73. Schulz TA, Choi MG, Raychaudhuri S, Mears JA, Ghirlando R, Hinshaw JE, Prinz WA (2009) Lipid-regulated sterol transfer between closely apposed membranes by oxysterol-binding protein homologues. J Cell Biol 187:889–903
74. Kim YE, Chen J, Chan JR, Langen R (2010) Engineering a polarity-sensitive biosensor for time-lapse imaging of apoptotic processes and degeneration. Nat Methods 7:67–73
75. Song Y, Hustedt EJ, Brandon S, Sanders CR (2013) Competition between homodimerization and cholesterol binding to the C99 domain of the amyloid precursor protein. Biochemistry 52:5051–5064
76. Löhr F, Tumulka F, Bock C, Abele R, Dötsch V (2015) An extended combinatorial ^{15}N, $^{13}C\alpha$, and $^{13}C'$ labeling approach to protein backbone resonance assignment. J Biomol NMR 62:263–279
77. Ikeya T, Takeda M, Yoshida H, Terauchi T, Jee JG, Kainosho M, Guntert P (2009) Automated NMR structure determination of stereo-array isotope labeled ubiquitin from minimal sets of spectra using the SAIL-FLYA system. J Biomol NMR 44:261–272
78. Hefke F, Bagaria A, Reckel S, Ullrich SJ, Dotsch V, Glaubitz C, Guntert P (2011) Optimization of amino acid type-specific ^{13}C and ^{15}N labeling for the backbone assignment of membrane proteins by solution- and solid-state NMR with the UPLABEL algorithm. J Biomol NMR 49:75–84
79. Opitz C, Isogai S, Grzesiek S (2015) An economic approach to efficient isotope labeling in insect cells using homemade ^{15}N-, ^{13}C- and ^{2}H-labeled yeast extracts. J Biomol NMR 62:373–385

80. Sattler M, Fesik SW (1996) Use of deuterium labeling in NMR: overcoming a sizeable problem. Structure 4:1245–1249
81. Tugarinov V, Muhandiram R, Ayed A, Kay LE (2002) Four-dimensional NMR spectroscopy of a 723-residue protein: chemical shift assignments and secondary structure of malate synthase g. J Am Chem Soc 124:10025–10035
82. Revington M, Zuiderweg ER (2004) TROSY-driven NMR backbone assignments of the 381-residue nucleotide-binding domain of the Thermus Thermophilus DnaK molecular chaperone. J Biomol NMR 30:113–114
83. Lescop E, Schanda P, Brutscher B (2007) A set of BEST triple-resonance experiments for time-optimized protein resonance assignment. J Magn Reson 187:163–169
84. Schanda P, Van Melckebeke H, Brutscher B (2006) Speeding up three-dimensional protein NMR experiments to a few minutes. J Am Chem Soc 128:9042–9043
85. Velyvis A, Ruschak AM, Kay LE (2012) An economical method for production of (2)H, (13)CH3-threonine for solution NMR studies of large protein complexes: application to the 670 kDa proteasome. PLoS One 7:e43725
86. Gennis RB (1989) Biomembranes: molecular structure and function. Springer, New York, NY
87. Palmier MO, Van Doren SR (2007) Rapid determination of enzyme kinetics from fluorescence: overcoming the inner filter effect. Anal Biochem 371:43–51
88. Li L, Shi X, Guo X, Li H, Xu C (2014) Ionic protein-lipid interaction at the plasma membrane: what can the charge do? Trends Biochem Sci 39:130–140

Chapter 6

Using Small Angle X-Ray Scattering (SAXS) to Characterize the Solution Conformation and Flexibility of Matrix Metalloproteinases (MMPs)

Louise E. Butt, Robert A. Holland, Nikul S. Khunti, Debra L. Quinn, and Andrew R. Pickford

Abstract

Small angle X-ray scattering (SAXS) provides information about the conformation and flexibility of proteins in solution, and hence provides complementary structural information to that obtained from X-ray crystallography and nuclear magnetic resonance spectroscopy. In this chapter, we describe the methods for the preparation of matrix metalloproteinase (MMP) samples for SAXS analyses, and for the acquisition, processing and interpretation of the SAXS data.

Key words Protein dynamics, Protein flexibility, Multi-domain, Oligomer, Structure–function relationship

1 Introduction

Matrix metalloproteinases (MMPs) are a family of zinc-dependent proteases that perform vital roles in essential processes such as embryogenesis, development, and wound healing [1]. Impaired regulation of MMP activity contributes to disease states such as aneurism, arthritis, cancer, and fibrosis [1]. Although primarily known for their degradation of extracellular matrix (ECM) components, MMPs are also involved in signaling through their liberation of ECM-bound growth factors and cytokines [2]. A more thorough understanding of the roles of MMPs in homeostasis and disease requires detailed knowledge of the relationship between their structure and function.

The consensus minimal architecture for an active MMP comprises a catalytic (CAT) domain—the location of the hydrolytic apparatus—connected to a hemopexin-like (HPX) domain by a polypeptide linker [3]. This holds true for all except MMP-7, which lacks a HPX domain. In many MMPs, the minimal

Charles A. Galea (ed.), *Matrix Metalloproteases: Methods and Protocols*, Methods in Molecular Biology, vol. 1579, DOI 10.1007/978-1-4939-6863-3_6, © Springer Science+Business Media LLC 2017

CAT-linker-HPX arrangement is supplemented by additional domains for substrate binding (e.g., the fibronectin type 2 domains in MMP-2 and -9), or cell surface localization, as seen in the membrane-type (MT) MMPs. The linker connecting the CAT and HPX domains varies in length from 14 residues in the prototypic collagenase MMP-1, to 68 residues in the gelatinase MMP-9. In their zymogenic proMMP forms, a small pro-peptide (PRO) domain docks into the active site of the CAT domain sterically blocking substrate access, and chemically blocking hydrolytic activity by chelation of the reactive Zn^{2+} ion with a Cys residue (the so-called "cysteine switch").

1.1 Structural Analyses of Intact MMPs

This domain architecture of MMPs makes them amenable to molecular dissection through recombinant expression of isolated domains for subsequent structural analysis by macromolecular X-ray crystallography (MX) and nuclear magnetic resonance (NMR) spectroscopy. To date, MX has been the predominant technique in such investigations, providing us with high-resolution structural information, not only of isolated MMP domains, but also of their complexes with small molecule inhibitors, TIMPs, and other MMPs (reviewed in [3]).

The first structure of an intact MMP, that of porcine MMP-1, was solved in 1995 [4]. It revealed a defined relative location and orientation of the HPX domain with respect to the CAT domain, despite the two being connected by what had been presumed to be a flexible linker. The belief that the domains interacted in a preferred manner was reinforced with the later MX structures of human MMP-1 [5], proMMP-1 [6] and proMMP-2 [7]. The recent MX structure of MMP-13 bound to peptide fragments of its own PRO domain revealed a similar arrangement, albeit with a 30° rotation of the HPX domain [8]. In each case, a consensus interface (nicknamed the "ball-and-socket" joint [9]) exists between the CAT and HPX domains; a cluster of three residues (the "ball") from the first blade of the HPX domain docks into a depression (the "socket") in the CAT domain surface comprised of residues from the Met-turn, the S1' wall forming segment and the specificity loop. The subsequent crystal structure of MMP-1 in complex with a collagen-like triple helical peptide suggested that this joint remains intact when the enzyme binds its substrate [10].

1.2 Flexibility in MMPs

Although these MX studies have revealed intricate intramolecular and intermolecular interactions with exquisite precision, we must keep in mind that the MMPs (or fragments thereof) are not in their native, solution state. Instead, the macromolecules are constrained within the crystal by protein–protein contacts with their neighbors, an enforced order that may restrict their dynamic behavior. Furthermore, the very process of crystallization requires the protein of interest to become more ordered, a more entropically challenging prospect for an inherently flexible macromolecule. In a solution of such a protein, there will be multiple conformations, and some of these will be more amenable to crystallization than others under a given set of conditions. Hence, those crystals that do form may be trapping a "snapshot" structure that is not fully representative of the ensemble population in the solution state.

It is well established that proteins are inherently flexible, that they exhibit conformational changes of varying magnitudes and on a variety of timescales, and that these molecular motions are inextricably linked with their function [11–13]. Therefore, in order to fully understand the structure–function relationship in (pro) MMPs, we must establish how their structure changes with time. In MX analyses, flexibility is often inferred from high B-factors or a lack of electron density for some stretches of the polypeptide. For example, the protease-sensitive "bait region" in the PRO domain of many MMP zymogens is believed to be flexible because its electron density is absent in the crystal structures of proMMP-1 [6], proMMP-3$^{\Delta HPX}$ [14] and proMMP-9$^{\Delta HPX}$ [15]. But does the presence of defined electron density in a crystal structure imply that the protein is rigid in solution?

In a landmark paper, Bertini and coworkers both solved the crystal structure of MMP-12, and characterized its solution conformation using small angle X-ray scattering (SAXS) and NMR [16]. The medium resolution MX study revealed an unambiguous and specific relative domain orientation quite different to the "ball-and-socket" joints seen in proMMP-1, MMP-1, and proMMP-2 (*see* Subheading 1.1, above); instead, in MMP-12, the fourth blade of the HPX domain contacts a point on the opposite side of the CAT domain to the above "socket". However, the authors showed that this MX structure is not fully representative of the enzyme in solution, as their data from both NMR and SAXS were inconsistent with that single static conformation. Furthermore, no single structure from a pool of possible MMP-12 conformations (with differing CAT and HPX domain locations/orientations generated by randomizing the linker connecting them) could, by itself, explain the SAXS data. Instead, using an ensemble optimization modeling (EOM) approach [17], the authors were able to select from that pool a family of conformers that, on average, did satisfy the SAXS data. Whilst approximately one-half of the conformers in solution were of similar size to the MX structure, the other half were significantly more extended [16].

These results on MMP-12, which reveal a far more dynamic solution conformation than the "snapshot" crystal structures would suggest, led to a reappraisal of the structural plasticity of MMP-1 using SAXS and NMR [18]. The combined study revealed that MMP-1 is far more dynamic in solution than the previous crystal structures suggest, with the population ensemble including extended conformers where the non-covalent interface between the

CAT and HPX domains is disrupted [18]. This transient dislocation of the "ball-and-socket" interface between the two domains may reveal a collagen binding exosite on the HPX domain that is otherwise concealed [9]. This structural flexibility in MMP-1 is important for function, as shown by a subsequent NMR study which revealed that the enzyme initially binds to a collagen-like triple helical peptide with its CAT-HPX interface dislocated, then condenses to the compact structure prior to collagenolysis [19, 20].

In contrast, separate SAXS analyses have shown that the crystal structure of proMMP-1 [6] is a good representation of the solution conformation [9]; the non-covalent PRO–HPX interface seen in the zymogen crystal structure appears to maintain a tightly compact domain arrangement in solution. This highlights our inability to predict the overall flexibility of multi-domain proteins purely from MX analyses alone, and emphasizes the importance of performing complementary studies on MMPs (and proMMPs) in solution by methods such as SAXS and/or NMR.

1.3 Small Angle X-Ray Scattering (SAXS)

In MX, diffraction of X-rays from well-ordered atoms in oriented molecules within the crystal lattice gives rise to discrete reflections of varying intensity on the two-dimensional (2D) detector; transformation of this diffraction pattern gives rise to the electron density map, which can precisely locate atomic positions within the protein. However, in small-angle X-ray scattering (SAXS), the randomized orientation of the isotropically tumbling molecules in the sample gives rise to a circularly symmetric pattern of elastically scattered X-rays [21, 22]. As with MX, a 2D detector is utilized, but the rotational symmetry means the SAXS data is, in effect, a one-dimensional profile, varying in intensity with angular deviation (2θ) out from the direction of the incident collimated X-ray beam (of wavelength λ). The resultant scattering profile, which is described as intensity $I(q)$ versus q, where $q = (4\pi \sin \theta)/\lambda$ (*see* **Note 1**), can be interpreted in terms of the protein conformation but the rotational averaging in solution means that both the information content and resolution (typically about 10 Å) are much lower than that obtained from MX [21, 22]. Nevertheless, SAXS provides an effective way not only of validating those protein structures solved by MX, but also of characterizing the mixed populations of the varying conformations that exist in solution.

1.4 Acquisition of SAXS Data

Protein solution SAXS is an inherently insensitive technique requiring a synchrotron X-ray radiation source and high concentrations of purified protein (typically 1-5 mg/mL) for satisfactory results [21–24]. Needless to say, protein solubility can be a limiting factor, and attempting SAXS at the limit of a protein's solubility is problematic since even minimal aggregation can cause disproportionately large distortions to the scattering profile. To alleviate this problem, many modern SAXS beamlines (e.g., B21 at Diamond

Light Source (DLS), BM29 at the European Synchrotron Radiation Facility (ESRF), and P12 at the Deutsches Elektronen-Synchrotron (DESY)) offer in-line size-exclusion chromatography (SEC) to allow removal of aggregated material immediately before the sample enters the X-ray beam. Typically, protein SAXS data is acquired over the range of q values from approximately 0.01–0.5 $Å^{-1}$, which corresponds to an interatomic spacing (d) range of 628–13 Å (since $d = 2\pi/q$). Even with a well-behaved, highly concentrated, monodisperse (i.e., non-aggregated) protein, most of the scattering arises from the buffer, and this contribution must be subtracted prior to interpretation of the protein signal. If more than one conformer is present in the sample, the scattering profile will be an average of their individual contributions, weighted by their relative abundance.

The high protein concentrations required for SAXS analyses render samples containing uninhibited CAT domains susceptible to self-hydrolysis (autolysis), most readily in the protease-susceptible CAT-HPX linker. Hence, studies on intact, mature MMPs (i.e., lacking the PRO domain) have utilized hydrolytically impaired enzymes in which the active site Glu residue is mutated to Ala (i.e., mutant E219A in both MMP-1 [9, 18, 20] and MMP-12 [16]). The equivalent mutation in MMP-7 reduces hydrolytic activity by approximately 99.9% [25]. In some studies, further autolytic inhibition is achieved through the addition of a hydroxamic acid-based inhibitor, for example *N*-isobutyl-*N*-[4-methoxyphenyl-sulfonyl] glycyl hydroxamic acid (NNGH), and by substitution of the active site Zn^{2+} ion by Cd^{2+} by exhaustive dialysis [26]. However, metal ion chelating agents, such as ethylenediaminetetraacetic acid (EDTA) or 1,10-phenanthroline should not be used, as this will strip Zn^{2+} and/or Ca^{2+} ions from the enzyme and thus disrupt its structure.

1.5 Interpretation of the Scattering Profile

Scattering at the lower end of the q range gives information about the size of the protein in solution, that is the apparent molecular weight (MW) and the approximate radius of gyration (R_g) [21–24]. The MW reveals the oligomeric state of the sample, and the R_g provides a quantitative measure of the protein's compactness. These parameters can be used to reveal any monomer-dimer transitions, for example as exhibited by the HPX domain of MMP-14 in solution [27]. Scattering at higher q values gives information about the molecular shape, albeit at low resolution compared to MX. A Kratky plot ($q^2I(q)$ vs. q), a mathematical manipulation of the scattering profile, can distinguish between folded, unfolded and flexible proteins.

An indirect Fourier transform of the reciprocal-space scattering profile ($I(q)$ vs. q) over a finite q range yields the real-space pairwise distance function ($P(r)$ vs. r), which is in effect a histogram of the possible interatomic distances (r) in the protein

[21–24]. From the $P(r)$, we can determine the maximum dimension of the protein (D_{max}) in solution, and a more accurate measurement of the R_g (often called the real-space R_g). A $P(r)$ distribution can be calculated from a particular atomic resolution structure (such as that derived from MX), but the reverse is not possible; the $P(r)$ distribution contains insufficient detail to be transformed into a unique atomic resolution structure. Previously solved MX/NMR structures can be validated by back-calculation of either reciprocal-space scattering data or real-space $P(r)$ from the atomic coordinates, and then comparing with the experimental SAXS data.

Methods for generating three-dimensional (3D) models from the SAXS data fall into two broad categories, namely volumetric modeling and atomistic modeling [21–24]. In volumetric (also known as ab initio) modeling, a molecular envelope is generated from the $P(r)$ distribution using dummy atoms (or "beads") of fixed volume with no assumptions made about the protein structure. Subsequently, a previously determined MX/NMR structure may be superimposed with this envelope by rigid body minimization. In the atomistic approach, all-atom models are generated that satisfy the SAXS data, and that use previously determined atomic resolution structures of the protein or fragments thereof (typically, individual domains). When studying flexible proteins, the scattering cannot be explained by a single atomistic structure, so an ensemble optimization modeling (EOM) approach [17] is commonplace.

1.6 Disclaimer

This chapter describes generalized methods for assessing the conformational state of (pro)MMPs in solution using SAXS. The strategies employed are adapted from those published previously [9, 16, 18, 20, 27]. However, they remain "first-shot" approaches, and may require some optimization (e.g., sample buffer conditions) in order to obtain the highest quality data. As with other structural methods, the quality of results will depend on the MMP (or proMMP) under study, and the purity of the sample. In recent years, the amount of available SAXS experiment time has grown with the expansion of dedicated synchrotron beamlines, and data acquisition is becoming more automated and routine. However, we strongly advise anyone interested in performing SAXS analyses to consult their local beamline scientist and identify any potential pitfalls prior to planning experiments and production of their MMP samples.

2 Materials

2.1 Preparation of Protein Samples and Buffers

1. *N*-isobutyl-*N*-[4methoxyphenylsulfonyl] glycyl hydroxamic acid (NNGH).
2. Acetohydroxamic acid (AHA).

3. Cadmium (II) chloride (toxic).
4. Bovine serum albumin (BSA).
5. HiLoad 26/60 Superdex 75 pg (GE Healthcare).
6. ÄKTAprime plus chromatography system (GE Healthcare).
7. Standard SAXS buffer: 50 mM Tris, pH 7.4, 150 mM NaCl, 10 mM $CaCl_2$, 0.02% (w/v) NaN_3.
8. Cd^{2+} SAXS buffer: 20 mM Tris, pH 7.2, 10 mM $CaCl_2$, 0.3 M NaCl, 0.2 M AHA, 0.3 mM $CdCl_2$.
9. Glycerol.
10. Ascorbic acid (vitamin C).
11. NanoDrop 2000 spectrophotometer (Thermo Scientific).
12. Refrigerated bench-top centrifuge.
13. Refrigerated microcentrifuge.
14. Vivacon-500 10,000 Da MW cutoff (MWCO) ultrafiltration device (Sartorius).
15. Vivacon-500 100,000 Da MW cutoff (MWCO) ultrafiltration device (Sartorius) (*see* **Note 2**).
16. Dialysis membrane (10,000 Da MWCO).

2.2 Data Acquisition Using the BioSAXS Robot

1. BioSAXS™ sample handling robot (Arinax).
2. BioSAXS™-compatible 96-well microplates (*see* **Note 3**).

2.3 Data Acquisition Using Inline SEC-SAXS

1. Agilent 1200 high-pressure liquid chromatography (HPLC) system.
2. Shodex KW402.5 4F (4.6 mm internal diameter × 300 mm length) HPLC SEC column (Showa Denko).

2.4 Processing of SAXS Data

1. Data reduction software (beamline-dependent).
2. *ATSAS* software package [28, 29], version 2.7.1 or later (www.embl-hamburg.de/biosaxs) (*see* **Note 4**).
3. *ScÅtter* software, version 2.3 or later (www.bioisis.net/scatter).
4. Access to the *EOM online* server (www.embl-hamburg.de/biosaxs/atsas-online/eom.php).

3 Methods

3.1 Procedures for Preparation of Protein Samples and Buffers

To date, all SAXS analyses on MMPs have been performed on recombinant proteins, expressed either from *E. coli* (proMMP-1 [9], MMP-1 [9, 18, 20], MMP-12 [16] and the HPX domain of MMP-14 [27]) or Sf9 insect cells (for proMMP-9 [30]). Recombinant expression more readily provides the substantial (milligram) amounts of protein required for SAXS analysis, and offers the potential of studying mutant enzymes. Use of higher eukaryotic host systems is required for native-like glycosylation on some MMPs, e.g., proMMP-9 [30]. Methodology for recombinant expression is beyond the scope of this chapter, but the interested reader is referred to suggested expert texts [31–33].

MMPs are generally amenable to SAXS analysis at near-neutral pH and physiological salt concentrations. Calcium is added to ensure occupancy of binding sites in the CAT and HPX domains. Addition of zinc is not considered necessary as it is retained by the folded enzymes with very high affinity (*see* **Note 5**). Extended exposure to X-rays damages proteins inducing aggregation and thus ruining the data. Addition of 3–5% (v/v) glycerol or 1–2 mM ascorbic acid to the buffer may reduce such damage (*see* **Note 6**).

If the (pro)MMP sample or the BSA MW standard contains aggregated protein then the interpretation of the scattering data will be extremely challenging. Therefore, we recommend using the following SEC scheme as the final step in sample purification prior to SAXS analysis. We use the same scheme regardless of whether the sample will subsequently be delivered into the beam by the BioSAXS™ robot (Subheading 3.2) or inline SEC (Subheading 3.3) (*see* **Note 7**). In addition, we always purify the BSA to be used as a MW standard in the same way as the (pro)MMP of interest.

1. Pre-rinse a 10,000 Da MWCO ultracentrifugation device with the buffer to be used for preparative SEC. The buffer composition will depend on the (pro)MMP under study.
2. Concentrate the protein using the rinsed device to a concentration of at least 5 mg/mL. Use multiple short spins (less than 3 min duration) to avoid excessively high protein concentration building up at the membrane surface (*see* **Note 8**).
3. Purify the protein by low-pressure SEC using the HiLoad 26/60 Superdex 75 pg column on an ÄKTAprime plus chromatography system (or equivalent) collecting 5 mL fractions across the elution range of void volume (v_0) to total volume (v_t) (*see* **Note 9**).
4. Assess the purity of the fractions by sodium dodecyl sulfate (SDS) polyacrylamide electrophoresis (PAGE), and pool the desired fractions containing the (pro)MMP of interest.
5. If required to prevent autolysis, add equimolar NNGH and/or an excess of AHA to the SEC-purified (pro)MMP sample and incubate for 1 h at 4 °C to equilibrate.

3.2 Procedure for Data Acquisition Using the BioSAXS Robot

1. Dialyse the sample overnight at 4 °C against the SAXS buffer. The precise buffer used will depend on the (pro)MMP under study. Do not discard the dialysate (*see* **Note 10**).
2. Clarify the dialysate using a 0.2 μm filter and keep as the SAXS reference buffer.

3. Measure the protein concentration of the dialysed sample by spectrophotometry using the Nanodrop 2000. If the protein concentration is below 5 mg/mL (or the solubility limit, if that is lower), concentrate the sample by ultrafiltration as in Subheading 3.1, **step 1** above (*see* **Note 11**).
4. Centrifuge the sample in the refrigerated bench-top microfuge for 5 min at 15,000 × *g* and 4 °C. Retain the supernatant.
5. Check that the sample is monodisperse. A variety of methods can be used for this, for example native PAGE, multi-angle laser light scattering or dynamic light scattering (*see* **Note 12**). Each of these techniques is beyond the scope of this chapter.
6. Prepare suitable dilutions of the concentrated stock sample using the clarified SAXS reference buffer. Typical protein concentrations for SAXS are 1.0, 2.0, and 5.0 mg/mL. Precise measurements of the protein concentration are needed (*see* **Note 13**) for accurate determination of MW from the forward scattering $I(0)$ (*see* Subheading 3.4.4).
7. Centrifuge the sample again, as in **step 4**.
8. Load 30 μL of each concentration of the sample and an excess of buffer reference into appropriate wells of the BioSAXS™-compatible microplate. The samples should be analysed in the order *buffer-sample-buffer*, so the volume of buffer used will be twice the total volume of sample.
9. Program the BioSAXS™ robot to deliver the sample into the X-ray beam once the shutter is open, collecting multiple tandem 10-s exposures (frames) of raw scattering data with the buffer/sample under flow (*see* **Note 14**).

3.3 Procedure for Data Acquisition Using SEC-SAXS

In SEC-SAXS, as the sample (or BSA standard) passes through the SEC column it becomes more dilute resulting in a lower scattering signal. Hence, the initial concentration required is slightly higher than that for the BioSAXS™ robot.

1. Measure the protein concentration of the dialysed sample by spectrophotometry using the Nanodrop 2000. If the concentration is below 7 mg/mL (or the solubility limit, if that is lower), concentrate the sample by ultrafiltration as in Subheading 3.1, **step 1** above.
2. Prior to the first analysis, pass at least five column volumes of SAXS buffer through the SEC column to ensure that it is thoroughly equilibrated (*see* **Note 9**).
3. Program the HPLC to inject 40-50 μL of (pro)MMP sample onto the SEC column and into the X-ray beam once the shutter is open. Start monitoring the A_{280} with the sample running onto the column at a flowrate of 0.16 mL/min, and begin acquiring tandem 3-s exposures (frames) of raw scattering data.

4. Use selected frames recorded prior to elution of the column void volume (v_0) as the reference frames for buffer subtraction.
5. Stop recording the A_{280} and scattering data once a complete column volume (v_t) of buffer has passed through the column.
6. With the recommended setup, each SEC-SAXS run will take approximately 30 min. Prepare the next protein sample for analysis during this time.

3.4 Procedures for Processing of SAXS Data

3.4.1 Data Normalization and Reduction

Parameters for data normalization and reduction will be beamline dependent, and some (e.g., beam center position) may need calibrating. Data reduction and normalization is usually performed at the synchrotron beamline using bespoke software (i.e., neither *ScÅtter* nor the *ATSAS* package); in many cases it will be an automated step with no user input necessary. If unsure, consult your local beamline scientist for assistance and advice with the following steps.

1. Apply a mask to all frames of interest (both buffer and sample) to eliminate those pixels behind the beamstop (and any supporting structures), and any pixels from inactive areas of the detector.
2. Radially average the datasets through an appropriate sector (typically between 60° and 120°) to reduce the 2D SAXS images to one-dimensional dataset (intensity vs. radial distance from beam center).
3. Normalize this dataset using the camera length (the distance from sample to detector) and the intensity of the transmitted beam (measured at the beamstop diode) to obtain the scattering curve ($I(q)$ vs. q).

3.4.2 Buffer Subtraction and Data Quality

Data reduction provides us with a one-dimensional profile of the scattering from both protein and buffer [21–24]. For both the (pro)MMP sample and the BSA standard, we obtain the protein's scattering profile by subtracting the contribution from the buffer (Fig. 1). Buffer subtraction and inspection of the data quality can be performed in either *PRIMUS* [34] or *ScÅtter*.

1. Check for consistency of the multiple buffer frames (either the pre- and post-sample buffers if using the BioSAXS™ robot setup, or the pre-void volume buffer if using SEC-SAXS) by overlaying their scattering profiles.
2. Average those buffer frames showing consistent scattering for use in buffer subtraction.
3. Subtract this buffer scattering profile from each of the sample frames in order to visualize the scattering from just the protein of interest.
4. If SEC-SAXS was not used, every buffer-subtracted sample frame must be checked for signs of aggregation (*see* **Note 15**). In each case, examine the low-q^2 ($<\sim 0.005$) region of the

Guinier plot of ln($I(q)$) vs. q^2. A nonlinear relationship between ln ($I(q)$) and q^2 with upward deviation at low q^2 (i.e., a positive second derivative) indicates the presence of aggregation which may have been X-ray induced.

5. Do not continue analysing any frames for which aggregation is suspected, since scattering from aggregates strongly influences the entire dataset.
6. Examine the data for signs of inter-particle repulsion; this is indicated by a downward deviation from linearity in the Guinier plot at low q^2 (i.e., a negative second derivative). Inter-particle repulsion may be alleviated by using more dilute samples or increasing the ionic strength of the buffer.
7. For a given sample, combine all good quality frames by averaging the scattering ($I(q)$) data over the entire q-range. If using SEC-SAXS, average those frames which correspond to an absorbance (A_{280}) greater than or equal to half the maximum peak intensity (*see* **Note 16**). If the BioSAXS™ robot setup was used, average those frames for a given sample concentration that show no obvious signs of aggregation.
8. Where multiple sample concentrations have been analysed using the BioSAXS™ robot, a single scattering profile should be generated by merging the data sets (*see* **Note 17**). Using a concentration-weighted average, combine the data at low angle (e.g., $q < 0.15$ Å^{-1}) from the lowest protein concentration (e.g., 1.0 mg/mL), and the data at higher angle (e.g., $q >$ 0.10 Å^{-1}) from the highest protein concentration (e.g., 5.0 mg/mL). Visually check for consistency between the two data sets in the overlapping region (i.e., 0.10 Å^{-1} < q < 0.15 Å^{-1}) before merging; a systematic error in this range is indicative of inaccurate protein concentration measurement.

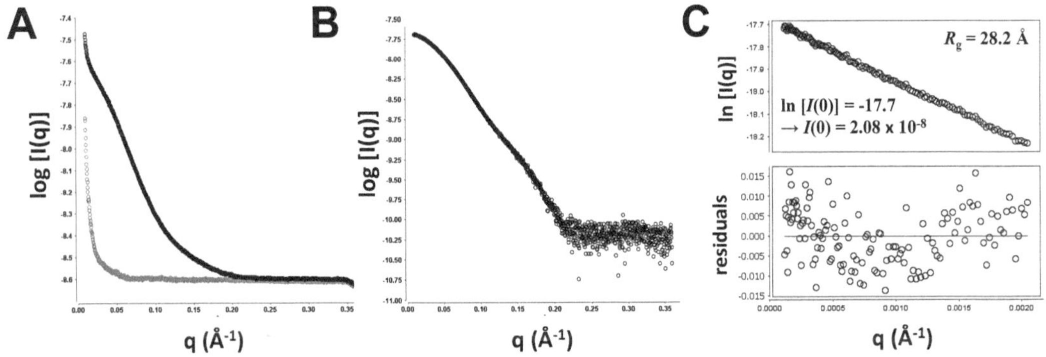

Fig. 1 Buffer subtraction and Guinier analysis. SAXS data were acquired on proMMP-1 at DLS beamline B21. (**a**) Overlay of buffer-only (*grey*) and proMMP-1 (*black*) scattering profiles; (**b**) Buffer-subtracted scattering profile for proMMP-1; (**c**) Guinier analysis of proMMP-1 (*above*) with residuals (*below*), showing the reciprocal-space approximation to R_g and extrapolation of $I(0)$. Graphs were produced using *ScÅtter*

3.4.3 Determination of Forward Scattering, I(0), and Reciprocal-Space $\mathbf{R}_g$ *by Guinier Analysis*

In the absence of aggregation, the forward scattering, $I(0)$, is one of the most accurate parameters that can be derived from SAXS data [21–24]. However, $I(0)$ is not measured directly (the beamstop prevents us from doing so), but instead must be obtained by backward extrapolation of the scattering profile to zero q. Guinier analysis (i.e., inspecting $\ln(I(q))$ vs. q^2 at low q^2) is a convenient way of obtaining $I(0)$. The same process yields the reciprocal-space R_g, an excellent approximation of the real-space R_g (*see* Subheading 3.4.5). For globular proteins, Guinier approximation of R_g is only robust when utilizing the range of q-values where $q.R_g \leq 1.3$ [21–24]. Interactive Guinier analysis can be performed in either *PRIMUS* [34] or *ScÅtter*.

1. Initiate an interactive Guinier analysis on the buffer-subtracted scattering profile (either the sample or BSA standard).
2. Truncate the low-q^2 datapoints where interference from the beamstop is suspected (*see* **Note 18**).
3. Truncate the high-q^2 datapoints until $q.R_g \leq 1.3$.
4. Note the q^2-range used, the resultant reciprocal-space R_g and the reverse-extrapolated $I(0)$ value which can used for determination of the apparent MW (*see* Subheading 3.3, **step 4**).

3.4.4 Determination of Apparent MW

The forward scattering, $I(0)$, has been shown to scale linearly with MW if one assumes a constant protein density [21–24]. It may be obtained from the Guinier region (*see* Subheading 3.4.3) or the $P(r)$ distribution (*see* Subheading 3.4.5). If using the *ATSAS* suite, use *PRIMUS* [34] for Guinier analysis, and *GNOM* [35] for $P(r)$ generation. *ScÅtter* can perform both of these functions.

1. Determine the $I(0)$ value for each concentration of BSA. The $I(0)$ value should scale linearly with BSA concentration. Choose one pair of BSA concentration and resultant $I(0)$ for the calculation of MW in **step 3** below.
2. Determine the $I(0)$ value for each concentration of the (pro) MMP sample.
3. For each sample concentration, calculate the apparent MW from the concentration and forward scattering using the following equation:

$$MW_{(pro)MMP} = MW_{BSA}\frac{I(0)_{(pro)MMP} / [(pro)MMP]}{I(0)_{BSA} / [BSA]}$$

The MW of BSA is 66,463 Da. All concentrations are in mg/mL.

4. If the apparent MW increases with increasing concentration, it indicates either aggregation (although this should be apparent from the nonlinearity of the Guinier plot—*see* Subheading 3.4.2) or changes in the multimerization state of the protein.

A similar increase in R_g value will also usually be observed if this is the case.

*3.4.5 Generating the **P**(r) Distribution*

The real-space pairwise distance function, $P(r)$, is generated by indirect Fourier transformation of the reciprocal-space scattering profile over a finite q range [21–24]. From the $P(r)$, we can deduce the D_{max} (the r where $P(r)$ returns to zero) and a more accurate (real-space) measurement of the R_g than that obtained from Guinier approximation (*see* Subheading 3.4.3). In addition, the $P(r)$ profile gives an indication of the global shape of the protein in solution, and can be used for volumetric modeling (*see* Subheading 3.4.8). D_{max} is actually a "soft" parameter that must be iteratively optimized by the user in the original *GNOM* [35] software, or by a recursive algorithm in the more recent *AUTOGNOM* program [28], and in *ScÅtter*. The following instructions apply to *GNOM* and *ScÅtter*:

1. Load your buffer-subtracted, merged scattering data (*see* Subheading 3.4.2) into the program.
2. Specify the desired data range, excluding data at low q that may be distorted by the beamstop, and data at high q that are excessively noisy.
3. Begin the iterative $P(r)$ analysis, starting with a D_{max} of approximately three times the reciprocal-space R_g that was obtained from Guinier analysis (*see* Subheading 3.4.3).
4. Optimize the $P(r)$ distribution by changing the D_{max} value. The $P(r)$ distribution should decay smoothly to the x-axis towards D_{max} (Fig. 2). If D_{max} is set too low, there will be a sharp drop in $P(r)$ as r approaches D_{max}. If the function drops below zero, then D_{max} has been set too high.
5. Note down the values of the real-space R_g and the forward-scattering, $I(0)$. Discrepancies between these values and those calculated from Guinier analysis (Subheading 3.4.3) typically arise from small amounts of aggregation that have a greater impact on the low-q data.
6. Assess the profile of the $P(r)$ distribution which informs the shape of the protein in solution; a single, near-symmetrical peak is indicative of a single globular domain, whilst multi-domain proteins give rise to multiple peaks or a shoulder on the main peak.

3.4.6 Back-Calculation of Scattering Data from Atomic Coordinates

Known high-resolution atomic structures can be compared to experimental intensity data using *CRYSOL* [36], a program which generates theoretical solution scattering profiles that can be subsequently fitted to experimental scattering data. Importantly, these simulated profiles are created using spherical harmonic reconstructions that take into account the hydration shell. Fitting is achieved by minimization of the χ^2 discrepancy between the back-calculated and experimental scattering profiles through varying the following parameters: (1) average displaced solvent volume per atomic group, (2) contrast of the hydration shell, and (3) relative background. Familiarity with command-line arguments is strongly advised, but the program may also be run in interactive mode allowing the user to easily define option parameters. The default values for these parameters are usually sufficient but can be adjusted accordingly by the user, if deemed appropriate (*see* the *CRYSOL* manual for further details).

1. Enter the required Protein Data Bank (PDB) file as the initial input file. *CRYSOL* will read the atomic coordinates and evaluate scattering profiles for all models and chains present, unless otherwise specified.
2. If comparison with experimental data is required, specify the buffer-subtracted experimental scattering data (*see* Subheading 3.4.2), and select the appropriate angular units (Å^{-1} or nm^{-1}) for this data.
3. After execution, multiple output files are generated (*see* **Note 19**). The file with extension ".log" contains the experimental and theoretical R_g values. The fit to the experimental scattering data and the χ^2 value are located in the file with extension ".fit".
4. Inspect the χ^2 values; low values indicate good agreement between the known atomic structure and the experimental SAXS data. However, for highly flexible proteins, one expects variations throughout the sampled q range. Examine the data for signs of any deviation; upward deviation of the experimental data at low q range (< 0.2 Å^{-1}) is suggestive of molecular elongation, whilst deviation within the medium q range (> 0.2 Å^{-1}) may indicate a difference in relative domain orientation/location between the two structures.

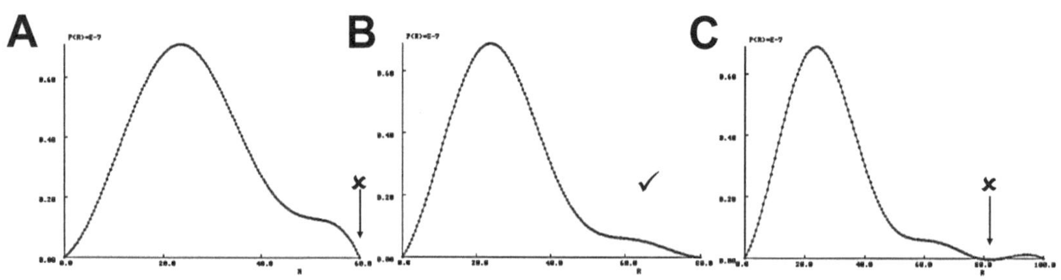

Fig. 2 Optimizing D_{max} in the $P(r)$ distribution. SAXS data were acquired on proMMP-1$^{\Delta HPX}$ at DLS beamline B21. Pairwise distance distributions are shown for (**a**) D_{max} = 60 Å; (**b**) D_{max} = 80 Å, the correct value; and (**c**) D_{max} = 100 Å. In (**a**), the D_{max} is too low, giving rise to a truncated $P(r)$ (*vertical arrow*). In (**c**), the D_{max} is too high, making the $P(r)$ descend below the *x*-axis (*vertical arrow*). Data displayed using GNOM

3.4.7 Volumetric Modeling

For generating three-dimensional ab initio reconstructions from 1D scattering data, we recommend using the program *DAMMIF* [37] which uses the *GNOM* output file (*see* Subheading 3.4.5) as input to produce low-resolution envelopes of dummy atoms (*see* **Note 20**). Thereafter, *DAMAVER* [38] is used to align, average, and filter these envelopes according to the frequency of dummy atom occupancy. The process in brief involves comparison of all reconstructed envelopes to find the most probable one, outliers (defined as models with two standard deviations from the mean) are removed, and the remaining models are aligned. A probability map is created from the average aligned models, then filtered and refined to generate a final averaged 'core' molecular envelope. If a high-resolution atomic structure is available, *SUPCOMB* [39] can be used to superimpose this structure onto the envelope by minimizing the normalized spatial discrepancy (NSD) between the envelope and crystal structure. Users are referred to the *DAMMIF*, *DAMAVER* and *SUPCOMB* manuals for detailed information regarding command-line arguments and output files (*see* **Note 21**).

1. In *DAMMIF*, enter the required *GNOM* file (*see* Subheading 3.4.5); *DAMMIF* will read the previously computed distance distribution function from this file.
2. Select an appropriate annealing configuration, either "slow" (which we recommend), "fast" or "interactive" (*see* **Note 22**).
3. Enforce the appropriate symmetry, typically P1 (*see* **Note 23**).
4. Define appropriate anisometry: prolate, oblate, or unknown. This may be inferred from previous experiments.
5. Once submitted, *DAMMIF* will output a runtime log (*see* **Note 24**) which includes step-by-step χ^2 values which indicate the level of agreement between the experimental scattering profile and one simulated from the reconstructed molecular envelope.
6. Generate multiple *DAMMIF* reconstructions (*see* **Note 25**) prior to submitting the modeled envelopes (contained in the −1.pdb output files) to *DAMAVER* [38] for processing in batch mode (*see* **Note 26**).
7. Check the *DAMAVER* damsel.log file for the NSD values. A mean NSD value of less than 0.9 indicates a unique model.
8. Where a high-resolution MX/NMR structure is available, compare it to the reconstructed envelope (the output damfilt.pdb file from *DAMAVER*) using *SUPCOMB*. A single .pdb output file (appended with "*r*," unless otherwise specified) is generated containing the successful superposition. Note the NSD value, in which a low value (< 1.0) indicates an acceptable fit. Higher NSD values may indicate that the (pro)MMP has disordered regions or conformational freedom.

3.4.8 Atomistic and Ensemble Optimization Modeling

Atomistic modeling involves the generation of all-atom conformers by assembling predetermined (by MX/NMR) domain structures in a particular arrangement (i.e., position and orientation) that satisfies the experimental SAXS data. For (pro)MMPs, where interdomain flexibility may be present, a particularly powerful approach is to generate a population of conformers which together satisfy the SAXS data using the *EOM* [17] program within the *ATSAS* suite. *EOM online* is a web portal-based access to an *EOM* server at EMBL, but has some restrictions on parameter selection (*see* **Note 27**). The following instructions apply to *EOM online*:

1. Identify the domains in your (pro)MMP of interest for which atomic resolution structures have been solved previously (*see* **Note 28**). Using a text editor, generate a separate PDB file for each chosen domain containing the atomic coordinate entries (lines starting with the "ATOM" keyword) (*see* **Note 29**).
2. In *EOM online*, provide a file containing the (pro)MMP amino acid sequence in one-letter code.
3. Select the overall symmetry for the protein, typically P1 (*see* **Note 23**).
4. Upload the PDB files for the specified domains, and specify whether each of these domains should be fixed in space or free to move (*see* **Note 30**).
5. Specify the chain type, either "random coil" or "native-like." This influences the rotamer distribution of the built polypeptides—"random coil" gives a torsional distribution typical of chemically denatured proteins, whilst "native-like" gives a (slightly more restrictive) distribution consistent with intrinsically disordered proteins.
6. Provide one or more scattering curve files ($I(q)$ vs. q), together with the q units (either nm^{-1} or $Å^{-1}$) used in those files.
7. When run, the program will generate a pool of 10,000 models by randomizing the linker(s) between the ordered domains, and then use a genetic algorithm to select a subset ensemble which, as a minimum population, gives the best fit to the experimental data.
8. Download the results files. These contain the R_g, D_{max}, and volume distributions for the initial pool and selected ensemble (Fig. 3b), the ensemble-averaged theoretical scattering curve (Fig. 3c), and PDB coordinate files for representative conformers.

4 Notes

1. Those new to SAXS should note that there are a variety of abbreviations used in the literature for the scattering angle: q (as used here), Q, s, and S.
2. High MW filtration device for removal of protein aggregates.

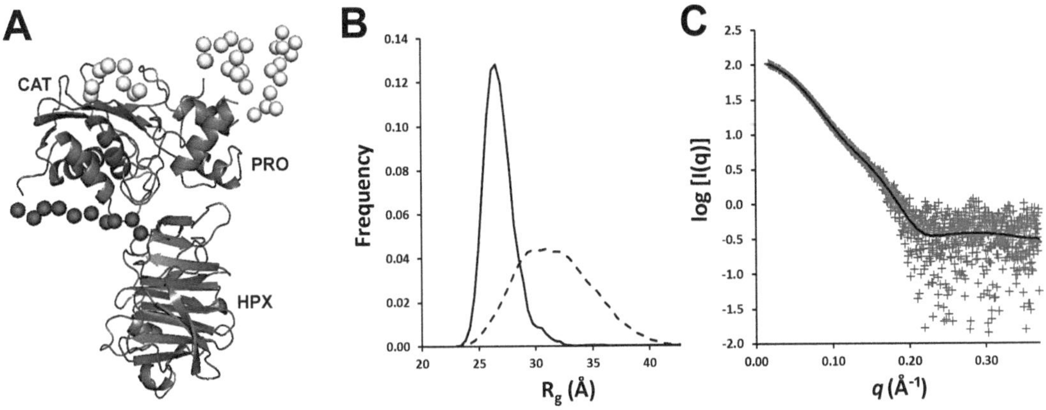

Fig. 3 Ensemble optimization modeling of proMMP-1. (**a**) Preparing the crystal structure for an *EOM* calculation. The PDB file was divided into four separate ordered "domains": the first helix of the PRO domain, the remainder of the structured PRO domain, the CAT domain and the HPX domain. All except the HPX domain are fixed in space during the *EOM* calculation. The program generates C_α atomic positions for the unstructured regions that connect these, i.e., the N-terminal tail, "bait" region, and PRO-CAT linker (*white spheres* in the diagram). Although precisely positioned in the crystal structure, the CAT-HPX linker (*black spheres*) was allowed to flex during the *EOM* calculation to accommodate varying positions/orientations of the HPX domain; (**b**) Results of the *EOM* calculation showing the R_g distribution for the starting pool of proMMP-1 conformers (*dashed line*) and the ensemble of selected conformers (*solid line*). Unlike the mature enzyme, proMMP-1 remains compact in solution [9]; (**c**) Superposition of the theoretical scattering from the ensemble of selected conformers (*black line*) over the experimental data (*grey crosses*)

3. We recommend Greiner Bio-One 96-well microplates.
4. The *ATSAS* software package incorporates numerous programs for SAXS data processing and analysis, such as *PRIMUS*, *GNOM*, *CRYSOL*, *DAMMIF*, *SUBCOMP*, and *EOM*
5. Adding excessive Zn^{2+} to the samples may promote protein cross-linking and aggregation.
6. If using SEC-SAXS, addition of glycerol increases the column back-pressure, so 3% (v/v) glycerol is the recommended maximum concentration. Although commonly used to protect against radiation damage in other protein SAXS studies [40], the reducing agents DTT and tris(2-carboxyethyl)phosphine (TCEP) are not recommended (even when studying MMP fragments that do not contain disulfide bonds) as each can chelate Zn^{2+} ions.
7. Although SEC purification of the sample may seem redundant if the SAXS analysis is to use inline SEC as the delivery method, the purification step serves to remove aggregates that will otherwise serve as nucleation points for further aggregation (and hence sample loss) during any concentrating step.
8. Between spins, ensure to mix the retentate by gentle pipetting. Using a NanoDrop 2000 spectrophotometer, measure the A_{280} of 1 μL aliquots of both the retentate and filtrate at

regular intervals (for example, every five spins) to ensure that the protein is being retained by the membrane and is increasing in concentration.

9. Refer to the SEC column manufacturer for the column's total volume (v_t) and void volume (v_0). Alternatively, the v_0 may be measured using dextran blue, a high-MW polysaccharide to which dye molecules are covalently bound. If MW calibration of the column is desired, we recommend using a mixture of aprotinin (6.5 kDa), ribonuclease A (13.7 kDa), carbonic anhydrase (29 kDa), ovalbumin (43 kDa), and conalbumin (75 kDa).
10. It is imperative that the chemical composition of the buffer exactly matches that of the sample. Even with the most careful pipetting, separately prepared buffers are usually insufficiently matched leading to erroneous buffer subtraction. In our experience with the BioSAXS™ robot, the best results are obtained when using the dialysate from the final dialysis step. However, it is also possible to use the filtrate from centrifugal ultrafiltration.
11. During protein concentration, low levels of aggregates can serve as nucleation sites for further aggregation, so it is advisable to remove these by passage through a high-MWCO (e.g., 100,000 Da) ultrafiltration device beforehand. If the sample does not aggregate on standing, pre-prepared high concentration stocks can be taken to the beamline. Otherwise, we recommend concentrating the sample on-site at the beamline immediately prior to SAXS analysis.
12. Although a convenient technique, observing a single band on an SDS-PAGE gel of the sample is not indicative of sample monodispersity.
13. For accurate apparent MW determination, it is imperative that the (pro)MMP concentrations are themselves accurate. We recommend using nanoliter-scale spectrophotometry. Unless they are particularly well standardized for the protein of interest, generic protein concentration assays (e.g., Bradford and BCA) are usually not sufficiently precise.
14. Having the sample under flow reduces X-ray induced damage (and resultant aggregation) as the irradiated protein is constantly moving out of the beam and being replaced.
15. In theory, it is possible to observe aggregation in SEC-SAXS, but only if the flow rate is too low (and hence X-ray damage is occurring), or if the protein being analysed is eluting at the void volume (i.e., it is aggregated even before it enters the X-ray beam). In principle, it is possible to reduce protein damage by attenuating the beam and hence lowering the X-ray flux (consult your local beamline scientist). However, the reduced

number of photons passing through the sample will also reduce the data quality by diminishing the signal–noise ratio.

16. Addition of further data below this cutoff will mostly add noise to the scattering profile.
17. $I(q)$ decays roughly exponentially as a function of q, so the data at wider angles has a much poorer signal–noise ratio. Hence, for good quality data at high q, a high protein concentration is required. However, this renders the sample more susceptible to X-ray induced aggregation, which has a greater effect on the low q data.
18. If present, beamstop interference will be apparent in both the buffer and sample scattering profiles (*see* Fig. 1a) as deviations at very low q (<0.01 Å^{-1}).
19. The following files will be saved as outputs after *CRYSOL* has run: .log (ASCII file—CRYSOL log), .sav (binary file—*CRYSOL* information), .flm (ASCII file—multipole coefficients), .int (ASCII file—scattering intensities), .fit (ASCII file—fit to experimental data), and .alm (binary file—net partial amplitudes).
20. *DAMMIF* [37] is an improvement over the earlier software *DAMMIN* [38], maintaining the key features but offering algorithm improvements which substantially increase the speed of model reconstruction.
21. Interactive configuration modules are available in *DAMMIF*, *DAMAVER*, and *SUBCOMP*, but note that some parameter changes will incur significant penalties in CPU usage.
22. In *DAMMIF*, the "slow" mode will generate a more accurate reconstruction of the envelope than in "fast" mode by increasing both the number of harmonics and iterations per step during the annealing process; this may be beneficial for elongated molecules. The "interactive" mode allows customization of the annealing procedures as required.
23. This will usually be "P1 (no symmetry)", but for (pro)MMP fragments that dimerize via the HPX domain (e.g., MMP-14 [27]), "P2 (dimer)" may be more appropriate.
24. The following files are saved as outputs after *DAMMIF* has run: a .log file containing the screen output; a −0.pdb file containing the search volume; a −1.pdb containing the modeled envelope; a .fit file of the simulated scattering profile compared to the smoothed experimental data; a .fir file of the simulated scattering profile compared to the experimental data; and a .in file, which can be used to rerun *DAMMIF* without changing parameters.
25. Ten independent reconstructions are recommended. However, it may be necessary to increase the number of *DAMMIF*

reconstructions to ensure a valid averaged "core" molecular envelope is generated.

26. In *DAMAVER*, the modeled envelope PDB input files are processed by a suite of functions, namely *damsel*, *damsup*, *damaver*, *damfilt*, and *damstart*. These may be run individually, but it is recommended to run them all automatically in batch mode.
27. *EOM online* will suit most (pro)MMP studies. However, if greater versatility is required (e.g., changing the size of the conformer pool, or simultaneous fitting to multiple SAXS datasets), we suggest using the downloadable *EOM* executable. For further details, see the *EOM* manual (http://www.embl-hamburg.de/biosaxs/manuals/eom.html).
28. If high-resolution (MX/NMR) structures are not available for the (pro)MMP of interest then approximate domain folds can be produced by homology modeling using the online SWISS-MODEL (swissmodel.expasy.org) [41].
29. The current version of the *EOM* program (2.0) is unable to handle PDB coordinate files that have a noncontiguous amino acid sequence. This is problematic for any domains containing unstructured regions, e.g., the PRO domain "bait" region in crystal structure of proMMP-1 [6], as their MX PDB coordinate files will have missing atomic entries. Our solution is to split these PDB files into two separate sub-domains.
30. If a (pro)MMP domain has been split into two (or more) sub-domains for *EOM* (*see* **Note 29**), then they must be fixed in space, otherwise they will dissociate during the calculation.

Acknowledgments

We gratefully acknowledge the financial support of IBBS and the BBSRC.

References

1. Murphy G, Nagase H (2008) Progress in matrix metalloproteinase research. Mol Aspects Med 29(5):290–308
2. Mott JD, Werb Z (2004) Regulation of matrix biology by matrix metalloproteinases. Curr Opin Cell Biol 16(5):558–564
3. Nagase H, Visse R, Murphy G (2006) Structure and function of matrix metalloproteinases and TIMPs. Cardiovasc Res 69(3):562–573
4. Li J, Brick P, O'Hare MC, Skarzynski T, Lloyd LF, Curry VA, Clark IM, Bigg HF, Hazleman BL, Cawston TE, Blow BM (1995) Structure of full-length porcine synovial collagenase reveals a C-terminal domain containing a calcium-linked, four-bladed beta-propeller. Structure 3:541–549
5. Iyer S, Visse R, Nagase H, Acharya KR (2006) Crystal structure of an active form of human MMP-1. J Mol Biol 362(1):78–88
6. Jozic D, Bourenkov G, Lim NH, Visse R, Nagase H, Bode W, Maskos K (2005) X-ray structure of human proMMP-1—new insights into procollagenase activation and collagen binding. J Biol Chem 280(10):9578–9585

7. Morgunova E, Tuuttila A, Bergmann U, Isupov M, Lindqvist Y, Schneider G, Tryggvason K (1999) Structure of human pro-matrix metalloproteinase-2: activation mechanism revealed. Science 284(5420):1667–1670
8. Stura EA, Visse R, Cuniasse P, Dive V, Nagase H (2013) Crystal structure of full-length human collagenase 3 (MMP-13) with peptides in the active site defines exosites in the catalytic domain. FASEB J 27(11):4395–4405
9. Arnold LH, Butt LE, Prior SH, Read CM, Fields GB, Pickford AR (2011) The interface between catalytic and hemopexin domains in matrix metalloproteinase-1 conceals a collagen binding exosite. J Biol Chem 286(52):45073–45082
10. Manka SW, Carafoli F, Visse R, Bihan D, Raynal N, Farndale RW, Murphy G, Enghild JJ, Hohenester E, Nagase H (2012) Structural insights into triple-helical collagen cleavage by matrix metalloproteinase 1. Proc Natl Acad Sci U S A 109(31):12461–12466
11. Pastor N, Amero C (2015) Information flow and protein dynamics: the interplay between nuclear magnetic resonance spectroscopy and molecular dynamics simulations. Front Plant Sci 6:306
12. Orozco M (2014) A theoretical view of protein dynamics. Chem Soc Rev 43(14):5051–5066
13. Schwartz SD (2013) Protein dynamics and the enzymatic reaction coordinate. Top Curr Chem 337:189–208
14. Becker JW, Marcy AI, Rokosz LL, Axel MG, Burbaum JJ, Fitzgerald PM, Cameron PM, Esser CK, Hagmann WK, Hermes JD et al (1995) Stromelysin-1: three-dimensional structure of the inhibited catalytic domain and of the C-truncated proenzyme. Protein Sci 4(10):1966–1976
15. Elkins PA, Ho YS, Smith WW, Janson CA, D'Alessio KJ, McQueney MS, Cummings MD, Romanic AM (2002) Structure of the C-terminally truncated human ProMMP9, a gelatin-binding matrix metalloproteinase. Acta Crystallogr D Biol Crystallogr 58(Pt 7):1182–1192
16. Bertini I, Calderone V, Fragai M, Jaiswal R, Luchinat C, Melikian M, Mylonas E, Svergun DI (2008) Evidence of reciprocal reorientation of the catalytic and hemopexin-like domains of full-length MMP-12. J Am Chem Soc 130(22):7011–7021
17. Bernado P, Mylonas E, Petoukhov MV, Blackledge M, Svergun DI (2007) Structural characterization of flexible proteins using small-angle X-ray scattering. J Am Chem Soc 129(17):5656–5664
18. Bertini I, Fragai M, Luchinat C, Melikian M, Mylonas E, Sarti N, Svergun DI (2009) Interdomain flexibility in full-length matrix metalloproteinase-1 (MMP-1). J Biol Chem 284(19):12821–12828
19. Bertini I, Fragai M, Luchinat C, Melikian M, Toccafondi M, Lauer JL, Fields GB (2012) Structural basis for matrix metalloproteinase 1-catalyzed collagenolysis. J Am Chem Soc 134(4):2100–2110
20. Cerofolini L, Fields GB, Fragai M, Geraldes CF, Luchinat C, Parigi G, Ravera E, Svergun DI, Teixeira JM (2013) Examination of matrix metalloproteinase-1 in solution: a preference for the pre-collagenolysis state. J Biol Chem 288(42):30659–30671
21. Mertens HD, Svergun DI (2010) Structural characterization of proteins and complexes using small-angle X-ray solution scattering. J Struct Biol 172(1):128–141
22. Rambo RP, Tainer JA (2015) Modeling macromolecular motions by X-ray-scattering-constrained molecular dynamics. Biophys J 108(10):2421–2423
23. Svergun DI, Koch MH (2002) Advances in structure analysis using small-angle scattering in solution. Curr Opin Struct Biol 12(5):654–660
24. Trewhella J, Hendrickson WA, Kleywegt GJ, Sali A, Sato M, Schwede T, Svergun DI, Tainer JA, Westbrook J, Berman HM (2013) Report of the wwPDB small-angle scattering task force: data requirements for biomolecular modeling and the PDB. Structure 21(6):875–881
25. Cha J, Auld DS (1997) Site-directed mutagenesis of the active site glutamate in human matrilysin: investigation of its role in catalysis. Biochemistry 36(50):16019–16024
26. Bertini I, Fragai M, Lee YM, Luchinat C, Terni B (2004) Paramagnetic metal ions in ligand screening: the Co(II) matrix metalloproteinase 12. Angew Chem Int Ed Engl 43(17):2254–2256
27. Tochowicz A, Goettig P, Evans R, Visse R, Shitomi Y, Palmisano R, Ito N, Richter K, Maskos K, Franke D, Svergun D, Nagase H, Bode W, Itoh Y (2011) The dimer interface of the membrane type 1 matrix metalloproteinase hemopexin domain: crystal structure and biological functions. J Biol Chem 286(9):7587–7600
28. Petoukhov MV, Konarev PV, Kikhney AG, Svergun DI (2007) *ATSAS* 2.1 – towards automated and web-supported small-angle scattering data analysis. J Appl Cryst 40:S223–S228
29. Petoukhov MV, Franke D, Shkumatov AV, Tria G, Kikhney AG, Gajda M, Gorba C, Mertens

HD, Konarev PV, Svergun DI (2012) New developments in the program package for small-angle scattering data analysis. J Appl Cryst 45(Pt 2):342–350

30. Rosenblum G, Van den Steen PE, Cohen SR, Grossmann JG, Frenkel J, Sertchook R, Slack N, Strange RW, Opdenakker G, Sagi I (2007) Insights into the structure and domain flexibility of full-length pro-matrix metalloproteinase-9/gelatinase B. Structure 15(10):1227–1236

31. Nie J, Pei D (2010) Expression and purification of membrane-type MMPs. Methods Mol Biol 622:99–110

32. Rodgers UR, Clark IM (2010) Expression of recombinant MMP-28 in mammalian cells. Methods Mol Biol 622:55–65

33. Windsor LJ, Steele DL (2010) Expression of recombinant matrix metalloproteinases in *Escherichia coli*. Methods Mol Biol 622:67–81

34. Konarev PV, Volkov VV, Sokolova AV, Koch MHJ, Svergun DI (2003) *PRIMUS*: a Windows PC-based system for small-angle scattering data analysis. J Appl Cryst 36:1277–1282

35. Svergun DI (1992) Determination of the regularization parameter in indirect-transform methods using perceptual criteria. J Appl Cryst 25:495–503

36. Svergun D, Barberato C, Koch MHJ (1995) *CRYSOL* – a program to evaluate x-ray solution scattering of biological macromolecules from atomic coordinates. J Appl Cryst 28:768–773

37. Franke D, Svergun DI (2009) *DAMMIF*, a program for rapid ab-initio shape determination in small-angle scattering. J Appl Cryst 42(2):342–346

38. Volkov VV, Svergun DI (2003) Uniqueness of ab initio shape determination in small-angle scattering. J Appl Crystallogr 36(3 Part 1):860–864

39. Kozin MB, Svergun DI (2001) Automated matching of high- and low-resolution structural models. J Appl Cryst 34(1):33–41

40. Jeffries CM, Graewert MA, Svergun DI, Blanchet CE (2015) Limiting radiation damage for high-brilliance biological solution scattering: practical experience at the EMBL P12 beamline PETRAIII. J Synchrotron Radiat 22(Pt 2):273–279

41. Biasini M, Bienert S, Waterhouse A, Arnold K, Studer G, Schmidt T, Kiefer F, Cassarino TG, Bertoni M, Bordoli L, Schwede T (2014) SWISS-MODEL: modelling protein tertiary and quaternary structure using evolutionary information. Nucleic Acids Res 42(W1):W252–W258

Part III

Computational Simulations of Matrix Metalloproteases

Chapter 7

Molecular Dynamics Studies of Matrix Metalloproteases

Natalia Díaz and Dimas Suárez

Abstract

Matrix metalloproteases are multidomain enzymes with a remarkable proteolytic activity located in the extracellular environment. Their catalytic activity and structural properties have been intensively studied during the last few decades using both experimental and theoretical approaches, but many open questions still remain. Extensive molecular dynamics simulations enable the sampling of the configurational space of a molecular system, thus contributing to the characterization of the structure, dynamics, and ligand binding properties of a particular MMP. Based on previous computational experience, we provide in this chapter technical and methodological guidelines that may be useful to and stimulate other researchers to perform molecular dynamics simulations to help address unresolved questions concerning the molecular mode of action of MMPs.

Key words Free energy calculations, Ligand binding, Molecular dynamics, Molecular modeling, Multidomain enzymes, Peptide hydrolysis

1 Introduction

Matrix metalloproteinases (MMPs), as their name implies, display hydrolytic activity against extracellular matrix structural elements like collagen. However, the zinc-dependent MMPs also cleave bioactive molecules thus acting as pivotal effectors in many physiological processes [1]. Moreover, MMPs contribute to the initiation and progression of prevalent pathologies like cancer and, accordingly, numerous laboratories have pursued the design of MMP inhibitors. Unfortunately, compounds developed over the last 20 years have failed in clinical trials probably due to their broad-spectrum activity (i.e., they inhibit several MMPs and other metallo-enzymes). Nowadays, MMPs are still considered *druggable* targets but it is now recognized that more specific inhibitors need to be developed [2]. Nevertheless, the design of specific MMP inhibitors, which could certainly benefit from computational aided drug design approaches, remains a challenge due to the inherent flexibility of these structurally similar macromolecules.

Charles A. Galea (ed.), *Matrix Metalloproteases: Methods and Protocols*, Methods in Molecular Biology, vol. 1579,
DOI 10.1007/978-1-4939-6863-3_7,

MMPs are multi-domain proteins that exhibit specific structural features which must be taken into account when planning molecular simulations [3]. NMR and X-ray crystal structures have shown that the ~165-residue catalytic domain displays a similar fold for all MMPs and consists of an upper *N*-terminal subdomain with a five-stranded β-sheet, two α-helices and several loops, and a lower *C*-terminal subdomain formed by one α-helix and a large Ω-loop. The narrow active site cleft separates both subdomains and binds the catalytic zinc ion (Zn_1) involved in the hydrolysis reaction. A second zinc (Zn_2) and several calcium ions are also observed in all the catalytic domains and these play a structural role in the folding of several loops. Peptide substrates bind within the active site via polar and hydrophobic contacts with the so-called S_3–S_3' binding subsites. The S_1' hydrophobic pocket has been suggested to be particularly important for increasing selectivity during inhibitor design, since it is partially delimited by the Ω-loop and displays some variability in size and composition across the MMPs. However, induced-fit effects have been observed at this site and it is likely that the dynamics of the S_1' pocket should be included in future drug design efforts [4]. Structures available for the mature catalytic domain also differ in the length and the position of the *N*-terminal coil which, in turn, affects the proteolytic activity of MMPs [5]. In the physiologically relevant form (i.e., *superactive* form) of the enzyme, an *N*-terminal tyrosine or phenylalanine remains packed against a concave hydrophobic surface of the molecule via a salt-bridge interaction with a strictly conserved aspartic acid residue.

Besides the catalytic domain, active MMPs (except MMP-7 and MMP-26) possess additional modules or domains. These noncatalytic domains mediate the interaction of MMPs with other macromolecules, thus determining the cellular location, activation, and proteolytic activity of the enzymes [6]. These include the hemopexin-like domain that is present in the majority of multidomain MMPs (except MMP23), the three fibronectin-like modules inserted in the catalytic domains of MMP-2 and MMP-9, and the transmembrane domain characteristic of the membrane-type MMPs (MT-MMPs). Interestingly, these noncatalytic domains provide secondary sites for substrate binding that could be targeted by MMP inhibitors. In fact, binding to the exosites (i.e., substrate-binding sites located outside the active-site cleft) has been proposed as a promising new route to more specific MMP inhibitors. Crystal structures for full-length MMPs are available for MMP-1, MMP-2, MMP-12, and MMP-13 and these structures suggest a conserved and well-defined spatial relationship between the catalytic and noncatalytic domains. However, NMR and spectroscopy analyses show that the catalytic and hemopexin-like domains of MMP-1, MMP-9, and MMP-12 are able to adopt both compact and significantly extended conformations in solution [7].

Molecular dynamics (MD) is a valuable simulation method for analyzing the structure and dynamics of complex macromolecules like MMPs. MD simulations usually employ atomic force field models to sample the configurational space of systems comprising many thousands of atoms over time scales of nanoseconds and beyond. Most MD studies of multidomain MMPs have focused on the isolated catalytic domains of MMP-1, MMP-2, MMP-3, or MMP-13 with the aim of computing relative binding free energies for active site inhibitors (e.g., carboxylate, hydroxamate, phosphonate, thiadiazole urea, alloxan, and pyrimidine dicarboxamide inhibitors) [8–15]. MD simulations have also been used to analyze the dynamic flexibility of each recognition pocket within the active site cleft and their substrate specificities [16–18]. The MMP-2 activation process (i.e., removal of the propeptide from the active site) and its interaction with collagen-like peptides have also been investigated with the aid of MD simulations [19, 20]. Our group has intensively applied MD-based methods to analyze various aspects of the structure and dynamics of MMPs such as the specific roles of the noncatalytic Zn_2 ion and several calcium ions on the structure of the catalytic domain [21], the effect of the conformation adopted by the *N*-terminal coil [22], interactions with small and macromolecular peptide substrates [22–26], the influence of noncatalytic domains [27–29], etc.

In this chapter, we will attempt to distill our previous experience into technical and methodological guidelines that could be useful to and encourage other researchers to perform future MD simulations aimed at addressing unresolved challenges regarding the molecular mode of action of MMPs. Subheading 2 discusses the appropriate choice of software and hardware, and highlights the AMBER package of molecular simulation programs as a reference tool [30, 31]. In Subheading 3, we review the various tasks and procedures required to set up, run, and analyze an extended MD simulation of a biomolecular system, placing emphasis on the points that are particularly relevant to MMPs. We comment on how to derive realistic parameters for metal-containing systems and what specific structural features of MMPs need special attention during the preparation of the system, differentiating between simulations of the MMP catalytic domain alone and the full-length enzyme in the presence of a peptide substrate, inhibitor, or collagen-like molecule.

2 Materials

The three main components required to run an MD simulation are the initial spatial coordinates of the system of interest (e.g., a particular MMP), software for implementing the various methods, and hardware for performing the calculations.

2.1 Initial Structure

To start an MD simulation, initial coordinates for all the atoms in the desired MMP should be provided to the minimization/simulation packages.

1. Check the Protein Data Bank for available structures and select a suitable X-ray crystal structure with a high resolution. Currently, structural information is only available for 12 of the 23 human MMPs: MMP-1 (catalytic domain and full-length enzyme), MMP-2 (catalytic domain, fibronectin modules, hemopexin domain, and full-length enzyme), MMP-3 (catalytic domain), MMP-7 (catalytic domain), MMP-8 (catalytic domain), MMP-9 (catalytic-fibronectin domains, hemopexin domain, and a model for the full-length enzyme), MMP-10 (catalytic domain), MMP-12 (catalytic domain, hemopexin domain, and full-length enzyme), MMP-13 (catalytic domain, hemopexin domain, and full-length enzyme), MMP-14 (catalytic and hemopexin domains), MMP-16 (catalytic domain), and MMP-20 (catalytic domain). For the remaining MMPs, homology modeling may provide a convenient initial structure provided the homology models are carefully validated (*see* **Note 1**).
2. Edit the initial structure by removing additional molecules, reverse mutations (e.g., the catalytic glutamic acid), and selecting appropriate protonation states for the titratable amino-acids (*see* **Note 2**). Take into account that the catalytic Zn_1 is coordinated by three histidines through their side chain Nε atom, but Zn_2 is bound to two Nε and one Nδ atoms of three different histidines.
3. Refine the structure by adding missing atoms/residues and optimize the conformational state of some side chains (i.e., His, Asn, Gln) (*see* **Note 3**).
4. The use of explicit solvent in extended molecular simulations of MMPs is highly recommended to realistically account for both entropic and enthalpic solvent effects, which can influence inter and intra-domain conformational changes as well as ligand binding. Therefore, the MMP molecule should be placed in an appropriate solvent box that extends at least 15 Å from the protein atoms. Neutralize the system by randomly adding Na^+/Cl^- counterions.

2.2 Software

Fortunately, a variety of programs are readily available for preparing and running MD simulations of biomolecules assuming an all atom representation. Our preferred choice for studying MMPs is the AMBER (Assisted Model Building with Energy Refinement) package, which integrates a collection of programs to set up, perform, and analyze MD simulations [32]. AMBER can be used with several types of molecular mechanics (MM) methods or *force fields* for calculating potential energies and forces for large biomolecules, small organic molecules, and typical solvents. For extended MD simulations of MMP systems, we recommend the family of *amber* force fields since previous experience shows that these potentials exhibit a favorable tradeoff between the quality of their structural and energetic predictions and their computational cost. Later versions of *amber* force fields, which use fixed partial charges centered on atoms, are *ff14SB* for proteins and the general *amber* force field (*GAFF*) for small organic molecules.

Among the numerous programs included in the AMBER package, the following tools are particularly important:

1. In the preparation toolbox, the program *antechamber* generates parameter files for nonstandard residues (e.g., zinc coordinated histidines) and ligands (e.g., MMP inhibitors), whereas *LEaP* creates the coordinate and topology files required for performing energy calculations and MD simulations.
2. The MD simulation programs comprise *sander* and *pmemd*. *Sander* implements many simulation techniques and potential-energy methods including hybrid quantum mechanical/molecular mechanical methods (QM/MM), while *pmemd* maximizes the performance of the most used simulation techniques combined with classical MM force fields.
3. Structure and dynamic analyses of MD trajectories can be carried out using *cpptraj*, a powerful tool available in the latest versions of AMBER. In addition, AMBER offers various tools (e.g., *MMPBSA.py*) to extract energetic information from MD trajectories using end-point free energy methods.
4. For those learning to use the software, various aids are available on the AMBER website (http://ambermd.org/) such as an easy to read but comprehensive manual, an updated mailing list, and numerous tutorials.

Other popular packages such as CHARMM (http://www.charmm.org/), NAMD (http://www.ks.uiuc.edu/Research/namd/), or GROMACS (http://www.gromacs.org/) are suitable alternatives in terms of functionality, efficiency, and license cost. Besides the adoption of a reference package like AMBER, a research project on molecular modeling of MMPs will likely demand the use of quantum mechanical (QM) methods and the execution of other miscellaneous tasks for the preparation/analysis of the simulations that will require further specialized software tools (*see* Subheading 3).

2.3 Hardware

Computational algorithms and hardware technologies have experienced impressive developments during the last years that bring high performance computing (HPC) to workstations affordable for a wide range of researchers. Recent developments in Graphic

Processing Units (GPUs) have led to a dramatic acceleration of MD simulations. In this respect, it should be noted that *pmemd* efficiently runs on GPUs and is currently one of the best programs for running extended MD simulations in a timely manner. Nevertheless, efficient editing and parameterization, as well as the postprocessing of MD trajectories for simulations run over extended periods, demands more computational resources in terms of conventional central processing units (CPUs), rapid access memory (RAM), high performance storage on multiple hard drives, graphic cards for visualization, etc.

The recommended starting configuration for a workstation or a computer server dedicated to perform molecular simulations of systems with the typical size of MMPs, would include a dual socket motherboard equipped with two multicore CPUs, 4 GB of RAM per CPU core and a distributed storage capacity greater than 2 TB constituted by SAS/SSD drives in a RAID device. The same computer should be equipped with at least two state-of-the-art NVIDIA GPU cards (currently Maxwell GPUs). Assuming that a Linux operating system (e.g., CENTOS, Scientific Linux, etc.) and the AMBER package are installed, such a workstation/server machine would be able to *simultaneously* carry out production runs on the GPUs with minimum CPU usage and other tasks for the preparation and analysis of MD trajectories using the remaining RAM and CPU cores. Further information pertaining to the recommended hardware can be found on the AMBER website (http://ambermd.org/gpus). Of course, all major HPC hardware vendors offer workstations and clusters suitable for performing MD simulations.

3 Methods

In this section, we comment on selected methodological and technical issues that should be addressed to set up and run a molecular simulation for a particular MMP either in its unbound form or in complex with inhibitor molecules, short peptide substrates, or triple-helical substrates mimicking collagen molecules. Consideration of specific points to be discussed here, which have been primarily chosen on the basis of our previous experience in the molecular modeling of MMPs, may be helpful in the design of future studies.

3.1 MM Parametrization of the Coordination Environment of Metal Ions

The MMP catalytic domain binds two zinc ions (Zn_1 catalytic and Zn_2 structural) and a number of calcium ions that are structurally and functionally relevant. To describe these metal ions in classical force fields for extended MD simulations, we assign these metal ions to either a nonbonded or a bonded representation.

1. In the nonbonded form, only the total charge (+2) and van der Waals parameters available in the literature are required to describe these ions. We suggest using the nonbonding parameters developed by Li *et al.* that reproduce experimental ion-oxygen distance values and coordination number for the first solvation shell [33]. This works well for calcium ions but it is not recommended for describing the coordination environment surrounding the zinc ions.
2. Different coordination spheres have been observed for zinc ions in MMPs (e.g., tetrahedral for the structural and the propeptide bound catalytic zincs, distorted trigonal bipyramid for the hydroxamate inhibited catalytic zinc, etc.) which, in principle, can be well represented in the bonded metal-ligand configuration (*see* **Note 4**).
3. If there is incomplete experimental data to define the protonation state of the zinc-binding groups (ZBGs) and the essential Glu residue, and/or the number and position of water molecules, preliminary QM/MM studies should be carried out to determine the most likely configuration of the zinc sites [34]. For example, by initially assigning a nonbonded conformation for the zinc ion in AMBER, it is relatively simple to define a QM region comprising the catalytic Zn_1 ion, closest residues, and ligand/water molecules for each configuration. The QM region is described by either one of the semiempirical QM methods implemented in *sander* or a more sophisticated method using an *external* QM code (Gaussian, ORCA, GAMESS; *see* **Note 5**). Subsequent energy minimization of the QM region followed by structural comparisons with the experimental structures and/or energy comparison allow for discrimination between various configurations.
4. Equilibrium values and force constants for all zinc-ligand bonds and angles should be provided in the bound configuration. In order to do this we usually carry out QM calculations on realistic cluster models of the zinc coordination spheres (Fig. 1). For example, the selected cluster model for the structural Zn_2 should comprise the side chains of the coordinating ligands (three histidines and an aspartic acid). On the other hand, for the active site Zn_1, it is advisable to include the catalytic glutamic acid located in the second coordination shell and other relevant ligands (e.g., water molecules, inhibitor ZBGs, etc.). Bond and angle reference values involving zinc can be directly taken from the optimized cluster models. To compute force constants, perturbed structures should be generated by selectively modifying all the bond distances and angles involving the zinc ion and fitting the relative energies to a second-order polynomial [35]. The torsion parameters associated with the zinc-ligand interactions can be set to zero.

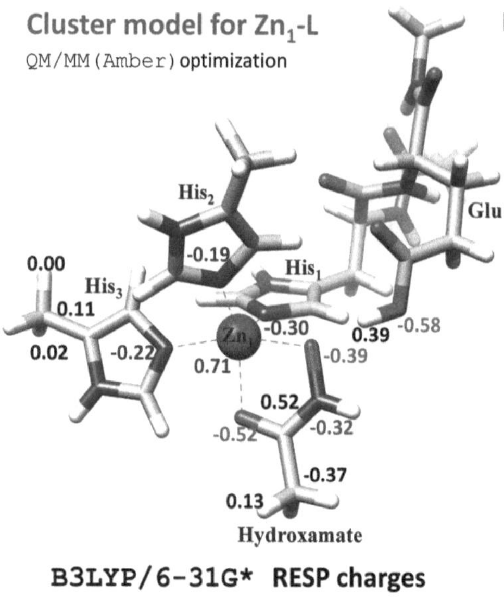

Equilibrium Geometry and **Force Constants**

(B3LYP/6-31G* geometry perturbations)

	K_r	r
His_1-Nε···Zn_1	101.0	2.06
His_2-Nε···Zn_1	93.0	2.06
His_3-Nε···Zn_1	68.0	2.13
Hydroxamate-O_1···Zn_1	99.0	2.13
Hydroxamate-O_2···Zn_1	105.0	2.08

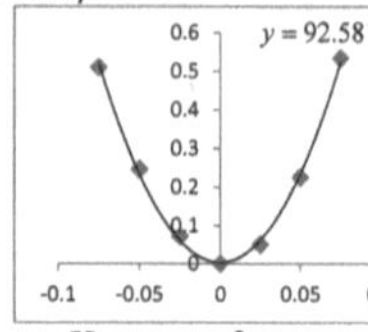

	K_θ	θ
His_1-Nε···Zn_1···Nε-His_2	38.0	119.0
His_1-Nε···Zn_1···Nε-His_3	96.0	88.0
His_2-Nε···Zn_1···Nε-His_3	71.0	94.0
His_1-Nε···Zn_1···O_1-Hydroxamate	80.0	96.0
His_1-Nε···Zn_1···O_2-Hydroxamate	68.0	113.0
His_2-Nε···Zn_1···O_1-Hydroxamate	97.0	90.0
His_2-Nε···Zn_1···O_2-Hydroxamate	54.0	128.0
His_3-Nε···Zn_1···O_1-Hydroxamate	107.0	163.0
His_3-Nε···Zn_1···O_2-Hydroxamate	144.0	87.0
Hydroxamate-O_1···Zn_1···O_2-Hydroxamate	69.0	100.0
Zn_1···His_1-Nε···His_1-Cε	268.0	128.0
Zn_1···His_1-Nε···His_1-Cδ	380.0	126.0

Fig. 1 Ball-and-stick representation of a cluster model used to parametrize the Zn_1 ion following the bonded approach. Selected atomic charges and MM parameters (*r*: reference bond distance, *θ*: reference bond angle, *Kr*: bond force constant, *Kθ*: angle force constant for Zn_1–L bonds and the corresponding angles) as obtained from quantum mechanical (B3LYP/6-31G*) calculations are included. An example of curve fitting to obtain a force constant is also shown for the His_2-Nε···Zn_1 bond (*y*: relative energy of the perturbed structure with respect to the unperturbed one; *x*: distance perturbation)

5. To properly describe metal-ligand charge transfer effects in the zinc-bound conformation, atomic charges from standard force fields should be avoided for the zinc ions and their zinc-binding groups. New partial charges must be derived by fitting to the QM electrostatic potential computed at a theoretical level compatible with the selected force field (e.g., a density functional theory method like B3LYP combined with a double-ζ basis set as 6-31G+*; HF-based charges are not recommended because they underestimate Zn-ligand charge-transfer effects). The electrostatic potential should be computed for the optimized QM cluster model used to derive metal-ligand force-constants or on the isolated QM region extracted from a QM/MM model. During the charge fitting procedure, the global charges of those capping residues (acetyl or *N*-methylamide) or H-atoms included in the cluster model to saturate open termini should be adjusted to zero or to the existing values in the chosen force field.

In line with the residue-based approach characteristic of MM methods, partial atomic charges and force constants for metal clusters should be transferable among MMPs. Thus, parameter sets cited for some MMPs in the literature may be applicable to simulations involving other MMPs. Alternatively, the latest version of AMBER provides a semi-automated procedure, named MCPB for

Metal Center Parameter Builder, for building a bound representation of a metal ion center within a metalloprotein (*see* **Note 6**).

3.2 Simulation of the Catalytic Domain

To date, most MD simulations of MMPs have focused on the catalytic domain, mainly examining the binding of active site inhibitors. We believe that future studies of the catalytic domain should carefully address the following points:

1. The *N*-terminal coil in most MMP catalytic domain crystal structures is either longer or shorter than the biologically relevant form because its length depends on the protocol used for the activation of the pro-MMP. This disparity prevents the binding of the *N*-terminal portion to the surface of the catalytic domain via the formation of a salt bridge with a conserved aspartic acid residue (Fig. 2) and leads to a reduction in the catalytic activity of the enzyme. Thus, for MD simulations of MMPs it is important to select a structure with the correct *N*-terminus, usually a tyrosine or a phenylalanine, and a properly positioned *N*-terminal coil. If this physiologically relevant (*superactivated*) structure is not available for a particular MMP, it can be easily generated by superposition with structures for other MMPs.
2. The number of bound calcium ions fluctuates between one and three, depending on the crystal structure chosen (Fig. 3). These calcium ions interact with several loops and turns and contribute toward generating a catalytically competent enzyme. For MMP-2, previous MD simulations followed by approximate free energy calculations have found that the binding of

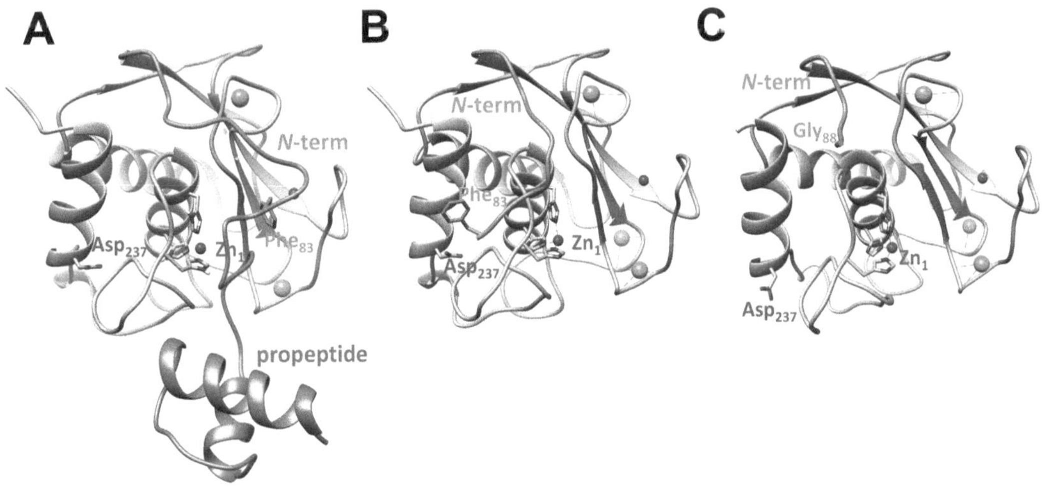

Fig. 2 Ribbon representation of the MMP-3 catalytic domain generated for different crystal structures with PDB codes 1SLM (**a**), 1CAQ (**b**), and 1HFS (**c**) illustrating the different positions of the *N*-terminal coil (in *orange*). The side chains of the conserved aspartic acid (Asp_{237}) and *N*-terminal residues are shown as *sticks*. Zinc and calcium ions are depicted as *spheres* (*magenta* and *green*, respectively) and the active site histidines as *sticks*

two calcium ions is energetically favored [21]. However, it is advisable to carefully determine the number and location of calcium ions to include in future simulations of other MMP catalytic domains. This can be done by using a similar approach to that mentioned above of extended MD simulations and energetic analyses.

3. The catalytic domain of the gelatinases MMP-2 and MMP-9 has an insertion of around 200 amino-acid residues corresponding to the three fibronectin-like modules. If needed, the fibronectin insertion can be replaced by a short connecting peptide to reduce the size of the molecule, without altering the structural and dynamic properties of the catalytic domain. In fact, crystal structures are available for both MMP-2 and MMP-9 with and without the fibronectin modules.

3.3 Binding of Small Ligands to the Active Site

Numerous MD simulations have previously analyzed the binding of small peptide substrates and inhibitors within the active site of various MMPs. Moreover, MD simulations of the Michaelis complex formed between the MMP catalytic domain and a substrate of interest must be considered a prerequisite prior to attempting to study the mechanism of catalysis (*see* **Note 7**).

1. The initial MMP/peptide complex can be built taking into account previously published data showing that small peptide substrates extend within the active site and make contact via a number of polar and hydrophobic interactions with the *S3–S3′* binding sites. Several crystal structures are available that can act as a template for initially binding a short peptide around the catalytic zinc ion (Fig. 4). For this purpose, use a program

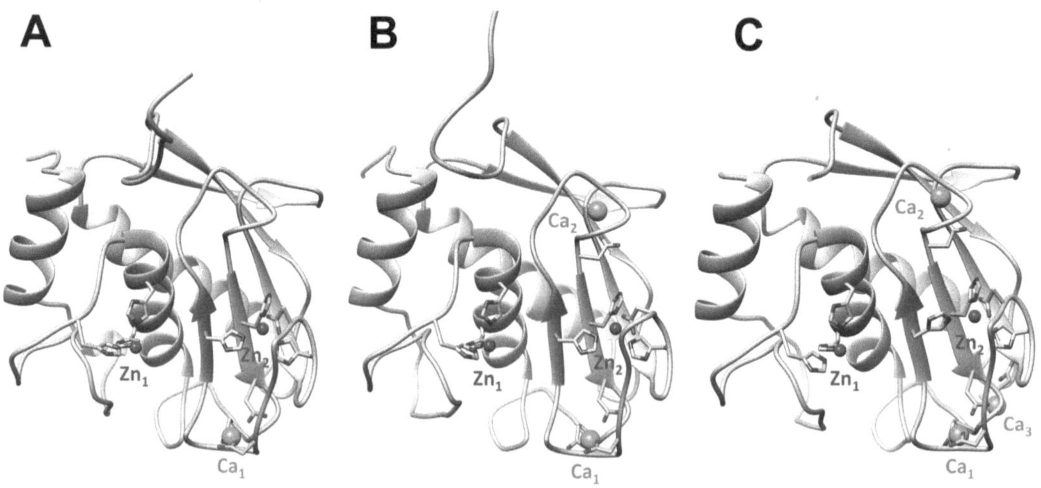

Fig. 3 Ribbon representation of the MMP-13 catalytic domain generated using different crystal structures with PDB codes 1YOU (**a**), 1XUC (**b**), and 1ZTQ (**c**). Zinc and calcium ions are depicted as *spheres* (*magenta* and *green*, respectively) and the side chains of peptide-binding residues as *sticks*

for interactive visualization and editing of molecular structures (e.g., *chimera:* http://www.cgl.ucsf.edu/chimera/). For larger substrates, the conformation of the *N*- and/or *C*-terminal portions of the peptide can be properly sampled using a conformational search algorithm prior to production MD simulations. For example, the Low MODe (LMOD) algorithm in *sander* performs a conformational search by applying structural perturbations to a molecular system derived from low frequency vibrational modes representing large amplitude concerted motions.

2. Automatic docking programs like *autodock4.0* (http://autodock.scripps.edu/wiki/AutoDock4) can be used to bind novel inhibitors within the active site. The parameters for these docking programs usually require some adjustment to take into account interactions between the zinc ion and the inhibitor, which largely control binding to the whole molecule. Normally, it is easy to recognize the ZBG of the inhibitor, but it is also important to realize that some ZBGs can modify their protonation state upon going from solution into the active site (e.g., the hydroxamate group transfers a proton to the active site glutamate residue when binding to MMP-2 and MMP-9). Make sure that the catalytic zinc ion is included in the docking calculations and select a proper configuration (i.e., atom types and partial charges) for the ZBG and other active sites. This is readily accomplished in *autodock4.0* since this program allows for defining different protonation states and it is possible to modify the partial charges of selected residues. Alternatively,

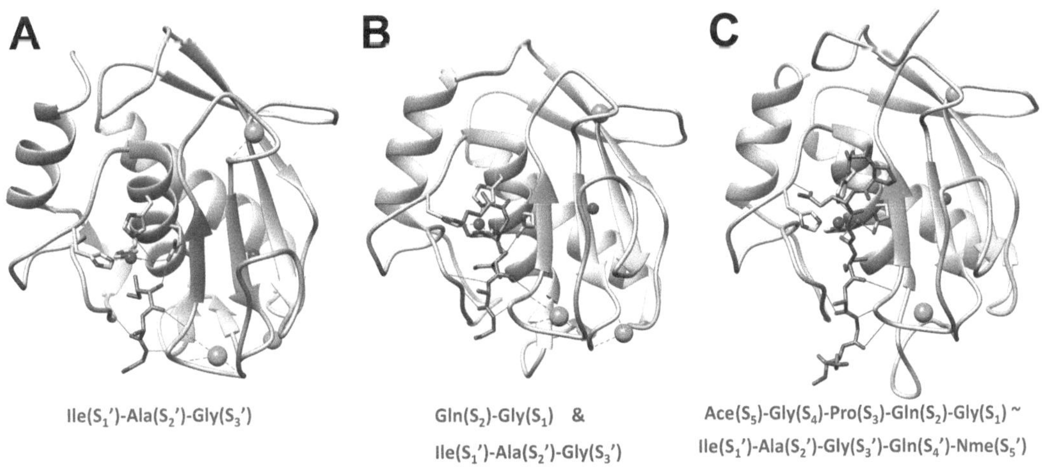

Fig. 4 Ribbon representation of the catalytic domains of MMP-12 (**a**, **b**) and MMP-2 (**c**) generated from crystal structures with PDB codes 2OXW, 2OXZ, and 1CK7, respectively. Ligands bound within the active site are illustrated as *sticks* with carbon atoms colored *green*. The ligand sequence and position within the active site is also indicated below each structure. Zinc and calcium ions are depicted as *spheres* (*magenta* and *green*, respectively) and the side chains of the Zn_1 binding protein residues as *sticks*

use the new force field *Autodock4Zn*, which includes a specialized potential that describes the interactions of zinc coordinating ligands [36].

3. Protein flexibility can be effectively included during docking calculations by considering different protein configurations. It is highly recommended to select a set of 25–50 snapshots from a previous MD simulation of the desired catalytic domain (preferably in complex with another ligand). Extract the coordinates of the catalytic domain alone and explore the docking of the inhibitor to the different protein frames. Due to the inherent plasticity of the Ω-loop and the induced fit effects observed in the *S1′* pocket, the use of different protein configurations can be indispensable for docking inhibitors with bulky side chains.

3.4 Binding of Collagen-like Substrates

Large and structurally complex biomolecules like those based on the collagen triple helix are among the natural substrates of the MMPs. For collagen-like substrates the complexity of both the substrate and the receptor (enzyme) increases the level of difficulty in the preparation and analysis of molecular simulations involving MMP complexes. Although such studies are particularly challenging, they are potentially more interesting. Therefore, in this subsection, we give some broad guidelines that will hopefully complement procedures available in the literature.

1. Initial coordinates for triple helical peptides (THPs) containing standard amino-acid residues can be obtained from the amino acid sequence using an interactive triple helical collagen building script termed *THe BuScr.* (http://structbio.biochem.dal.ca/jrainey/THeBuScr.html). Nevertheless, an initial MD simulation of the solvated helix is required to properly relax the system, especially if nonstandard amino-acid residues need to be included in the THP model.

2. The triple helix must locally unfold to accommodate one of its chains within the narrow active site of a MMP. Targeted MD simulations can be used to locally force one of the three chains into a more extended configuration and relax the whole triple helix. This transition can be smoothly driven along the targeted MD simulation by defining a harmonic penalty based on the root mean squared deviation of the backbone residues located around the scissile peptide bond. The reference structure used for the constraint can be taken from a previous MD simulation of the complex formed between the MMP and a small linear peptide substrate with the appropriate sequence.

3. We recommend implementing a semiflexible docking protocol combining various software tools (e.g., *cpptraj*, *sander*, *Mmpbsa.py*) using linux shell scripts to dock the distorted triple

helix within the active site of a MMP originally in complex with a smaller ligand as well as to partially relax the most severe steric clashes between the MMP and the THP (Fig. 5). First, MMP and THP coordinates, respectively, are retrieved from equally spaced snapshots from the targeted MD simulation of the distorted triple helix and from a previous MMP/peptide trajectory. The two sets of coordinates are combined by superposing the backbone coordinates of the triple helix around the scissile peptide bond onto the corresponding ones for the linear peptide substrate, thus generating a large set of MMP/THP models. Energy minimization and short MD simulations at high temperature (500 K) are used to relax steric clashes in the MMP/THP models by promoting uphill moves of bulky side chains (at this stage only residues involved in steric clashes are allowed to move). Finally, the relaxed MMP/THP complexes are scored by calculating the MMP···THP interaction energies using the MM-PBSA method.

4. Alternatively, several MMP/THP complexes have previously been published and can be used as templates in future studies examining other MMPs and/or triple helices. To date, several structures are available, including a crystal structure for the interaction of full-length MMP-1 with collagen in a nonreac-

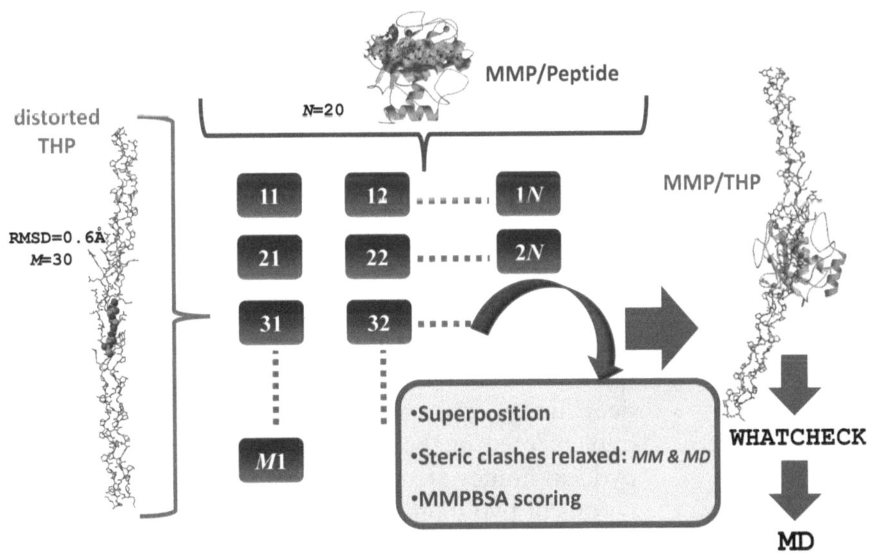

Fig. 5 Schematic representation of the semiflexible protocol proposed for docking a THP molecule within the active site of an MMP. Initial coordinates for the MMP/THP complexes (11, 12, ...,1*N*,..., *MN*) come from the superposition of *M* snapshots obtained from the MD simulation of the distorted THP onto *N* snapshots extracted from the trajectory computed for the MMP bound to a short linear peptide. After relaxing the steric clashes and scoring the complexes with MMPBSA energies, the most stable MMP/THP complexes are structurally evaluated using the *whatcheck* software tool (http://swift.cmbi.ru.nl/gv/whatcheck/) prior to run extended MD simulations

tive complex (4AUO) and computational models for several MMP-2/THP pre-reactive complexes [25, 26, 37]. Starting from these templates, molecular editing software can also be used to build initial structures for other complexes.

3.5 Modeling of the Full-Length Enzyme

MD simulations of full-length MMPs can be used to gain detailed insights into the nature and location of *exosites*, the relative mobility of the different domains, or the interaction of MMPs with macromolecular substrates. Preparation of simulations to address these questions is not trivial and the following points should be taken into account:

1. Crystal structures for full-length enzymes are only available for MMP-1, MMP-2, MMP-12, and MMP-13. For other MMPs, molecular models could be generated from crystal structures that are available for the isolated domains by assuming a relative orientation similar to that observed for the full-length enzyme. However, the linker segment connecting the catalytic and hemopexin-like domains is quite variable in length (from 18 amino-acid residues in MMP-21 to 67 in MMP-15) and composition. The secondary structure of this linker segment can be predicted using tools such as BioAssemblyModeler BAM (http://dunbrack.fccc.edu/BAM/). It is worth noting that this is also the case for MMP-2 since the linker in the crystal structures for this enzyme has missing residues (8–11 depending on the selected structure).
2. In general, full-length MMPs are assumed to experience some degree of relative inter-domain motion, so sufficiently large solvent boxes should be employed for these MD simulations (*see* **Note 8**). Moreover, extensive simulations (at least hundreds of nanoseconds) and/or the use of enhanced sampling techniques are mandatory to investigate inter-domain conformational freedom.
3. Protein-protein docking calculations may be used to generate alternative conformations of the full-length enzyme. By considering the catalytic and hemopexin-like domains as independent units, it is possible to use programs like pyDock implemented in the pyDockWeb (http://life.bsc.es/servlet/pydock/home/) to provide the best rigid-body docking orientations of two fragments. Alternatively, pyDockTET has been developed to obtain domain-domain assemblies by scoring the domain-domain poses in terms of binding energies and a pseudo-energy term based on restraints derived from linker end-to-end distances. The energetically most favored poses should then be sampled using extended MD simulations to relax the system and analyze the strength of the inter-domain interactions.

3.6 Running the MD Simulations

To properly run an MD simulation for MMP systems in explicit solvent, the following steps should be sequentially performed to relax and thermalize the system:

1. Solvent boxes around macromolecules are usually generated by replicating the coordinates of a small and pre-equilibrated box in three directions. As a result, the large solvent box surrounding an MMP molecule contains vacuum bubbles at the corners of the interfaces formed by the smaller boxes. Thus, before starting the simulation for the entire system we first relax the large solvent box. This can be done by performing an initial energy minimization, a short MD simulation at constant pressure, and a final minimization of the solvent molecules (counterions can be also included in these steps). In this way, the solvent molecules are allowed to adapt to the solute (i.e., the protein) and the volume of the box changes to the correct density.
2. The whole system should be minimized to remove potentially bad contacts in the initial MMP structure. If needed, this can be done in several steps relaxing first those parts of the system with more severe steric clashes.
3. The system is then gradually heated from 0 K to the desired simulation temperature, usually 300 K. To avoid any problem, heating is best done by running a number of short MD simulations to progressively increase the target temperature of the system. We usually employ six constant volume simulations of 10 ps each to increase the temperature according to the following sequence 0 K → 50 K → 100 K → 150 K → 200 K → 250 K → 300 K.
4. After thermalization, the actual MD simulation can be started by choosing either constant volume (NVT) or constant pressure (NPT) conditions. If using NVT conditions, we recommend running a preliminary pressurization job under NPT conditions (1–2 ns).
5. During the MD simulation of the MMP, it is necessary to discriminate between *equilibration* and *production* phases. In general simulations begin in a state (edited X-ray structure, homology model, docking model, etc.) that can be structurally different from the one that you wish to sample and, therefore, the first part of the MD trajectory is rather sensitive to the initial conditions and to singular transitions occurring during the structural relaxation of the system. Thus, during the equilibration phase, which can extend for tens or even hundreds of *ns* in the case of full-length MMPs, the structure and energy of the system normally evolve to reach a relatively stable plateau. Sampling during the production phase of the simulation occurs along this plateau where the system is in equilibrium (Fig. 6).

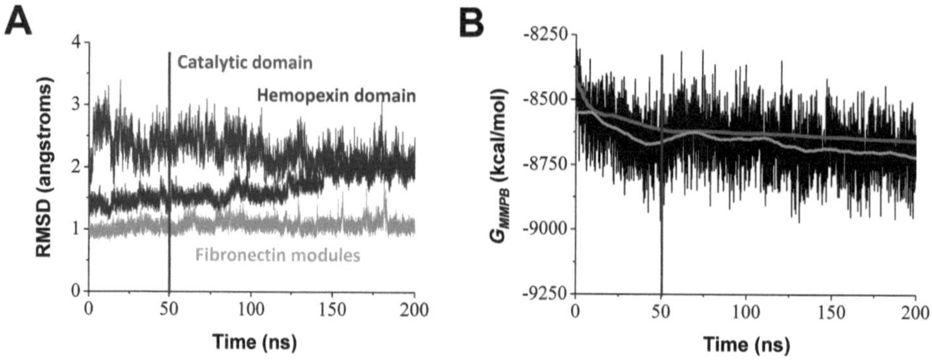

Fig. 6 Time evolution of (**a**) RMSD and (**b**) MM-PB energy for the MD simulation of full-length MMP-2 (PDB ID: 1CK7). The *vertical blue line* separates the equilibration and production phases. The *red* and *green curves* represent the evolution of the accumulated average and the adjacent average, respectively

6. From our experience MD simulations involving the MMP catalytic domain alone require at least tens of *ns* to sample the flexibility of the important Ω-loop, while simulations of full-length MMPs unbound or in complex with THP molecules should extend to hundreds of *ns* to properly explore the conformational landscape of slow inter-domain motions.

3.7 Structural Analysis of MD Simulations

Structural analyses of MD simulations can be used to characterize different aspects of MMP structure and flexibility. Depending on the particular objectives, different kinds of analyses can be performed using a wide variety of tools. But, in general, the following approaches may be useful for extracting relevant information from MMP simulations:

1. Along with the conventional structural analysis, inspection of the time evolution of the root mean squared deviation (RMSD) with respect to a reference structure (usually the initial structure) allows for the selection of the production phase for further analysis. Besides computing RMSDs for the whole protein it is advisable to segregate it into secondary (helices, β-sheets, loops) and tertiary (domains) structural elements.
2. Relevant structural elements within the MMP catalytic domain include the long Ω-loop and the hydrophobic S_1' binding site. To characterize the structure and flexibility of the Ω-loop, clustering analysis based on RMSD values for Cα and Cβ Cartesian coordinates or backbone torsion angles can be performed on selected snapshots extracted from the production phase of the MD trajectory. This can be achieved using the *cpptraj* program in the AMBER package or other programs like the MMTSB-tool set (http://blue11.bch.msu.edu/mmtsb/Cluster.pl).
3. Accessibility to different binding sites can be characterized by computing the *radius of accessibility* for selected backbone and

side chain groups located within the active site cleft. The radius of accessibility [21] is defined as the maximum radius of a spherical ligand that can touch the desired target and its determination is based on the rapid calculation of the molecular surface using programs such as MSMS (http://mgl.scripps.edu/people/sanner/html/msms_home.html).

4. The evolution of the size and shape of cavities in the MMPs like the *S1′* pocket during the MD simulation can be monitored using a program like *mdpocket* (http://fpocket.sourceforge.net/).
5. Interdomain and enzyme-ligand contacts include H-bonds and hydrophobic interactions. H-bonds can be characterized on the basis of geometrical criteria (e.g., X···Y distance <3.5 Å and X–H···Y angle >120°; *cpptraj* has different commands to search for and characterize H-bond contacts). In addition, nonpolar interactions can be detected by measuring the distance between the centers of mass of two interacting groups. We have also used in-house developed software (available upon request) to evaluate the empirical dispersion attraction between pairs of atoms belonging to two hydrophobic groups.

3.8 Energetic Analysis of MD Trajectories

Relative free energies can be *estimated* from conventional MD simulations using end-point free energy methods [38, 39] to evaluate the inhibitory ability of structurally dissimilar compounds toward MMPs and/or to discriminate among different configurations or inter-domain arrangements of full-length MMPs. In particular, the MM-PB(GB)SA (molecular mechanics-Poisson Boltzmann or generalized Born surface area) method offers a good balance between speed and accuracy and has been widely used to compute relative binding affinities [11, 40, 41]. However, there are different variants of this method and various protocols describing its application. We comment below on options that particularly apply to MMPs.

1. Extract a minimum of 500 snapshots separated by 25–50 ps from the production phase. Compute solvation energies using the Poisson-Boltzman (PB) method (this method is especially applicable for large systems). Both linear and nonlinear solvers of the PB equation are implemented in the program *pbsa* that is included in the AMBER package.
2. The calculation of binding energies for ligands bearing ZBGs with a *strong* affinity for the catalytic Zn ion can be problematic. First, a nonbonded approach for the Zn environment should be used for the MM-PBSA calculations. However, the purely MM representation of the Zn···ZBG interaction can result in unbalanced interaction energies and, consequently, a careful validation of the relative MM-PBSA energies would be

necessary and/or alternative approaches using semiempirical QM methods (e.g., QM-GBSA) could be employed.

3. In the case of small and *rigid* ligands bound to MMPs, it can be acceptable to approximate the full binding energy ($\Delta_{\text{bind}}G$) by the interaction energy ($\Delta_{\text{int}}G$) evaluated using the distorted geometries of the protein and ligand moieties extracted from the MD simulation of the complex (i.e., the so-called one-trajectory approximation). This approach minimizes the statistical uncertainty of the average energy differences and facilitates the estimation of relative binding affinities for a series of ligand molecules. Entropic corrections may be omitted.
4. For complexes involving the MMP catalytic domain and a flexible peptide or peptide-mimetic ligands, conventional MM-PBSA calculations will suffice provided that a mixed bonded and nonbonded parameterization of the catalytic Zn site is used (e.g., the direct or water-mediated Zn···O=C(peptide) can be reasonably described by nonbonding MM terms; *see* **Note 4**). However, due to the intrinsic flexibility of peptides, it is necessary to estimate the configurational entropy [42] change associated with the binding of the ligand. To do this we recommend performing energy minimizations and normal mode calculations for a subset of MD snapshots when using the AMBER MM-PBSA tools. These calculations yield absolute entropy terms based on the rigid-rotor and harmonic oscillator approximations that can be improved by adding the purely conformational entropy of the flexible ligand computed with the program *cencalc* (http://sourceforge.net/projects/cencalc/). Relative binding energies must be obtained by combining the MM-PBSA energies from *independent* trajectories of the MMP/peptide and isolated peptide molecules (e.g., Pep1 + MMP/Pep2 → Pep2 + MMP/Pep1) [24].
5. The MM-PBSA method is probably the only viable approach to assess the relative stability and binding affinity of various configurations of full-length MMPs and MMPs complexes with large THP substrates [26]. As mentioned earlier, relative MM-PBSA binding energies must be estimated using independent trajectories because interaction energies from the one-trajectory approximation ignore distortion effects, which are important in large biomolecular systems. Configurational entropy can be estimated using the quasi-harmonic method as implemented in the program *cpptraj*. This method is very computationally efficient and facilitates an estimation of the absolute entropy for large macromolecules. However, it largely overestimates the configurational entropy and provides only an insight into the entropy effects at work.

6. For extended MD simulations of full-length MMPs, the MM-PBSA energies oscillate widely through fast sub-nanosecond motions and can also exhibit lower-amplitude oscillations on longer timescales (~ns), as shown by the evolution of the adjacent average energies in Fig. 6. In this case, it is recommended to assess the statistical uncertainty of the MM-PBSA values by reporting both the standard error and the estimated *block-averaging* error.

The application of rigorous free energy methods, such as Thermodynamic integration (TI) and free energy perturbation (FEP), is beyond the scope of this chapter. Nevertheless, we note that these methods are of limited applicability due to their high computational cost and convergence problems. In principle, they can compute free energy changes for any chemical transformation occurring within the system. But in practice, only very limited changes in the structure of an MMP-bound inhibitor can be tolerated to achieve convergence of the relative binding affinities within an affordable sampling [43]. Furthermore, Ω-loop flexibility and induced-fit effects that are experimentally observed for different MMP/inhibitor complexes require extended simulation times for TI energy calculations [43].

4 Supporting Information

Readers interested in practicing the preparation and execution of an MD simulation can download a tar file (*mini_howto_mmp12_hae_MD.tar*) containing a tutorial for MD simulations of the catalytic domain of the human MMP-12 enzyme complexed with acetohydroxamic acid (PDB code: 1Y93).

5 Notes

1. Due to the structural similarity of the catalytic and hemopexin domains of various MMPs, it is possible to build an initial model using homology modeling for MMPs with no known three-dimensional structure. The amino acid sequences for the MMP of interest can be obtained from the UniProt database (http://www.uniprot.org/) and aligned against sequences in the PDB database using BLAST to find proteins with similar sequences and known structure. Based on sequence similarity, structure coverage, and crystal structure resolution, three or four templates should be selected to perform the comparative modeling using a software tool such as MODELLER (https://salilab.org/modeller/).

2. Ionization states for titratable residues can be selected based on approximate pK_a calculations using the H++ web server (http://biophysics.cs.vt.edu/H++). Only reasonably high resolution (i.e., <2.5 Å) crystal structures should be used in these calculations. Besides the structure, the user should provide a value for the dielectric constant within the protein. An interior dielectric constant of 10 is usually recommended though lower values of 4 and 6 seem to provide better results for buried side chains. Thus, it may be a good strategy to examine the consistency of the predicted pK_a values by comparing the results obtained with two or three high quality structures and a few values (i.e., 4, 6, 10, and 20) for the interior dielectric constant. To include the zinc and calcium ions in the pK_a calculations, edit the input PDB file to change "HETATM" to "ATOM" and use "CA" as residue name and atom type for calcium, and "ZN" as residue name and atom type for zinc. Alternatively, "manually" include the lines for zinc and calcium in the PQR file (a PDB file including atomic charges and bondi radii) which is generated by a preliminary H++ execution retrieving the structure from the protein data bank. In this case, charges (+2.0) and radii (1.00 for Zn^{2+} and 1.70 for Ca^{2+}) should be provided.
3. Building and refinement of side chain conformations in proteins can be easily done using the program SCWRL4 developed in the Dunbrack Lab (http://dunbrack.fccc.edu/scwrl4/). This refinement is particularly important for the asparagine and glutamine side chains because they may be flipped with respect to their most favored conformation in crystal structures. The program also evaluates the protonation of Nδ or Nε atoms for histidine residues.
4. A hybrid bonded and nonbonded representation has been previously employed to model the catalytic zinc in MMP-2 bound to different peptide substrates [22]. In this hybrid model, explicit Zn-L bonds were defined to the three coordinating histidines and to the hydrolytic water molecule, but no explicit bond was defined to the carbonyl group of the scissile peptide bond that acts as the fifth zinc ligand. This hybrid description results in a stable coordination sphere around the metal and, at the same time, is flexible enough to allow for some fluctuations over important reactive distances (e.g., Zn–O···C-peptide for the nucleophilic attack and Glu–COOH···N-peptide for the protonation of the leaving group).
5. The flexible binding environment of the MMP catalytic Zn ion, which can accommodate a large variety of ZBGs in potential inhibitors as well as to bind and hydrolyze peptide substrates, explains why both QM and QM/MM methods play a prominent role in molecular simulation studies of MMPs. As described in Subheading 3, state-of-the-art QM methods are

used to obtain the reference data necessary for deriving the MM parameters for the Zn_1 site in the presence of different ligands and ZBGs using the bonded approach. Parameterization using these QM methods can be computationally costly due to the levels of theory used in these methods (i.e., the still popular B3LYP method combined with a double-ζ basis set). Of course, specialized QM codes like *ORCA* (https://orcaforum.cec.mpg.de/), *Gaussian* (http://www.gaussian.com/), or *GAMESS* (http://www.msg.ameslab.gov/gamess/) are needed. For beginners, we highly recommend *ORCA* since this program offers a wide variety of QM methods and is very efficient. It is also distributed freely for academic purposes and comes with an excellent manual. Importantly, *ORCA* and other programs are readily recognized by the QM interface implemented in the *sander* program, thus facilitating the execution of high level QM/MM calculations (if desired, the *sander* interface can be used to drive full QM single-point or geometry optimizations).

Researchers interested in MMP modeling should be aware of *approximated* QM methods like the Density Functional based Tight Binding method using a self-consistent redistribution of Mulliken charges (this is the so-called SCC-DFTB method; http://www.dftb.org), which can yield reasonably accurate energies and forces much faster than conventional QM methods provided that pairwise atomic parameters are available for the system of interest. Over the last decade many SCC-DFTB applications on Zn enzymes have been reported and perform remarkably well [44]. This has prompted software developers to implement SCC-DFTB/MM simulation methods to carry out energy minimization, molecular dynamics in explicit solvent and potential of mean force calculations. Thus, latter versions of the *sander* program include a *second-order* SCC-DFTB implementation that can be useful for studying both ligand binding and chemical reactivity in MMPs (we recommend, however, validating the structural and energetic SCC-DFTB data by comparison with conventional QM/MM calculations). The current SCC-DFTB implementation can be used to produce MD trajectories on the *ps-ns* time scales rather than the *ns-μs* scales that can be reached by classical MM jobs on GPUs. In addition, the ability of SCC-DFTB to model flexible metal coordination environments is a major achievement that eliminates the need for obtaining MM parameters for the metal sites using the rather cumbersome bonded approach. Nevertheless, the choice between SCC-DFTB and purely MM methods still depends on the required amount of sampling and the goal of the simulation. Note also that a *third-order* DFTB methodology having improved parameters for biological and organic molecules containing the following elements, Br-C-Ca-Cl-F--

H-I-K-Mg-N-Na-O-P-S-Zn, is currently available and will presumably be accessible in upcoming revisions of the AMBER package.

It is envisaged that with software tools like AMBER, integration of QM and SCC-DFTB methods and continuous improvement of hardware technologies will lead to even more realistic and useful simulations of metallo-enzymes including MMPs.

6. The Metal Center Parameter Builder (MCPB) computes bond and angle parameters in a sidechain cluster model (i.e., includes the sidechains of the ligating residues terminated with a methyl group) and charges in an extended cluster model also including the backbone (capped with acetyl and *N*-methyl groups). It uses the *Gaussian* program to perform geometry optimization and frequency calculations at the B3LYP/6-31G* level of theory.

7. Although computational studies of enzymatic catalysis are beyond the scope of this chapter, it may be interesting to note that to study the molecular details of MMP-mediated peptide hydrolysis and/or reaction with ZBGs, it is essential to build a realistic model of the pre-reactive complex and explore its conformational landscape by means of MD simulations. However, QM calculations on cluster models or *static* QM/MM calculations starting from a single geometry provide an incomplete picture of the reaction mechanism. Such calculations can miss many subtle details of the reaction mechanism that render them useless for determining specificity for various substrates. Therefore, QM/MM studies of reaction profiles should be started from at least several snapshots selected from the reference MD trajectory using geometrical criteria.

8. For linear systems like THPs or those that may experience expansion during an MD simulation, octahedral solvent boxes are preferred over the rectangular ones to avoid any direct contact between different copies of the protein.

References

1. Butler GS, Overall CM (2009) Updated biological roles for matrix metalloproteinases and new "intracellular" substrates revealed by degradomics. Biochemistry 48:10830–10845
2. Fingleton B (2008) MMPs as therapeutic targets—still a viable option? Semin Cell Dev Biol 19:61–68
3. Bode W, Maskos K (2003) Structural basis of the matrix metalloproteinases and their physiological inhibitors, the tissue inhibitors of metalloproteinases. Biol Chem 384:863–872
4. Fabre B, Ramos A, de Pascual-Teresa B (2014) Targeting matrix metalloproteinases: exploring the dynamics of the S_1' pocket in the design of selective, small molecule inhibitors. J Med Chem 57:10205–10219
5. Gioia M, Fasciglione GF, Marini S, D'Alessio S, de Sanctis G, Diekmann O, Pieper M, Politi V, Tschesche H, Coletta M (2002) Modulation of the catalytic activity of neutrophil collagenase MMP-8 on bovine collagen I. J Biol Chem 277:23123–23130

6. Overall CM (2002) Molecular determinants of metalloproteinase substrate specificity. Mol Biotechnol 22:51–86
7. Bertini I, Fragai M, Luchinat C, Melikian M, Mylonas E, Sarti N, Svergun DI (2009) Interdomain flexibility in full-length matrix metalloproteinase-1 (MMP-1). J Biol Chem 284:12821–12828
8. Donini OAT, Kollman PA (2000) Calculation and prediction of binding free energies for the matrix metalloproteinases. J Med Chem 43:4180–4188
9. Hou TJ, Zhang W, Xu XJ (2001) Binding affinities for a series of selective inhibitors of gelatinase-A using molecular dynamics with a linear interaction energy approach. J Phys Chem B 105:5304–5315
10. Aschi M, Roccatano D, di Nola A, Gallina C, Gavuzzo E, Pochetti G, Piepere M, Tscheschee H, Mazza F (2002) Computational study of the catalytic domain of human neutrophil collagenase. Specific role of the S_3 and S_3' subsites in the interaction with a phosphonate inhibitor. J Comput Aided Mol Des 16:213–225
11. Hou T, Guo S, Xu X (2002) Predictions of binding of a diverse set of ligands to gelatinase-A by a combination of molecular dynamics and continuum solvent models. J Phys Chem B 106:5527–5535
12. Rizzo RC, Toba S, Kuntz ID (2004) A molecular basis for the selectivity of thiadiazole urea inhibitors with stromelysin-1 and gelatinase-A from generalized Born molecular dynamics simulations. J Med Chem 47:3065–3074
13. Khandelwal A, Lukacova V, Comez D, Kroll DM, Raha S, Balaz S (2005) A combination of docking, QM/MM methods, and MD simulation for binding affinity estimation of metalloprotein ligands. J Med Chem 48:5437–5447
14. Durrant JD, de Oliveira CAF, McCammon JA (2010) Including receptor flexibility and induced fit effects into the design of MMP-2 inhibitors. J Mol Recognit 23:173–182
15. Giangreco I, Lattanzi G, Nicolotti O, Catto M, Laghezza A, Leonetti F, Stefanachi A, Carotti A (2011) Insights into the complex formed by matrix metalloproteinase-2 and alloxan inhibitors: molecular dynamics simulations and free energy calculations. PLoS One 6:e25597
16. de Oliveira CAF, Zissen M, Mongon J, McCammon JA (2007) Molecular dynamics simulations of metalloproteinases types 2 and 3 reveal differences in the dynamic behavior of the S_1' binding pocket. Curr Pharm Des 13:3471–3475
17. Falconi M, Altobelli G, Iovino MC, Politi V, Desideri A (2003) Molecular dynamics simulation of Matrix Metalloproteinase 2: fluctuations and time evolution of recognition pockets. J Comput Aided Mol Des 17:837–848
18. Manzetti S, McCulloch DR, Herington AC, van der Spoel D (2003) Modeling of enzyme–substrate complexes for the metalloproteases MMP-3, ADAM-9 and ADAM-10. J Comput Aided Mol Des 17:551–565
19. Azhagiya Singam ER, Rajapandian V, Subramanian V (2014) Molecular dynamics simulation study on the interaction of collagen-like peptides with gelatinase-A (MMP-2). Biopolymers 101:779–794
20. Kotra LP, Cross JB, Shimura Y, Fridman R, Schlegel HB, Mobashery S (2001) Insight into the complex and dynamic process of activation of matrix metalloproteinases. J Am Chem Soc 123:3108–3113
21. Díaz N, Suárez D (2007) Molecular dynamics simulations of matrix metalloproteinase 2: role of the structural metal ions. Biochemistry 46:8943–8952
22. Díaz N, Suárez D (2008) Molecular dynamics simulations of the active matrix metalloproteinase-2: positioning of the N-terminal fragment and binding of a small peptide substrate. Proteins 72:50–61
23. Díaz N, Suárez D (2008) Peptide hydrolysis catalyzed by matrix metalloproteinase 2: a computational study. J Phys Chem B 112:8412–8424
24. Díaz N, Suárez D, Suárez E (2010) Kinetic and binding effects in peptidesubstrate selectivity of matrix metalloproteinase-2: molecular dynamics and QM/MM calculations. Proteins 78:1–11
25. Díaz N, Suárez D, Valdés H (2013) Unraveling the molecular structure of the catalytic domain of matrix metalloproteinase-2 in complex with a triple-helical peptide by means of molecular dynamics simulations. Biochemistry 52:8556–8569
26. Díaz N, Suárez D (2015) Extensive simulations of the full-length matrix metalloproteinase-2 enzyme in a prereactive complex with a collagen triple-helical peptide. Biochemistry 54:1243–1258
27. Díaz N, Suárez D, Valdés H (2008) From the X-ray compact structure to the elongated form of the full-length MMP-2 enzyme in solution: a molecular dynamics study. J Am Chem Soc 130:14070–14071
28. Valdés H, Díaz N, Suárez D, Fernández-Recio J (2010) Interdomain conformations in the full-length MMP-2 enzyme explored by protein-protein docking calculations using

pyDock. J Chem Theory Comput 6:2204–2213
29. Díaz N, Suárez D (2012) Alternative interdomain configurations of the full-length MMP-2 enzyme explored by molecular dynamics simulations. J Phys Chem B 116:2677–2686
30. Salomon-Ferrer R, Case DA, Walker RC (2013) An overview of the Amber biomolecular simulation package. WIREs Comput Mol Sci 3:198–210
31. Case DA, Berryman JT, Betz RM, Cerutti DS, Cheatham TEI, Darden TA, Duke RE, Giese TJ, Gohlke H, Goetz AW, Homeyer N, Izadi S, Janowski P, Kaus J, Kovalenko A, Lee TS, LeGrand S, Li P, Luchko T, Luo R, Madej B, Merz KM, Monard G, Needham P, Nguyen H, Nguyen HT, Omelyan I, Onufriev A, Roe DR, Roitberg A, Salomon-Ferrer R, Simmerling CL, Smith W, Swails J, Walker RC, Wang J, Wolf RM, Wu X, Kollman PA (2015) AMBER15. University of California, San Francisco
32. Case DA, Babin V, Berryman JT, Betz RM, Cai Q, Cerutti DS, Cheatham TEI, Darden TA, Duke RE, Gohlke H, Goetz AW, Gusarov S, Homeyer N, Janowski P, Kaus J, Kolossváry I, Kovalenko A, Lee TS, LeGrand S, Luchko T, Luo R, Madej B, Merz KM, Paesani F, Roe DR, Roitberg A, Sagui C, Salomon-Ferrer R, Seabra G, Simmerling CL, Smith W, Swails J, Walker RC, Wang J, Wolf RM, Wu X, Kollman PA (2014) AMBER 14. University of California, San Francisco
33. Li P, Roberts BP, Chakravorty DK, Merz KM (2013) Rational design of particle mesh Ewald compatible Lennard-Jones parameters for +2 metal cations in explicit solvent. J Chem Theory Comput 9:2733–2748
34. Díaz N, Suárez D, Sordo TL (2006) Quantum chemical study on the coordination environment of the catalytic zinc ion in matrix metalloproteinases. J Phys Chem B 110:24222–24230
35. Sousa SF, Fernandes PA, Ramos MJ (2007) Effective tailor-made force field parameterization of the several Zn coordination environments in the puzzling FTase enzyme: opening the door to the full understanding of its elusive catalytic mechanism. Theor Chem Acc 117:171–181
36. Santos-Martins D, Forli S, Ramos MJ, Olson AJ (2014) AutoDock4Zn: an improved AutoDock force field for small-molecule docking to zinc metalloproteins. J Chem Inf Model 54:2371–2379
37. Manka SW, Carafoli F, Visse R, Bihan D, Raynal N, Farndale RW, Murphy G, Enghild JJ, Hohenester E, Nagase H (2012) Structural insights into triple-helical collagen cleavage by matrix metalloproteinase. Proc Natl Acad Sci U S A 109:12461–12466
38. Gohlke H, Case DA (2004) Converging free energy estimates: MM-PB(GB)SA studies on the protein-protein complex Ras-Raf. J Comput Chem 25:238–250
39. Homeyer N, Gohlke H (2012) Free energy calculations by the molecular mechanics Poisson-Boltzmann surface area method. Mol Inform 31:114–122
40. Zhou Z-G, Yao Q-Z, Lei D, Zhang Q-Q, Zhang J (2014) Investigations on the mechanisms of interactions between matrix metalloproteinase 9 and its flavonoid inhibitors using a combination of molecular docking, hybrid quantum mechanical/molecular mechanical calculations, and molecular dynamics simulations. Can J Chem 92:821–830
41. Marcial BL, Sousa SF, Barbosa IL, Dos Santos HF, Ramos MJ (2012) Chemically modified tetracyclines as inhibitors of MMP-2 matrix metalloproteinase: a molecular and structural study. J Phys Chem B 116:13644–13654
42. Suárez D, Díaz N (2015) Direct methods for computing single-molecule entropies from molecular simulations. WIREs Comput Mol Sci 5:1–26
43. Fabre B, Filipiak K, Díaz N, Zapico JM, Suárez D, Ramos A, de Pascual-Teresa B (2014) An integrated computational and experimental approach to gaining selectivity for MMP-2 within the gelatinase subfamily. Chembiochem 15:399–412
44. Xu D, Cui Q, Guo H (2014) Quantum mechanical/molecular mechanical studies of zinc hydrolases. Int Rev Phys Chem 33:1–41

Part IV

Determining Matrix Metalloprotease Substrate Specificity

Chapter 8

Determining the Substrate Specificity of Matrix Metalloproteases using Fluorogenic Peptide Substrates

Maciej J. Stawikowski, Anna M. Knapinska, and Gregg B. Fields

Abstract

A continuous assay method, such as the one that utilizes an increase in fluorescence upon hydrolysis, allows for rapid and convenient kinetic evaluation of proteases. To better understand MMP behaviors toward native substrates, a variety of fluorescence resonance energy transfer (FRET)/intramolecular fluorescence energy transfer (IFET) triple-helical substrates have been constructed to examine the collagenolytic activity of MMP family members. Results of these studies have been valuable for providing insights into (a) the relative triple-helical peptidase activities of the various collagenolytic MMPs, (b) the collagen preferences of these MMPs, and (c) the relative roles of MMP domains and specific residues in efficient collagenolysis. The present chapter provides an overview of MMP FRET triple-helical substrates and describes how to construct and utilize these substrates.

Key words Collagen, FRET substrates, Matrix metalloproteinases, Triple-helical substrates, Fluorogenic substrates

1 Introduction

1.1 Development of Peptide Substrates for Matrix Metalloproteinases

A continuous assay method, such as the one that utilizes an increase in fluorescence upon hydrolysis, allows for rapid and convenient kinetic evaluation of proteases, both in solution and cell surface bound. For this reason, significant research efforts have focused on designing substrates for matrix metalloproteinase (MMP) family members [1]. Sequence specificity studies for MMPs have been performed with peptides based on protein sequences surrounding MMP cleavage sites, and the results of these studies have been comprehensively reviewed [2, 3]. In addition, phage display and peptide libraries have been utilized to identify MMP-selective substrate sequences [4–10]. Results of the aforementioned studies were subsequently used to design fluorogenic MMP substrates.

Charles A. Galea (ed.), *Matrix Metalloproteases: Methods and Protocols*, Methods in Molecular Biology, vol. 1579,
DOI 10.1007/978-1-4939-6863-3_8, © Springer Science+Business Media LLC 2017

1.2 General Characteristics of Fluorogenic Substrates

Fluorogenic substrates provide a particularly convenient enzyme assay method, as they can be monitored continuously and utilized at reasonably low concentration ranges. There are three types of fluorogenic substrates: (1) aromatic amines, (2) contact-quenched, and (3) resonance energy transfer quenched [11]. MMP fluorogenic substrates have been developed using resonance energy transfer quenching, which are sometimes referred to as intramolecular fluorescence energy transfer substrates (IFETS) [12].

1.2.1 Resonance Energy Transfer

Förster resonance energy transfer (FRET) may occur between a fluorescent donor group and a quenching acceptor group when the fluorophore has a high quantum yield (Φ_F) and a fluorescence emission spectrum that is exactly overlapped by a strongly absorbing acceptor (quencher) [11]. The efficiency of transfer will also depend upon the distance between the donor and acceptor [11, 12]. Resonance energy transfer quenched fluorogenic substrates are thus created by incorporating the donor and acceptor on opposite sides of the scissile bond, at a distance that allows for highly efficient energy transfer. When the substrate is cleaved by an enzyme, diffusion of the donor-containing substrate fragment away from the acceptor-containing substrate fragment results in loss of energy transfer and subsequent fluorescence. The creation of an optimal MMP fluorogenic substrate thus depends upon (1) incorporation of a fluorescent donor that has a high quantum yield, (2) incorporation of an acceptor that absorbs at the donor fluorescence emission wavelength (λ_{em}), and (3) the number of amino acid residues between the donor and the acceptor.

1.2.2 Chemical Moieties Used for Donor/Acceptor Pairs

There are a wide variety of donor and acceptor groups that have been used in fluorogenic substrates [11, 12]. MMP substrates have used one of eight different fluorophores. The first was Trp [13], which has an ε_{280} value of 5600 M^{-1} cm^{-1} and Φ_F of 0.2. The quencher group was *N*-2,4-dinitrophenyl (Dnp) [13, 14], which has an absorption maximum at $\lambda = 363$ nm and a prominent shoulder at $\lambda = 410$ nm [15]. Alternatively, a fluorogenic substrate for the *Leishmania* surface metalloproteinase used a dansyl group to quench Trp fluorescence [16]. It should be noted that dansyl quenching of Trp decreases rapidly with intervening residue distance [11]. Knight and colleagues proposed the use of (7-methoxycoumarin-4-yl)acetyl (Mca; ε_{325} = 14,500 M^{-1} cm^{-1} and $\Phi_F = 0.49$) as a fluorophore for MMP substrates [15]. Mca is efficiently quenched by Dnp moieties [15, 17], as the shoulder in the Dnp absorption spectrum overlaps the Mca fluorescence emission spectrum [15]. Dnp is also used as a quencher for MMP substrates containing the fluorophore *N*-methylanthranilic acid (Nma) [18]. 5-[(2-Aminoethyl)amino]naphthalene-1-sulfonic acid (Edans; ε_{336} = 5400 M^{-1} cm^{-1} and $\Phi_F = 0.13$) has been used as a donor group for an MMP substrate [19]. Edans fluorescence is quenched by the 4-(4-dimethylaminophenylazo)benzoic acid (Dabcyl) group [19]. MMP substrates have been developed with Lucifer Yellow (LY) as a fluorophore [20, 21]. LY has an ε_{420} value of 12,000 M^{-1} cm^{-1} and is quenched by 5-carboxytetramethylrhodamine (Ctmr) [20]. MMP and ADAM (a disintegrin and metalloproteinase domain) substrates have utilized 5-carboxyfluorescein (5-Fam) as a fluorophore [22, 23]. Fam has an ε_{492} of 65,000 M^{-1} cm^{-1} and Φ_F of 0.92. Dabcyl was used for quenching 5-Fam in ADAM substrates [23], while QXL 520 was used to quench 5-Fam fluorescence in MMP substrates [22]. 2-Aminobenzoyl (Abz)/anthranilamide has been applied as a fluorophore for solution and solid-phase MMP-9 and MMP-12 substrates used for screening potential phosphinic peptide inhibitors [24, 25]. Tyr(NO_2) was used to quench Abz fluorescence that possesses an ε_{316} value of 2300 M^{-1} cm^{-1} and an Φ_F of 0.60 [26–28]. Lastly, Cy3 (ε_{550} = 150,000 M^{-1} cm^{-1} and Φ_F = 0.15) was used as fluorophore and Cy5Q as quencher for an MMP-3 high-throughput screening (HTS) substrate [29].

1.2.3 Practical Aspects of Donor/Acceptor Groups

Trp and Tyr(NO_2) can be incorporated easily by solid-phase peptide synthesis and Dnp or dansyl acylated to the *N*-terminus of the peptide. Other moieties require additional synthetic steps. For example, to use the Mca/Dnp pair, a Dnp derivative such as *N*-3-(2,4-dinitrophenyl)-L-2,3-diaminopropionic acid (Dpa) [15] or Lys(Dnp) [17] must be synthesized (see below) and incorporated into a peptide by solid-phase methods, while Mca is acylated to the peptide *N*-terminus. Alternatively, the Mca derivative L-2-amino-3-(7-methoxy-4-coumaryl)propionic acid (Amp) can be synthesized (see below) and incorporated by solid-phase methods [30]. For Dabcyl/Edans substrates, derivatives of Glu(Edans) must be prepared (see below). Substrates containing Nma, Fam, or Cy3 can be synthesized using "on-resin" reactions to create Lys(Nma), Lys(Fam), or Lys(Cy3) moieties (see below).

In contrast to solid-phase assembly of fluorogenic substrates, "post-synthesis" approaches have been used for the preparation of peptides containing LY and Ctmr [20, 21] or Cy5Q [29]. Although they are more laborious, post-synthesis approaches do alleviate one potential purification problem of fluorogenic peptides. Typically, fluorophores and quenchers are hydrophobic, and thus the reverse-phase (RP) HPLC separation of the desired peptide from deletion peptides containing these groups can be difficult.

1.3 Collagen-Model Triple-Helical Substrates

Collagen is the most abundant protein in animals, and is the major structural protein found in the connective tissues such as basement membranes, tendons, ligaments, cartilage, bone, and skin. The most defining feature of collagen is the supersecondary structure, composed of three parallel extended left-handed polyproline type II alpha chains of primarily repeating Gly-Xxx-Yyy triplets. Three left-handed strands intertwine in a right-handed fashioned around a common axis to form a triple-helix. In the collagen Gly-Xxx-Yyy triplet, the residue in the Xxx position is often L-proline (Pro) and the residue in the Yyy position is often 4(R)-hydroxy-L-proline (Hyp), accounting for 20% of the total amino acid composition in collagen [31]. The other commonly found amino acids are Ala, Lys, Arg, Leu, Val, Ser, and Thr [31]. The packing of the triple-helical coiled-coil structure requires Gly in every third position.

For several decades triple-helical peptides (THPs) consisting of collagen-model sequences and their three-dimensional folds have been constructed and studied to fully investigate the structural and biological roles of collagenous proteins [32–39]. THP models of the collagenase cleavage site in type I collagen were utilized to examine the substrate specificity of MMP family members. The α1(I)772–786 THP was hydrolyzed by MMP-1, MMP-1($\Delta_{243-450}$), MMP-2, and MMP-13 but not by MMP-3 or MMP-3($\Delta_{248-460}$) [40, 41]. Sequence and mass spectrometric analysis of the products generated by the MMPs indicated that MMP hydrolysis occurred primarily at the Gly-Ile bond, which is analogous to the bond cleaved in native type I collagen. Thus, the α1(I)772–786 THP contained all necessary information to direct MMP-1, MMP-2, and MMP-13 binding and proteolysis. Similar results were reported by Ottl et al. [42–44], who found that a heterotrimeric THP incorporating the $[\alpha1(I)]_2\alpha2(I)$772–783 sequence was efficiently hydrolyzed by MMP-8. The α1(I)772–786 THP was subsequently used to dissect the mechanism of the initial steps of MMP-1 catalyzed collagneolysis [45].

To measure collagenolytic MMP activities, our laboratory has described a number of FRET triple-helical peptide (fTHP) substrates that are either suitable for most collagenolytic MMPs or selective for different collagenolytic MMPs (Table 1) [46–52]. These triple-helical substrates utilize FRET primarily via incorporation of Mca as the fluorophore and the Dnp as the quencher within the same peptide chain [46, 53, 54]. Typically, the P_5' subsite of the fTHP accommodates Lys(Dnp) while the P_5 subsite accommodates Lys(Mca) (Fig. 1).

Designing fTHP substrates poses many challenges. Several key factors have to be taken into account: (1) the distance between a donor and acceptor allowing for efficient FRET; (2) suitable location of the donor/acceptor pair such that they do not impair substrate–enzyme interactions; and (3) suitable location of the donor/acceptor pair so as not to significantly influence triple-helical peptide stability. The situation is furthermore complicated by the fact that each MMP domain makes contact with different fTHP strands. For example, the leading strand of the triple-helix forms most of the contact with the catalytic (CAT) domain while the middle strand forms significant contact with the hemopexin-like (HPX) domain [45]. Quite often meeting all of the above criteria is not

Table 1
Hydrolysis of fTHPs by MMPs

Enzyme	Substrate abbreviation	Substrate sequence	k_{cat} (s^{-1})	K_M (μM)	k_{cat}/K_M[a] (s^{-1} M^{-1})	Reference
MMP-1	fTHP-15	$(Gly-Pro-Hyp)_5$-Gly-Pro-Lys(Mca)-Gly-Pro-Gln-Gly~Leu-Arg-Gly-Gln-Lys(Dnp)-Gly-Val-Arg-$(Gly-Pro-Hyp)_5$-NH_2	0.085 ± 0.019	15 ± 2.3	5680[b]	[52]
MMP-1(R291A)	fTHP-15	$(Gly-Pro-Hyp)_5$-Gly-Pro-Lys(Mca)-Gly-Pro-Gln-Gly~Leu-Arg-Gly-Gln-Lys(Dnp)-Gly-Val-Arg-$(Gly-Pro-Hyp)_5$-NH_2	0.13 ± 0.001	38 ± 14	3340[b]	[52]
MMP-1(I290A,R291A)	fTHP-15	$(Gly-Pro-Hyp)_5$-Gly-Pro-Lys(Mca)-Gly-Pro-Gln-Gly~Leu-Arg-Gly-Gln-Lys(Dnp)-Gly-Val-Arg-$(Gly-Pro-Hyp)_5$-NH_2	0.085 ± 0.017	24 ± 7.1	3540[b]	[52]
MMP-1(Δ243–450)	fTHP-15	$(Gly-Pro-Hyp)_5$-Gly-Pro-Lys(Mca)-Gly-Pro-Gln-Gly~Leu-Arg-Gly-Gln-Lys(Dnp)-Gly-Val-Arg-$(Gly-Pro-Hyp)_5$-NH_2	0.02 ± 0.010	13 ± 1.9	1500[b]	[52]
MMP-1	fTHP-16	$(Gly-Pro-Hyp)_5$-Gly-Pro-Ser-Gly-Ala-Glu-Gly-Pro-Lys(Mca)-Gly-Pro-Gln-Gly~Leu-Arg-Gly-Gln-Lys(Dnp)-Gly-Val-Arg-$(Gly-Pro-Hyp)_5$-NH_2	0.314 ± 0.012	20 ± 4.3	15,490[b]	[52]
MMP-1(R291A)	fTHP-16	$(Gly-Pro-Hyp)_5$-Gly-Pro-Ser-Gly-Ala-Glu-Gly-Pro-Lys(Mca)-Gly-Pro-Gln-Gly~Leu-Arg-Gly-Gln-Lys(Dnp)-Gly-Val-Arg-$(Gly-Pro-Hyp)_5$-NH_2	0.712 ± 0.008	166 ± 16	4273[b]	[52]
MMP-1(I290A,R291A)	fTHP-16	$(Gly-Pro-Hyp)_5$-Gly-Pro-Ser-Gly-Ala-Glu-Gly-Pro-Lys(Mca)-Gly-Pro-Gln-Gly~Leu-Arg-Gly-Gln-Lys(Dnp)-Gly-Val-Arg-$(Gly-Pro-Hyp)_5$-NH_2	0.524 ± 0.016	122 ± 2.9	4298[b]	[52]
MMP-1(Δ243–450)	fTHP-16	$(Gly-Pro-Hyp)_5$-Gly-Pro-Ser-Gly-Ala-Glu-Gly-Pro-Lys(Mca)-Gly-Pro-Gln-Gly~Leu-Arg-Gly-Gln-Lys(Dnp)-Gly-Val-Arg-$(Gly-Pro-Hyp)_5$-NH_2	0.085 ± 0.027	79 ± 17	1073[b]	[52]

(continued)

Table 1 (continued)

Enzyme	Substrate abbreviation	Substrate sequence	k_{cat} (s^{-1})	K_M (μM)	k_{cat}/K_M[a] (s^{-1} M^{-1})	Reference
MMP-1	fTHP-17	$(Gly\text{-}Pro\text{-}Hyp)_4$-Gly-Pro-Lys(Mca)-Gly-Pro-Gln-Gly~Leu-Arg-Gly-Gln-Lys(Dnp)-Gly-Val-Arg-Gly-Leu-Hyp-Gly-Gln-Arg-Gly-Glu-Arg-$(Gly\text{-}Pro\text{-}Hyp)_4$-$NH_2$	0.303 ± 0.003	6.8 ± 1.7	44,350[b]	[52]
MMP-1(R291A)	fTHP-17	$(Gly\text{-}Pro\text{-}Hyp)_4$-Gly-Pro-Lys(Mca)-Gly-Pro-Gln-Gly~Leu-Arg-Gly-Gln-Lys(Dnp)-Gly-Val-Arg-Gly-Leu-Hyp-Gly-Gln-Arg-Gly-Glu-Arg-$(Gly\text{-}Pro\text{-}Hyp)_4$-$NH_2$	0.105 ± 0.005	14 ± 2.3	7515[b]	[52]
MMP-1(I290A,R291A)	fTHP-17	$(Gly\text{-}Pro\text{-}Hyp)_4$-Gly-Pro-Lys(Mca)-Gly-Pro-Gln-Gly~Leu-Arg-Gly-Gln-Lys(Dnp)-Gly-Val-Arg-Gly-Leu-Hyp-Gly-Gln-Arg-Gly-Glu-Arg-$(Gly\text{-}Pro\text{-}Hyp)_4$-$NH_2$	0.144 ± 0.003	17 ± 0.8	8527[b]	[52]
MMP-1(Δ243–450)	fTHP-17	$(Gly\text{-}Pro\text{-}Hyp)_4$-Gly-Pro-Lys(Mca)-Gly-Pro-Gln-Gly~Leu-Arg-Gly-Gln-Lys(Dnp)-Gly-Val-Arg-Gly-Leu-Hyp-Gly-Gln-Arg-Gly-Glu-Arg-$(Gly\text{-}Pro\text{-}Hyp)_4$-$NH_2$	0.016 ± 0.003	32 ± 9.6	506[b]	[52]
MMP-1	fTHP-15	$(Gly\text{-}Pro\text{-}Hyp)_5$-Gly-Pro-Lys(Mca)-Gly-Pro-Gln-Gly~Leu-Arg-Gly-Gln-Lys(Dnp)-Gly-Val-Arg-$(Gly\text{-}Pro\text{-}Hyp)_5$-$NH_2$	0.0071 ± 0.0011	1.8 ± 0.30	4100 ± 800[c]	[66]
MMP-1	fTHP-16	$(Gly\text{-}Pro\text{-}Hyp)_5$-Gly-Pro-Ser-Gly-Ala-Glu-Gly-Pro-Lys(Mca)-Gly-Pro-Gln-Gly~Leu-Arg-Gly-Gln-Lys(Dnp)-Gly-Val-Arg-$(Gly\text{-}Pro\text{-}Hyp)_5$-$NH_2$	0.046 ± 0.0093	13 ± 2.2	3500 ± 580[c]	[66]
MMP-1	fTHP-17	$(Gly\text{-}Pro\text{-}Hyp)_4$-Gly-Pro-Lys(Mca)-Gly-Pro-Gln-Gly~Leu-Arg-Gly-Gln-Lys(Dnp)-Gly-Val-Arg-Gly-Leu-Hyp-Gly-Gln-Arg-Gly-Glu-Arg-$(Gly\text{-}Pro\text{-}Hyp)_4$-$NH_2$	0.105 ± 0.0046	2.9 ± 0.077	37,000 ± 4700[c]	[66]

(continued)

Table 1 (continued)

Enzyme	Substrate abbreviation	Substrate sequence	k_{cat} (s^{-1})	K_M (μM)	k_{cat}/K_M[a] (s^{-1} M^{-1})	Reference
MMP-2	fTHP-15	$(Gly\text{-}Pro\text{-}Hyp)_5$-Gly-Pro-Lys(Mca)-Gly-Pro-Gln-Gly~Leu-Arg-Gly-Gln-Lys(Dnp)-Gly-Val-Arg-$(Gly\text{-}Pro\text{-}Hyp)_5\text{-}NH_2$	1.2 ± 0.12	5.9 ± 0.57	210,000 ± 17,000[c]	[66]
MMP-2	fTHP-16	$(Gly\text{-}Pro\text{-}Hyp)_5$-Gly-Pro-Ser-Gly-Ala-Glu-Gly-Pro-Lys(Mca)-Gly-Pro-Gln-Gly~Leu-Arg-Gly-Gln-Lys(Dnp)-Gly-Val-Arg-$(Gly\text{-}Pro\text{-}Hyp)_5\text{-}NH_2$	1.2 ± 0.05	5.2 ± 0.84	230,000 ± 43,000[c]	[66]
MMP-2	fTHP-17	$(Gly\text{-}Pro\text{-}Hyp)_4$-Gly-Pro-Lys(Mca)-Gly-Pro-Gln-Gly~Leu-Arg-Gly-Gln-Lys(Dnp)-Gly-Val-Arg-Gly-Leu-Hyp-Gly-Gln-Arg-Gly-Glu-Arg-$(Gly\text{-}Pro\text{-}Hyp)_4\text{-}NH_2$	0.042 ± 0.0022	0.73 ± 0.062	57,000 ± 12,000[c]	[66]
MMP-8	fTHP-15	$(Gly\text{-}Pro\text{-}Hyp)_5$-Gly-Pro-Lys(Mca)-Gly-Pro-Gln-Gly~Leu-Arg-Gly-Gln-Lys(Dnp)-Gly-Val-Arg-$(Gly\text{-}Pro\text{-}Hyp)_5\text{-}NH_2$	0.088 ± 0.012	9.2 ± 2.1	9600 ± 1200[c]	[66]
MMP-8	fTHP-16	$(Gly\text{-}Pro\text{-}Hyp)_5$-Gly-Pro-Ser-Gly-Ala-Glu-Gly-Pro-Lys(Mca)-Gly-Pro-Gln-Gly~Leu-Arg-Gly-Gln-Lys(Dnp)-Gly-Val-Arg-$(Gly\text{-}Pro\text{-}Hyp)_5\text{-}NH_2$	0.24 ± 0.009	11 ± 0.44	22,000 ± 1800[c]	[66]
MMP-8	fTHP-17	$(Gly\text{-}Pro\text{-}Hyp)_4$-Gly-Pro-Lys(Mca)-Gly-Pro-Gln-Gly~Leu-Arg-Gly-Gln-Lys(Dnp)-Gly-Val-Arg-Gly-Leu-Hyp-Gly-Gln-Arg-Gly-Glu-Arg-$(Gly\text{-}Pro\text{-}Hyp)_4\text{-}NH_2$	0.30 ± 0.035	29 ± 2.4	10,000 ± 900[c]	[66]
MMP-13	fTHP-15	$(Gly\text{-}Pro\text{-}Hyp)_5$-Gly-Pro-Lys(Mca)-Gly-Pro-Gln-Gly~Leu-Arg-Gly-Gln-Lys(Dnp)-Gly-Val-Arg-$(Gly\text{-}Pro\text{-}Hyp)_5\text{-}NH_2$	4.2 ± 0.25	67 ± 3.9	63,000 ± 1200[c]	[66]

(continued)

Table 1 (continued)

Enzyme	Substrate abbreviation	Substrate sequence	k_{cat} (s^{-1})	K_M (μM)	k_{cat}/K_M[a] (s^{-1} M^{-1})	Reference
MMP-13	fTHP-16	$(Gly-Pro-Hyp)_5$-Gly-Pro-Ser-Gly-Ala-Glu-Gly-Pro-Lys(Mca)-Gly-Pro-Gln-Gly~Leu-Arg-Gly-Gln-Lys(Dnp)-Gly-Val-Arg-$(Gly-Pro-Hyp)_5$-NH_2	2.7 ± 0.12	25 ± 2.7	107,000 ± 8400[c]	[66]
MMP-13	fTHP-17	$(Gly-Pro-Hyp)_4$-Gly-Pro-Lys(Mca)-Gly-Pro-Gln-Gly~Leu-Arg-Gly-Gln-Lys(Dnp)-Gly-Val-Arg-Gly-Leu-Hyp-Gly-Gln-Arg-Gly-Glu-Arg-$(Gly-Pro-Hyp)_4$-NH_2	0.0058 ± 0.00054	0.15 ± 0.022	39,000 ± 12,000[c]	[66]
MT1-MMP	fTHP-15	$(Gly-Pro-Hyp)_5$-Gly-Pro-Lys(Mca)-Gly-Pro-Gln-Gly~Leu-Arg-Gly-Gln-Lys(Dnp)-Gly-Val-Arg-$(Gly-Pro-Hyp)_5$-NH_2	0.034 ± 0.0036	11 ± 1.9	3,000 ± 120[c]	[66]
MT1-MMP	fTHP-16	$(Gly-Pro-Hyp)_5$-Gly-Pro-Ser-Gly-Ala-Glu-Gly-Pro-Lys(Mca)-Gly-Pro-Gln-Gly~Leu-Arg-Gly-Gln-Lys(Dnp)-Gly-Val-Arg-$(Gly-Pro-Hyp)_5$-NH_2	0.102 ± 0.011	20 ± 5.8	5200 ± 120[c]	[66]
MT1-MMP	fTHP-17	$(Gly-Pro-Hyp)_4$-Gly-Pro-Lys(Mca)-Gly-Pro-Gln-Gly~Leu-Arg-Gly-Gln-Lys(Dnp)-Gly-Val-Arg-Gly-Leu-Hyp-Gly-Gln-Arg-Gly-Glu-Arg-$(Gly-Pro-Hyp)_4$-NH_2	0.017 ± 0.00037	5.2 ± 0.58	3300 ± 510[c]	[66]
MMP-8	fTHP-4[d]	$(Gly-Pro-Hyp)_5$-Gly-Pro-Lys(Mca)-Gly-Pro-Gln-Gly~Leu-Arg-Gly-Gln-Lys(Dnp)-Gly-Val-Arg-$(Gly-Pro-Hyp)_5$-NH_2	0.035	7.6	4500[e]	[50]
MMP-8	C_{10}-fTHP-4[d]	C_{10}-$(Gly-Pro-Hyp)_5$-Gly-Pro-Lys(Mca)-Gly-Pro-Gln-Gly~Leu-Arg-Gly-Gln-Lys(Dnp)-Gly-Val-Arg-$(Gly-Pro-Hyp)_5$-NH_2	0.089	14.4	6200[e]	[50]

(continued)

Table 1 (continued)

Enzyme	Substrate abbreviation	Substrate sequence	k_{cat} (s^{-1})	K_M (μM)	k_{cat}/K_M[a] (s^{-1} M^{-1})	Reference
MMP-8	C_{16}-fTHP-4[d]	C_{16}-(Gly-Pro-Hyp)$_5$-Gly-Pro-Lys(Mca)-Gly-Pro-Gln-Gly~Leu-Arg-Gly-Gln-Lys(Dnp)-Gly-Val-Arg-(Gly-Pro-Hyp)$_5$-NH_2	0.016	9.95	1600[e]	[50]
MMP-8	fTHP-9	(Gly-Pro-Hyp)$_5$-Gly-Pro-Lys(Mca)-Gly-Pro-Gln-Gly~**Cys(Mob)**-Arg-Gly-Gln-Lys(Dnp)-Gly-Val-Arg-(Gly-Pro-Hyp)$_5$-NH_2[f]	0.94	5.4	175,000[e]	[50]
MMP-8	C_{10}-fTHP-9	C_{10}-(Gly-Pro-Hyp)$_5$-Gly-Pro-Lys(Mca)-Gly-Pro-Gln-Gly~**Cys(Mob)**-Arg-Gly-Gln-Lys(Dnp)-Gly-Val-Arg-(Gly-Pro-Hyp)$_5$-NH_2[f]	0.64	8.8	73,000[e]	[50]
MMP-8	C_{16}-fTHP-9	C_{16}-(Gly-Pro-Hyp)$_5$-Gly-Pro-Lys(Mca)-Gly-Pro-Gln-Gly~**Cys(Mob)**-Arg-Gly-Gln-Lys(Dnp)-Gly-Val-Arg-(Gly-Pro-Hyp)$_5$-NH_2[f]	0.13	9.6	14,000[e]	[50]
MMP-8	fTHP-10	(Gly-Pro-Hyp)$_5$-Gly-Pro-Lys(Mca)-Gly-Pro-**Orn**-Gly~Leu-Arg-Gly-Gln-Lys(Dnp)-Gly-Val-Arg-(Gly-Pro-Hyp)$_5$-NH_2[f]	0.035	8.15	4300[e]	[50]
MMP-8	fTHP-11	(Gly-Pro-Hyp)$_5$-Gly-Pro-Lys(Mca)-Gly-Pro-**Orn**-Gly~**Cys(Mob)**-Arg-Gly-Gln-Lys(Dnp)-Gly-Val-Arg-(Gly-Pro-Hyp)$_5$-NH_2[f]	0.38	19.7	19,000[e]	[50]
MMP-8	C_{10}-fTHP-11	C_{10}-(Gly-Pro-Hyp)$_5$-Gly-Pro-Lys(Mca)-Gly-Pro-**Orn**-Gly~**Cys(Mob)**-Arg-Gly-Gln-Lys(Dnp)-Gly-Val-Arg-(Gly-Pro-Hyp)$_5$-NH_2[f]	0.007	3.60	1900[e]	[50]

(continued)

Table 1 (continued)

Enzyme	Substrate abbreviation	Substrate sequence	k_{cat} (s^{-1})	K_M (μM)	k_{cat}/K_M[a] (s^{-1} M^{-1})	Reference
MMP-8	C_{16}-fTHP-11	C_{16}-(Gly-Pro-Hyp)$_5$-Gly-Pro-Lys(Mca)-Gly-Pro-**Orn**-Gly~**Cys(Mob)**-Arg-Gly-Gln-Lys(Dnp)-Gly-Val-Arg-(Gly-Pro-Hyp)$_5$-NH_2[f]	0.007	3.50	2100[e]	[50]
MMP-8	C_6-fTHP-12	C_6-(Gly-Pro-Hyp)$_5$-Gly-Pro-Lys(Mca)-Gly-Pro-**Leu**-Gly~**Met**-Arg-Gly-Gln-Lys(Dnp)-Gly-Val-Arg-(Gly-Pro-Hyp)$_5$-NH_2[f]	0.070	14.5	5090[e]	[51]
MMP-8	fTHP-13	(Gly-Pro-Hyp)$_5$-Gly-Pro-Lys(Mca)-Gly-Pro-**Val-Asn**~**Phe**-Arg-Gly-Gln-Lys(Dnp)-Gly-Val-Arg-(Gly-Pro-Hyp)$_5$-NH_2[f]	0.028	9.30	3010[e]	[51]
MMP-8	C_6-fTHP-13	C_6-(Gly-Pro-Hyp)$_5$-Gly-Pro-Lys(Mca)-Gly-Pro-**Val-Asn**~**Phe**-Arg-Gly-Gln-Lys(Dnp)-Gly-Val-Arg-(Gly-Pro-Hyp)$_5$-NH_2[f]	0.020	3.90	3870[e]	[51]
MMP-13	fTHP-4[d]	(Gly-Pro-Hyp)$_5$-Gly-Pro-Lys(Mca)-Gly-Pro-Gln-Gly~Leu-Arg-Gly-Gln-Lys(Dnp)-Gly-Val-Arg-(Gly-Pro-Hyp)$_5$-NH_2	0.015	8.6	1600[e]	[50]
MMP-13	C_{10}-fTHP-4[d]	C_{10}-(Gly-Pro-Hyp)$_5$-Gly-Pro-Lys(Mca)-Gly-Pro-Gln-Gly~Leu-Arg-Gly-Gln-Lys(Dnp)-Gly-Val-Arg-(Gly-Pro-Hyp)$_5$-NH_2	0.013	3.4	3800[e]	[50]
MMP-13	C_{16}-fTHP-4[d]	C_{16}-(Gly-Pro-Hyp)$_5$-Gly-Pro-Lys(Mca)-Gly-Pro-Gln-Gly~Leu-Arg-Gly-Gln-Lys(Dnp)-Gly-Val-Arg-(Gly-Pro-Hyp)$_5$-NH_2	0.0020	1.4	1300[e]	[50]
MMP-13	fTHP-9	(Gly-Pro-Hyp)$_5$-Gly-Pro-Lys(Mca)-Gly-Pro-Gln-Gly~**Cys(Mob)**-Arg-Gly-Gln-Lys(Dnp)-Gly-Val-Arg-(Gly-Pro-Hyp)$_5$-NH_2[f]	0.16	6.25	25,000[e]	[50]

(continued)

Table 1
(continued)

Enzyme	Substrate abbreviation	Substrate sequence	k_{cat} (s^{-1})	K_M (μM)	k_{cat}/K_M[a] (s^{-1} M^{-1})	Reference
MMP-13	C_{10}-fTHP-9	C_{10}-(Gly-Pro-Hyp)$_5$-Gly-Pro-Lys(Mca)-Gly-Pro-Gln-Gly~**Cys(Mob)**-Arg-Gly-Gln-Lys(Dnp)-Gly-Val-Arg-(Gly-Pro-Hyp)$_5$-NH$_2$[f]	0.060	3.45	18,000[e]	[50]
MMP-13	C_{16}-fTHP-9	C_{16}-(Gly-Pro-Hyp)$_5$-Gly-Pro-Lys(Mca)-Gly-Pro-Gln-Gly~**Cys(Mob)**-Arg-Gly-Gln-Lys(Dnp)-Gly-Val-Arg-(Gly-Pro-Hyp)$_5$-NH$_2$[e]	0.028	2.50	11,000[e]	[50]
MMP-13	fTHP-10	(Gly-Pro-Hyp)$_5$-Gly-Pro-Lys(Mca)-Gly-Pro-**Orn**-Gly~Leu-Arg-Gly-Gln-Lys(Dnp)-Gly-Val-Arg-(Gly-Pro-Hyp)$_5$-NH$_2$[f]	0.58	30.1	19,000[e]	[50]
MMP-13	C_{10}-fTHP-10	C_{10}-(Gly-Pro-Hyp)$_5$-Gly-Pro-Lys(Mca)-Gly-Pro-**Orn**-Gly~Leu-Arg-Gly-Gln-Lys(Dnp)-Gly-Val-Arg-(Gly-Pro-Hyp)$_5$-NH$_2$[f]	0.31	17.4	18,000[e]	[50]
MMP-13	fTHP-11	(Gly-Pro-Hyp)$_5$-Gly-Pro-Lys(Mca)-Gly-Pro-**Orn**-Gly~**Cys(Mob)**-Arg-Gly-Gln-Lys(Dnp)-Gly-Val-Arg-(Gly-Pro-Hyp)$_5$-NH$_2$[f]	0.18	13.7	13,000[e]	[50]
MMP-13	C_{10}-fTHP-11	C_{10}-(Gly-Pro-Hyp)$_5$-Gly-Pro-Lys(Mca)-Gly-Pro-**Orn**-Gly~**Cys(Mob)**-Arg-Gly-Gln-Lys(Dnp)-Gly-Val-Arg-(Gly-Pro-Hyp)$_5$-NH$_2$[f]	0.013	7.10	1900[e]	[50]
MMP-13	C_{16}-fTHP-11	C_{16}-(Gly-Pro-Hyp)$_5$-Gly-Pro-Lys(Mca)-Gly-Pro-**Orn**-Gly~**Cys(Mob)**-Arg-Gly-Gln-Lys(Dnp)-Gly-Val-Arg-(Gly-Pro-Hyp)$_5$-NH$_2$[f]	0.007	7.68	940[e]	[50]
MMP-13	C_6-fTHP-12	C_6-(Gly-Pro-Hyp)$_5$-Gly-Pro-Lys(Mca)-Gly-Pro-**Leu**-Gly~**Met**-Arg-Gly-Gln-Lys(Dnp)-Gly-Val-Arg-(Gly-Pro-Hyp)$_5$-NH$_2$[f]	0.060	6.35	9710[e]	[51]

(continued)

Table 1 (continued)

Enzyme	Substrate abbreviation	Substrate sequence	k_{cat} (s^{-1})	K_M (μM)	k_{cat}/K_M[a] (s^{-1} M^{-1})	Reference
MMP-13	fTHP-13	(Gly-Pro-Hyp)$_5$-Gly-Pro-Lys(Mca)-Gly-Pro-**Val**-**Asn**~**Phe**-Arg-Gly-Gln-Lys(Dnp)-Gly-Val-Arg-(Gly-Pro-Hyp)$_5$-NH_2[f]	0.127	20.5	6116[e]	[51]
MMP-13	C_6-fTHP-13	C_6-(Gly-Pro-Hyp)$_5$-Gly-Pro-Lys(Mca)-Gly-Pro-**Val**-**Asn**~**Phe**-Arg-Gly-Gln-Lys(Dnp)-Gly-Val-Arg-(Gly-Pro-Hyp)$_5$-NH_2[f]	0.060	9.95	5800[e]	[51]
MT1-MMP(Δ279–523)	fTHP-4[d]	(Gly-Pro-Hyp)$_5$-Gly-Pro-Lys(Mca)-Gly-Pro-Gln-Gly~Leu-Arg-Gly-Gln-Lys(Dnp)-Gly-Val-Arg-(Gly-Pro-Hyp)$_5$-NH_2	1.37	23.1	59,000[e]	[50]
MT1-MMP(Δ279–523)	C_{10}-fTHP-4[d]	C_{10}-(Gly-Pro-Hyp)$_5$-Gly-Pro-Lys(Mca)-Gly-Pro-Gln-Gly~Leu-Arg-Gly-Gln-Lys(Dnp)-Gly-Val-Arg-(Gly-Pro-Hyp)$_5$-NH_2	0.0033	0.80	4100[e]	[50]
MT1-MMP(Δ279–523)	C_{16}-fTHP-4[d]	C_{16}-(Gly-Pro-Hyp)$_5$-Gly-Pro-Lys(Mca)-Gly-Pro-Gln-Gly~Leu-Arg-Gly-Gln-Lys(Dnp)-Gly-Val-Arg-(Gly-Pro-Hyp)$_5$-NH_2	0.0021	1.80	1200[e]	[50]
MT1-MMP(Δ279–523)	fTHP-9	(Gly-Pro-Hyp)$_5$-Gly-Pro-Lys(Mca)-Gly-Pro-Gln-Gly~**Cys(Mob)**-Arg-Gly-Gln-Lys(Dnp)-Gly-Val-Arg-(Gly-Pro-Hyp)$_5$-NH_2[f]	1.34	7.95	168,000[e]	[50]
MT1-MMP(Δ279–523)	C_{10}-fTHP-9	C_{10}-(Gly-Pro-Hyp)$_5$-Gly-Pro-Lys(Mca)-Gly-Pro-Gln-Gly~**Cys(Mob)**-Arg-Gly-Gln-Lys(Dnp)-Gly-Val-Arg-(Gly-Pro-Hyp)$_5$-NH_2[f]	0.080	2.80	29,000[e]	[50]
MT1-MMP(Δ279–523)	fTHP-10	(Gly-Pro-Hyp)$_5$-Gly-Pro-Lys(Mca)-Gly-Pro-**Orn**-Gly~Leu-Arg-Gly-Gln-Lys(Dnp)-Gly-Val-Arg-(Gly-Pro-Hyp)$_5$-NH_2[f]	0.25	10.2	25,000[e]	[50]

(continued)

Table 1
(continued)

Enzyme	Substrate abbreviation	Substrate sequence	k_{cat} (s^{-1})	K_M (μM)	k_{cat}/k_M[a] (s^{-1} M^{-1})	Reference
MT1-MMP(Δ279–523)	C_{10}-fTHP-10	C_{10}-(Gly-Pro-Hyp)$_5$-Gly-Pro-Lys(Mca)-Gly-Pro-**Orn**-Gly~Leu-Arg-Gly-Gln-Lys(Dnp)-Gly-Val-Arg-(Gly-Pro-Hyp)$_5$-NH_2[f]	0.091	4.40	20,000[e]	[50]
MT1-MMP(Δ279–523)	fTHP-11	(Gly-Pro-Hyp)$_5$-Gly-Pro-Lys(Mca)-Gly-Pro-**Orn**-Gly~**Cys(Mob)**-Arg-Gly-Gln-Lys(Dnp)-Gly-Val-Arg-(Gly-Pro-Hyp)$_5$-NH_2[f]	1.30	16.6	78,000[e]	[50]
MT1-MMP(Δ279–523)	C_{10}-fTHP-11	C_{10}-(Gly-Pro-Hyp)$_5$-Gly-Pro-Lys(Mca)-Gly-Pro-**Orn**-Gly~**Cys(Mob)**-Arg-Gly-Gln-Lys(Dnp)-Gly-Val-Arg-(Gly-Pro-Hyp)$_5$-NH_2[f]	0.036	3.50	11,000[e]	[50]
MT2-MMP/MMP-15(Δ268–628)	fTHP-11	(Gly-Pro-Hyp)$_5$-Gly-Pro-Lys(Mca)-Gly-Pro-**Orn**-Gly~**Cys(Mob)**-Arg-Gly-Gln-Lys(Dnp)-Gly-Val-Arg-(Gly-Pro-Hyp)$_5$-NH_2[f]	0.41	14.3	29,000[e]	[50]
MMP-8	fTHP-4[d]	(Gly-Pro-Hyp)$_5$-Gly-Pro-Lys(Mca)-Gly-Pro-Gln-Gly~Leu-Arg-Gly-Gln-Lys(Dnp)-Gly-Val-Arg-(Gly-Pro-Hyp)$_5$-NH_2	0.035 ± 0.016[a]	7.60 ± 2.40[a]	4500	[51]
MMP-8	C_6-fTHP-12	C_6-(Gly-Pro-Hyp)$_5$-Gly-Pro-Lys(Mca)-Gly-Pro-**Leu**-Gly~**Met**-Arg-Gly-Gln-Lys(Dnp)-Gly-Val-Arg-(Gly-Pro-Hyp)$_5$-NH_2[f]	0.070 ± 0.020	14.50 ± 4.80	5090	[51]
MMP-8	fTHP-13	(Gly-Pro-Hyp)$_5$-Gly-Pro-Lys(Mca)-Gly-Pro-**Val**-**Asn**~**Phe**-Arg-Gly-Gln-Lys(Dnp)-Gly-Val-Arg-(Gly-Pro-Hyp)$_5$-NH_2[f]	0.028 ± 0.004	9.30 ± 1.84	3010	[51]
MMP-8	C_6-fTHP-13	C_6-(Gly-Pro-Hyp)$_5$-Gly-Pro-Lys(Mca)-Gly-Pro-**Val**-**Asn**~**Phe**-Arg-Gly-Gln-Lys(Dnp)-Gly-Val-Arg-(Gly-Pro-Hyp)$_5$-NH_2[f]	0.020 ± 0.010	3.90 ± 2.30	3870	[51]

(continued)

Table 1 (continued)

Enzyme	Substrate abbreviation	Substrate sequence	k_{cat} (s^{-1})	K_M (μM)	k_{cat}/K_M[a] (s^{-1} M^{-1})	Reference
MMP-13	fTHP-4[d]	(Gly-Pro-Hyp)$_5$-Gly-Pro-Lys(Mca)-Gly-Pro-Gln-Gly~Leu-Arg-Gly-Gln-Lys(Dnp)-Gly-Val-Arg-(Gly-Pro-Hyp)$_5$-NH$_2$	0.015 ± 0.005[a]	8.60 ± 1.70[a]	1600	[51]
MMP-13	C$_6$-fTHP-12	C$_6$-(Gly-Pro-Hyp)$_5$-Gly-Pro-Lys(Mca)-Gly-Pro-**Leu**-Gly~**Met**-Arg-Gly-Gln-Lys(Dnp)-Gly-Val-Arg-(Gly-Pro-Hyp)$_5$-NH$_2$[f]	0.060 ± 0.030	6.35 ± 3.0	9710	[51]
MMP-13	fTHP-13	(Gly-Pro-Hyp)$_5$-Gly-Pro-Lys(Mca)-Gly-Pro-**Val**-**Asn**~**Phe**-Arg-Gly-Gln-Lys(Dnp)-Gly-Val-Arg-(Gly-Pro-Hyp)$_5$-NH$_2$[f]	0.127 ± 0.046	20.50 ± 6.22	6116	[51]
MMP-13	C$_6$-fTHP-13	C$_6$-(Gly-Pro-Hyp)$_5$-Gly-Pro-Lys(Mca)-Gly-Pro-**Val**-**Asn**~**Phe**-Arg-Gly-Gln-Lys(Dnp)-Gly-Val-Arg-(Gly-Pro-Hyp)$_5$-NH$_2$[f]	0.060 ± 0.040	9.95 ± 0.35	5800	[51]
MMP-9	fTHP-15	(Gly-Pro-Hyp)$_5$-Gly-Pro-Lys(Mca)-Gly-Pro-Gln-Gly~Leu-Arg-Gly-Gln-Lys(Dnp)-Gly-Val-Arg-(Gly-Pro-Hyp)$_5$-NH$_2$	0.78 ± 0.13	12.1 ± 0.6	64,851 ± 10,698	[52, 64]
MMP-9 CAT	fTHP-15	(Gly-Pro-Hyp)$_5$-Gly-Pro-Lys(Mca)-Gly-Pro-Gln-Gly~Leu-Arg-Gly-Gln-Lys(Dnp)-Gly-Val-Arg-(Gly-Pro-Hyp)$_5$-NH$_2$	1.05 ± 0.14	15.1 ± 1.4	69,566 ± 9138	[52, 64]
MMP-9 CAT-FN	fTHP-15	(Gly-Pro-Hyp)$_5$-Gly-Pro-Lys(Mca)-Gly-Pro-Gln-Gly~Leu-Arg-Gly-Gln-Lys(Dnp)-Gly-Val-Arg-(Gly-Pro-Hyp)$_5$-NH$_2$	1.75 ± 0.15	143.0 ± 13.9	12,227 ± 1077	[52, 64]
MMP-2	fTHP-15	(Gly-Pro-Hyp)$_5$-Gly-Pro-Lys(Mca)-Gly-Pro-Gln-Gly~Leu-Arg-Gly-Gln-Lys(Dnp)-Gly-Val-Arg-(Gly-Pro-Hyp)$_5$-NH$_2$	0.654	5.4 ± 0.97	120,423[b]	[67]
MMP-2	GDKA	Gly-Asp-Lys-(Gly-Pro-Hyp)$_4$-Gly-Pro-Lys(Mca)-Gly-Pro-Gln-Gly~Leu-Arg-Gly-Gln-Lys(Dnp)-Gly-Val-Arg-(Gly-Pro-Hyp)$_4$-Gly-Asp-Lys-NH$_2$	1.485	5.7 ± 0.79	262,004[b]	[67]

(continued)

Table 1
(continued)

Enzyme	Substrate abbreviation	Substrate sequence	k_{cat} (s^{-1})	K_M (μM)	k_{cat}/K_M[a] (s^{-1} M^{-1})	Reference
MMP-2	GDK B	Gly-Pro-Hyp-Gly-Asp-Lys-(Gly-Pro-Hyp)$_3$-Gly-Pro-Lys(Mca)-Gly-Pro-Gln-Gly~Leu-Arg-Gly-Gln-Lys(Dnp)-Gly-Val-Arg-(Gly-Pro-Hyp)$_3$-Gly-Asp-Lys-Gly-Pro-Hyp-NH$_2$	1.424	4.5 ± 0.60	314,950[b]	[67]
MMP-2	GDK C	(Gly-Pro-Hyp)$_2$-Gly-Asp-Lys-(Gly-Pro-Hyp)$_2$-Gly-Pro-Lys(Mca)-Gly-Pro-Gln-Gly~Leu-Arg-Gly-Gln-Lys(Dnp)-Gly-Val-Arg-(Gly-Pro-Hyp)$_2$-Gly-Asp-Lys-(Gly-Pro-Hyp)$_2$-NH$_2$	2.374	3.3 ± 0.35	728,309[b]	[67]
MMP-2	GDK D	(Gly-Pro-Hyp)$_3$-Gly-Asp-Lys-Gly-Pro-Hyp-Gly-Pro-Lys(Mca)-Gly-Pro-Gln-Gly~Leu-Arg-Gly-Gln-Lys(Dnp)-Gly-Val-Arg-Gly-Pro-Hyp-Gly-Asp-Lys-(Gly-Pro-Hyp)$_3$-NH$_2$	2.635	2.7 ± 0.62	986,155[b]	[67]
MMP-2	GDK E	(Gly-Pro-Hyp)$_4$-Gly-Asp-Lys-Gly-Pro-Lys(Mca)-Gly-Pro-Gln-Gly~Leu-Arg-Gly-Gln-Lys(Dnp)-Gly-Val-Arg-Gly-Asp-Lys-(Gly-Pro-Hyp)$_4$-NH$_2$	2.804	2.4 ± 0.42	1,162,216[b]	[67]
MMP-2	KGD A	Lys-Gly-Asp-Hyp-(Gly-Pro-Hyp)$_4$-Gly-Pro-Lys(Mca)-Gly-Pro-Gln-Gly~Leu-Arg-Gly-Gln-Lys(Dnp)-Gly-Val-Arg-(Gly-Pro-Hyp)$_3$-Gly-Pro-Lys-Gly-Asp-Hyp-NH$_2$	1.987	5.9 ± 0.94	337,988[b]	[67]
MMP-2	KGD B	Gly-Pro-Lys-Gly-Asp-Hyp-(Gly-Pro-Hyp)$_3$-Gly-Pro-Lys(Mca)-Gly-Pro-Gln-Gly~Leu-Arg-Gly-Gln-Lys(Dnp)-Gly-Val-Arg-(Gly-Pro-Hyp)$_2$-Gly-Pro-Lys-Gly-Asp-Hyp-Gly-Pro-Hyp-NH$_2$	2.296	7.1 ± 0.72	324,215[b]	[67]

(continued)

Table 1 (continued)

Enzyme	Substrate abbreviation	Substrate sequence	k_{cat} (s^{-1})	K_M (μM)	k_{cat}/K_M[a] (s^{-1} M^{-1})	Reference
MMP-2	KGD C	Gly-Pro-Hyp-Gly-Pro-Lys-Gly-Asp-Hyp-(Gly-Pro-Hyp)$_2$-Gly-Pro-Lys(Mca)-Gly-Pro-Gln-Gly~Leu-Arg-Gly-Gln-Lys(Dnp)-Gly-Val-Arg-Gly-Pro-Hyp-Gly-Pro-Lys-Gly-Asp-Hyp-(Gly-Pro-Hyp)$_2$-NH_2	2.627	4.8 ± 0.49	543,854[b]	[67]
MMP-2	KGD D	(Gly-Pro-Hyp)$_2$-Gly-Pro-Lys-Gly-Asp-Hyp-Gly-Pro-Hyp-Gly-Pro-Lys(Mca)-Gly-Pro-Gln-Gly~Leu-Arg-Gly-Gln-Lys(Dnp)-Gly-Val-Arg-Gly-Pro-Lys-Gly-Asp-Hyp-(Gly-Pro-Hyp)$_3$-NH_2	2.234	9.8 ± 0.97	227,927[b]	[67]
MMP-2	KGD E	(Gly-Pro-Hyp)$_3$-Gly-Pro-Lys-Gly-Asp-Hyp-Gly-Pro-Lys(Mca)-Gly-Pro-Gln-Gly~Leu-Arg-Gly-Gln-Lys(Dnp)-Gly-Val-Lys-Gly-Asp-Hyp-(Gly-Pro-Hyp)$_4$-NH_2	1.29	4.6 ± 0.60	279,035[b]	[67]
MMP-8	fTHP-15	(Gly-Pro-Hyp)$_5$-Gly-Pro-Lys(Mca)-Gly-Pro-Gln-Gly~Leu-Arg-Gly-Gln-Lys(Dnp)-Gly-Val-Arg-(Gly-Pro-Hyp)$_5$-NH_2	0.5	23.0 ± 9.4	21,753[b]	[67]
MMP-8	GDK A	Gly-Asp-Lys-(Gly-Pro-Hyp)$_4$-Gly-Pro-Lys(Mca)-Gly-Pro-Gln-Gly~Leu-Arg-Gly-Gln-Lys(Dnp)-Gly-Val-Arg-(Gly-Pro-Hyp)$_4$-Gly-Asp-Lys-NH_2	0.559	19.4 ± 0.75	28,837[b]	[67]
MMP-8	GDK B	Gly-Pro-Hyp-Gly-Asp-Lys-(Gly-Pro-Hyp)$_3$-Gly-Pro-Lys(Mca)-Gly-Pro-Gln-Gly~Leu-Arg-Gly-Gln-Lys(Dnp)-Gly-Val-Arg-(Gly-Pro-Hyp)$_3$-Gly-Asp-Lys-Gly-Pro-Hyp-NH_2	1.18	25.1 ± 2.1	47,008[b]	[67]

(continued)

Table 1
(continued)

Enzyme	Substrate abbreviation	Substrate sequence	k_{cat} (s^{-1})	K_M (μM)	k_{cat}/K_M[a] (s^{-1} M^{-1})	Reference
MMP-8	GDK C	$(Gly\text{-}Pro\text{-}Hyp)_2$-Gly-Asp-Lys-$(Gly\text{-}Pro\text{-}Hyp)_2$-Gly-Pro-Lys(Mca)-Gly-Pro-Gln-Gly~Leu-Arg-Gly-Gln-Lys(Dnp)-Gly-Val-Arg-$(Gly\text{-}Pro\text{-}Hyp)_2$-Gly-Asp-Lys-$(Gly\text{-}Pro\text{-}Hyp)_2$-$NH_2$	1.681	18.6 ± 0.74	90,357[b]	[67]
MMP-8	GDK D	$(Gly\text{-}Pro\text{-}Hyp)_3$-Gly-Asp-Lys-Gly-Pro-Hyp-Gly-Pro-Lys(Mca)-Gly-Pro-Gln-Gly~Leu-Arg-Gly-Gln-Lys(Dnp)-Gly-Val-Arg-Gly-Pro-Hyp-Gly-Asp-Lys-$(Gly\text{-}Pro\text{-}Hyp)_3$-$NH_2$	5.758	44.4 ± 2.4	129,685[b]	[67]
MMP-8	GDK E	$(Gly\text{-}Pro\text{-}Hyp)_4$-Gly-Asp-Lys-Gly-Pro-Lys(Mca)-Gly-Pro-Gln-Gly~Leu-Arg-Gly-Gln-Lys(Dnp)-Gly-Val-Arg-Gly-Asp-Lys-$(Gly\text{-}Pro\text{-}Hyp)_4$-$NH_2$	8.233	58.5 ± 6.7	140,727[b]	[67]
MMP-8	KGD A	Lys-Gly-Asp-Hyp-$(Gly\text{-}Pro\text{-}Hyp)_4$-Gly-Pro-Lys(Mca)-Gly-Pro-Gln-Gly~Leu-Arg-Gly-Gln-Lys(Dnp)-Gly-Val-Arg-$(Gly\text{-}Pro\text{-}Hyp)_3$-Gly-Pro-Lys-Gly-Asp-Hyp-NH_2	0.3	15.8 ± 6.5	18,624[b]	[67]
MMP-8	KGD B	Gly-Pro-Lys-Gly-Asp-Hyp-$(Gly\text{-}Pro\text{-}Hyp)_3$-Gly-Pro-Lys(Mca)-Gly-Pro-Gln-Gly~Leu-Arg-Gly-Gln-Lys(Dnp)-Gly-Val-Arg-$(Gly\text{-}Pro\text{-}Hyp)_2$-Gly-Pro-Lys-Gly-Asp-Hyp-Gly-Pro-Hyp-NH_2	0.947	44.7 ± 26.3	21,194[b]	[67]
MMP-8	KGD C	Gly-Pro-Hyp-Gly-Pro-Lys-Gly-Asp-Hyp-$(Gly\text{-}Pro\text{-}Hyp)_2$-Gly-Pro-Lys(Mca)-Gly-Pro-Gln-Gly~Leu-Arg-Gly-Gln-Lys(Dnp)-Gly-Val-Arg-Gly-Pro-Hyp-Gly-Pro-Lys-Gly-Asp-Hyp-$(Gly\text{-}Pro\text{-}Hyp)_2$-$NH_2$	0.663	30.8 ± 18.5	21,551[b]	[67]

(continued)

Table 1 (continued)

Enzyme	Substrate abbreviation	Substrate sequence	k_{cat} (s^{-1})	K_M (μM)	k_{cat}/K_M[a] (s^{-1} M^{-1})	Reference
MMP-8	KGD D	$(Gly\text{-}Pro\text{-}Hyp)_2$-Gly-Pro-Lys-Gly-Asp-Hyp-Gly-Pro-Hyp-Gly-Pro-Lys(Mca)-Gly-Pro-Gln-Gly~Leu-Arg-Gly-Gln-Lys(Dnp)-Gly-Val-Arg-Gly-Pro-Lys-Gly-Asp-Hyp-$(Gly\text{-}Pro\text{-}Hyp)_3$-$NH_2$	0.623	41.3 ± 1.5	15,100[b]	[67]
MMP-8	KGD E	$(Gly\text{-}Pro\text{-}Hyp)_3$-Gly-Pro-Lys-Gly-Asp-Hyp-Gly-Pro-Lys(Mca)-Gly-Pro-Gln-Gly~Leu-Arg-Gly-Gln-Lys(Dnp)-Gly-Val-Lys-Gly-Asp-Hyp-$(Gly\text{-}Pro\text{-}Hyp)_4$-$NH_2$	0.597	26.7 ± 3.5	22,389[b]	[67]
MMP-9	fTHP-15	$(Gly\text{-}Pro\text{-}Hyp)_5$-Gly-Pro-Lys(Mca)-Gly-Pro-Gln-Gly~Leu-Arg-Gly-Gln-Lys(Dnp)-Gly-Val-Arg-$(Gly\text{-}Pro\text{-}Hyp)_5$-$NH_2$	0.449	6.4 ± 0.79	69,771[b]	[67]
MMP-9	GDK A	Gly-Asp-Lys-$(Gly\text{-}Pro\text{-}Hyp)_4$-Gly-Pro-Lys(Mca)-Gly-Pro-Gln-Gly~Leu-Arg-Gly-Gln-Lys(Dnp)-Gly-Val-Arg-$(Gly\text{-}Pro\text{-}Hyp)_4$-Gly-Asp-Lys-$NH_2$	2.363	2.9 ± 0.52	803,827[b]	[67]
MMP-9	GDK B	Gly-Pro-Hyp-Gly-Asp-Lys-$(Gly\text{-}Pro\text{-}Hyp)_3$-Gly-Pro-Lys(Mca)-Gly-Pro-Gln-Gly~Leu-Arg-Gly-Gln-Lys(Dnp)-Gly-Val-Arg-$(Gly\text{-}Pro\text{-}Hyp)_3$-Gly-Asp-Lys-Gly-Pro-Hyp-NH_2	2.7	2.3 ± 0.14	1,171,153[b]	[67]
MMP-9	GDK C	$(Gly\text{-}Pro\text{-}Hyp)_2$-Gly-Asp-Lys-$(Gly\text{-}Pro\text{-}Hyp)_2$-Gly-Pro-Lys(Mca)-Gly-Pro-Gln-Gly~Leu-Arg-Gly-Gln-Lys(Dnp)-Gly-Val-Arg-$(Gly\text{-}Pro\text{-}Hyp)_2$-Gly-Asp-Lys-$(Gly\text{-}Pro\text{-}Hyp)_2$-$NH_2$	2.621	1.3 ± 0.22	1,948,364[b]	[67]

(continued)

Table 1
(continued)

Enzyme	Substrate abbreviation	Substrate sequence	k_{cat} (s^{-1})	K_M (μM)	k_{cat}/K_M[a] (s^{-1} M^{-1})	Reference
MMP-9	GDK D	$(Gly\text{-}Pro\text{-}Hyp)_3$-Gly-Asp-Lys-Gly-Pro-Hyp-Gly-Pro-Lys(Mca)-Gly-Pro-Gln-Gly~Leu-Arg-Gly-Gln-Lys(Dnp)-Gly-Val-Arg-Gly-Pro-Hyp-Gly-Asp-Lys-$(Gly\text{-}Pro\text{-}Hyp)_3$-$NH_2$	4.071	1.7 ± 0.21	2,443,634[b]	[67]
MMP-9	GDK E	$(Gly\text{-}Pro\text{-}Hyp)_4$-Gly-Asp-Lys-Gly-Pro-Lys(Mca)-Gly-Pro-Gln-Gly~Leu-Arg-Gly-Gln-Lys(Dnp)-Gly-Val-Arg-Gly-Asp-Lys-$(Gly\text{-}Pro\text{-}Hyp)_4$-$NH_2$	2.679	1.4 ± 0.15	1,982,689[b]	[67]
MMP-9	KGD A	Lys-Gly-Asp-Hyp-$(Gly\text{-}Pro\text{-}Hyp)_4$-Gly-Pro-Lys(Mca)-Gly-Pro-Gln-Gly~Leu-Arg-Gly-Gln-Lys(Dnp)-Gly-Val-Arg-$(Gly\text{-}Pro\text{-}Hyp)_3$-Gly-Pro-Lys-Gly-Asp-Hyp-NH_2	3.324	6.5 ± 0.63	511,077[b]	[67]
MMP-9	KGD B	Gly-Pro-Lys-Gly-Asp-Hyp-$(Gly\text{-}Pro\text{-}Hyp)_3$-Gly-Pro-Lys(Mca)-Gly-Pro-Gln-Gly~Leu-Arg-Gly-Gln-Lys(Dnp)-Gly-Val-Arg-$(Gly\text{-}Pro\text{-}Hyp)_2$-Gly-Pro-Lys-Gly-Asp-Hyp-Gly-Pro-Hyp-NH_2	4.51	4.5 ± 0.66	994,009[b]	[67]
MMP-9	KGD C	Gly-Pro-Hyp-Gly-Pro-Lys-Gly-Asp-Hyp-$(Gly\text{-}Pro\text{-}Hyp)_2$-Gly-Pro-Lys(Mca)-Gly-Pro-Gln-Gly~Leu-Arg-Gly-Gln-Lys(Dnp)-Gly-Val-Arg-Gly-Pro-Hyp-Gly-Pro-Lys-Gly-Asp-Hyp-$(Gly\text{-}Pro\text{-}Hyp)_2$-$NH_2$	5.335	4.5 ± 0.96	1,174,706[b]	[67]

(continued)

Table 1 (continued)

Enzyme	Substrate abbreviation	Substrate sequence	k_{cat} (s^{-1})	K_M (μM)	k_{cat}/K_M[a] (s^{-1} M^{-1})	Reference
MMP-9	KGD D	(Gly-Pro-Hyp)$_2$-Gly-Pro-Lys-Gly-Asp-Hyp-Gly-Pro-Hyp-Gly-Pro-Lys(Mca)-Gly-Pro-Gln-Gly~Leu-Arg-Gly-Gln-Lys(Dnp)-Gly-Val-Arg-Gly-Pro-Lys-Gly-Asp-Hyp-(Gly-Pro-Hyp)$_3$-NH$_2$	3.909	6.6 ± 1.3	588,329[b]	[67]
MMP-9	KGD E	(Gly-Pro-Hyp)$_3$-Gly-Pro-Lys-Gly-Asp-Hyp-Gly-Pro-Lys(Mca)-Gly-Pro-Gln-Gly~Leu-Arg-Gly-Gln-Lys(Dnp)-Gly-Val-Lys-Gly-Asp-Hyp-(Gly-Pro-Hyp)$_4$-NH$_2$	2.288	3.1 ± 0.60	729,451[b]	[67]
MMP-12	fTHP-15	(Gly-Pro-Hyp)$_5$-Gly-Pro-Lys(Mca)-Gly-Pro-Gln-Gly~Leu-Arg-Gly-Gln-Lys(Dnp)-Gly-Val-Arg-(Gly-Pro-Hyp)$_5$-NH$_2$	0.111	56.6 ± 0.50	1953[b]	[67]
MMP-12	GDK A	Gly-Asp-Lys-(Gly-Pro-Hyp)$_4$-Gly-Pro-Lys(Mca)-Gly-Pro-Gln-Gly~Leu-Arg-Gly-Gln-Lys(Dnp)-Gly-Val-Arg-(Gly-Pro-Hyp)$_4$-Gly-Asp-Lys-NH$_2$	0.153	26.3 ± 2.2	5791[b]	[67]
MMP-12	GDK B	Gly-Pro-Hyp-Gly-Asp-Lys-(Gly-Pro-Hyp)$_3$-Gly-Pro-Lys(Mca)-Gly-Pro-Gln-Gly~Leu-Arg-Gly-Gln-Lys(Dnp)-Gly-Val-Arg-(Gly-Pro-Hyp)$_3$-Gly-Asp-Lys-Gly-Pro-Hyp-NH$_2$	0.554	72.8 ± 6.3	7600[b]	[67]
MMP-12	GDK C	(Gly-Pro-Hyp)$_2$-Gly-Asp-Lys-(Gly-Pro-Hyp)$_2$-Gly-Pro-Lys(Mca)-Gly-Pro-Gln-Gly~Leu-Arg-Gly-Gln-Lys(Dnp)-Gly-Val-Arg-(Gly-Pro-Hyp)$_2$-Gly-Asp-Lys-(Gly-Pro-Hyp)$_2$-NH$_2$	0.892	57.4 ± 2.4	15,556[b]	[67]

(continued)

Table 1
(continued)

Enzyme	Substrate abbreviation	Substrate sequence	k_{cat} (s^{-1})	K_M (μM)	k_{cat}/K_M[a] (s^{-1} M^{-1})	Reference
MMP-12	GDK D	(Gly-Pro-Hyp)$_3$-Gly-Asp-Lys-Gly-Pro-Hyp-Gly-Pro-Lys(Mca)-Gly-Pro-Gln-Gly~Leu-Arg-Gly-Gln-Lys(Dnp)-Gly-Val-Arg-Gly-Pro-Hyp-Gly-Asp-Lys-(Gly-Pro-Hyp)$_3$-NH_2	0.361	19.9 ± 3.4	18,171[b]	[67]
MMP-12	GDK E	(Gly-Pro-Hyp)$_4$-Gly-Asp-Lys-Gly-Pro-Lys(Mca)-Gly-Pro-Gln-Gly~Leu-Arg Gly-Gln-Lys(Dnp)-Gly-Val-Arg-Gly-Asp-Lys-(Gly-Pro-Hyp)$_4$-NH_2	1.047	32.1 ± 4.3	32,557[b]	[67]
MMP-12	KGD A	Lys-Gly-Asp-Hyp-(Gly-Pro-Hyp)$_4$-Gly-Pro-Lys(Mca)-Gly-Pro-Gln-Gly~Leu-Arg-Gly-Gln-Lys(Dnp)-Gly-Val-Arg-(Gly-Pro-Hyp)$_3$-Gly-Pro-Lys-Gly-Asp-Hyp-NH_2	0.408	91.3 ± 23.6	4467[b]	[67]
MMP-12	KGD B	Gly-Pro-Lys-Gly-Asp-Hyp-(Gly-Pro-Hyp)$_3$-Gly-Pro-Lys(Mca)-Gly-Pro-Gln-Gly~Leu-Arg-Gly-Gln-Lys(Dnp)-Gly-Val-Arg-(Gly-Pro-Hyp)$_2$-Gly-Pro-Lys-Gly-Asp-Hyp-Gly-Pro-Hyp-NH_2	0.483	99.7 ± 10.8	4846[b]	[67]
MMP-12	KGD C	Gly-Pro-Hyp-Gly-Pro-Lys-Gly-Asp-Hyp-(Gly-Pro-Hyp)$_2$-Gly-Pro-Lys(Mca)-Gly-Pro-Gln-Gly~Leu-Arg-Gly-Gln-Lys(Dnp)-Gly-Val-Arg-Gly-Pro-Hyp-Gly-Pro-Lys-Gly-Asp-Hyp-(Gly-Pro-Hyp)$_2$-NH_2	0.618	79.0 ± 5.2	7827[b]	[67]

(continued)

Table 1
(continued)

Enzyme	Substrate abbreviation	Substrate sequence	k_{cat} (s^{-1})	K_M (μM)	k_{cat}/K_M[a] (s^{-1} M^{-1})	Reference
MMP-12	KGD D	$(Gly-Pro-Hyp)_2$-Gly-Pro-Lys-Gly-Asp-Hyp-Gly-Pro-Hyp-Gly-Pro-Lys(Mca)-Gly-Pro-Gln-Gly~Leu-Arg-Gly-Gln-Lys(Dnp)-Gly-Val-Arg-Gly-Pro-Lys-Gly-Asp-Hyp-$(Gly-Pro-Hyp)_3$-NH_2	0.606	148.6 ± 13.8	4077[b]	[67]
MMP-12	KGD E	$(Gly-Pro-Hyp)_3$-Gly-Pro-Lys-Gly-Asp-Hyp-Gly-Pro-Lys(Mca)-Gly-Pro-Gln-Gly~Leu-Arg-Gly-Gln-Lys(Dnp)-Gly-Val-Lys-Gly-Asp-Hyp-$(Gly-Pro-Hyp)_4$-NH_2	0.437	83.0 ± 20.7	5258[b]	[67]
MMP-13	fTHP-15	$(Gly-Pro-Hyp)_5$-Gly-Pro-Lys(Mca)-Gly-Pro-Gln-Gly~Leu-Arg-Gly-Gln-Lys(Dnp)-Gly-Val-Arg-$(Gly-Pro-Hyp)_5$-NH_2	0.677	21.8 ± 2.2	31,039[b]	[67]
MMP-13	GDK A	Gly-Asp-Lys-$(Gly-Pro-Hyp)_4$-Gly-Pro-Lys(Mca)-Gly-Pro-Gln-Gly~Leu-Arg-Gly-Gln-Lys(Dnp)-Gly-Val-Arg-$(Gly-Pro-Hyp)_4$-Gly-Asp-Lys-NH_2	0.289	8.5 ± 1.12	34,067[b]	[67]
MMP-13	GDK B	Gly-Pro-Hyp-Gly-Asp-Lys-$(Gly-Pro-Hyp)_3$-Gly-Pro-Lys(Mca)-Gly-Pro-Gln-Gly~Leu-Arg-Gly-Gln-Lys(Dnp)-Gly-Val-Arg-$(Gly-Pro-Hyp)_3$-Gly-Asp-Lys-Gly-Pro-Hyp-NH_2	0.33	5.9 ± 0.67	55,973[b]	[67]
MMP-13	GDK C	$(Gly-Pro-Hyp)_2$-Gly-Asp-Lys-$(Gly-Pro-Hyp)_2$-Gly-Pro-Lys(Mca)-Gly-Pro-Gln-Gly~Leu-Arg-Gly-Gln-Lys(Dnp)-Gly-Val-Arg-$(Gly-Pro-Hyp)_2$-Gly-Asp-Lys-$(Gly-Pro-Hyp)_2$-NH_2	1.213	11.4 ± 0.59	106,861[b]	[67]

(continued)

Table 1
(continued)

Enzyme	Substrate abbreviation	Substrate sequence	k_{cat} (s^{-1})	K_M (μM)	k_{cat}/K_M[a] (s^{-1} M^{-1})	Reference
MMP-13	GDK D	(Gly-Pro-Hyp)$_3$-Gly-Asp-Lys-Gly-Pro-Hyp-Gly-Pro-Lys(Mca)-Gly-Pro-Gln-Gly~Leu-Arg-Gly-Gln-Lys(Dnp)-Gly-Val-Arg-Gly-Pro-Hyp-Gly-Asp-Lys-(Gly-Pro-Hyp)$_3$-NH_2	3.402	25.6 ± 2.16	133,115[b]	[67]
MMP-13	GDK E	(Gly-Pro-Hyp)$_4$-Gly-Asp-Lys-Gly-Pro-Lys(Mca)-Gly-Pro-Gln-Gly~Leu-Arg Gly Gln-Lys(Dnp)-Gly-Val-Arg-Gly-Asp-Lys-(Gly-Pro-Hyp)$_4$-NH_2	4.197	19.9 ± 2.0	211,231[b]	[67]
MMP-13	KGD A	Lys-Gly-Asp-Hyp-(Gly-Pro-Hyp)$_4$-Gly-Pro-Lys(Mca)-Gly-Pro-Gln-Gly~Leu-Arg-Gly-Gln-Lys(Dnp)-Gly-Val-Arg-(Gly-Pro-Hyp)$_3$-Gly-Pro-Lys-Gly-Asp-Hyp-NH_2	1.803	32.8 ± 2.3	54,890[b]	[67]
MMP-13	KGD B	Gly-Pro-Lys-Gly-Asp-Hyp-(Gly-Pro-Hyp)$_3$-Gly-Pro-Lys(Mca)-Gly-Pro-Gln-Gly~Leu-Arg-Gly-Gln-Lys(Dnp)-Gly-Val-Arg-(Gly-Pro-Hyp)$_2$-Gly-Pro-Lys-Gly-Asp-Hyp-Gly-Pro-Hyp-NH_2	2.011	33.8 ± 1.4	59,460[b]	[67]
MMP-13	KGD C	Gly-Pro-Hyp-Gly-Pro-Lys-Gly-Asp-Hyp-(Gly-Pro-Hyp)$_2$-Gly-Pro-Lys(Mca)-Gly-Pro-Gln-Gly~Leu-Arg-Gly-Gln-Lys(Dnp)-Gly-Val-Arg-Gly-Pro-Hyp-Gly-Pro-Lys-Gly-Asp-Hyp-(Gly-Pro-Hyp)$_2$-NH_2	5.632	56.9 ± 3.6	98,946[b]	[67]

(continued)

Table 1 (continued)

Enzyme	Substrate abbreviation	Substrate sequence	k_{cat} (s^{-1})	K_M (μM)	k_{cat}/K_M[a] (s^{-1} M^{-1})	Reference
MMP-13	KGD D	$(Gly\text{-}Pro\text{-}Hyp)_2$-Gly-Pro-Lys-Gly-Asp-Hyp-Gly-Pro-Hyp-Gly-Pro-Lys(Mca)-Gly-Pro-Gln-Gly~Leu-Arg-Gly-Gln-Lys(Dnp)-Gly-Val-Arg-Gly-Pro-Lys-Gly-Asp-Hyp-$(Gly\text{-}Pro\text{-}Hyp)_3$-$NH_2$	5.819	120.8 ± 2.4	48,172[b]	[67]
MMP-13	KGD E	$(Gly\text{-}Pro\text{-}Hyp)_3$-Gly-Pro-Lys-Gly-Asp-Hyp-Gly-Pro-Lys(Mca)-Gly-Pro-Gln-Gly~Leu-Arg-Gly-Gln-Lys(Dnp)-Gly-Val-Lys-Gly-Asp-Hyp-$(Gly\text{-}Pro\text{-}Hyp)_4$-$NH_2$	5.005	72.9 ± 12.4	68,652[b]	[67]
MT1-MMP	fTHP-15	$(Gly\text{-}Pro\text{-}Hyp)_5$-Gly-Pro-Lys(Mca)-Gly-Pro-Gln-Gly~Leu-Arg-Gly-Gln-Lys(Dnp)-Gly-Val-Arg-$(Gly\text{-}Pro\text{-}Hyp)_5$-$NH_2$	0.3	13.1 ± 1.6	20,784[b]	[67]
MT1-MMP	GDK A	Gly-Asp-Lys-$(Gly\text{-}Pro\text{-}Hyp)_4$-Gly-Pro-Lys(Mca)-Gly-Pro-Gln-Gly~Leu-Arg-Gly-Gln-Lys(Dnp)-Gly-Val-Arg-$(Gly\text{-}Pro\text{-}Hyp)_4$-Gly-Asp-Lys-$NH_2$	0.136	6.9 ± 0.89	19,779[b]	[67]
MT1-MMP	GDK B	Gly-Pro-Hyp-Gly-Asp-Lys-$(Gly\text{-}Pro\text{-}Hyp)_3$-Gly-Pro-Lys(Mca)-Gly-Pro-Gln-Gly~Leu-Arg-Gly-Gln-Lys(Dnp)-Gly-Val-Arg-$(Gly\text{-}Pro\text{-}Hyp)_3$-Gly-Asp-Lys-Gly-Pro-Hyp-NH_2	0.305	9.1 ± 0.86	33,502[b]	[67]
MT1-MMP	GDK C	$(Gly\text{-}Pro\text{-}Hyp)_2$-Gly-Asp-Lys-$(Gly\text{-}Pro\text{-}Hyp)_2$-Gly-Pro-Lys(Mca)-Gly-Pro-Gln-Gly~Leu-Arg-Gly-Gln-Lys(Dnp)-Gly-Val-Arg-$(Gly\text{-}Pro\text{-}Hyp)_2$-Gly-Asp-Lys-$(Gly\text{-}Pro\text{-}Hyp)_2$-$NH_2$	0.589	8.3 ± 0.86	71,005[b]	[67]

(continued)

Table 1 (continued)

Enzyme	Substrate abbreviation	Substrate sequence	k_{cat} (s^{-1})	K_M (μM)	k_{cat}/K_M[a] (s^{-1} M^{-1})	Reference
MT1-MMP	GDK D	(Gly-Pro-Hyp)$_3$-Gly-Asp-Lys-Gly-Pro-Hyp-Gly-Pro-Lys(Mca)-Gly-Pro-Gln-Gly~Leu-Arg-Gly-Gln-Lys(Dnp)-Gly-Val-Arg-Gly-Pro-Hyp-Gly-Asp-Lys-(Gly-Pro-Hyp)$_3$-NH$_2$	1.026	9.5 ± 0.29	108,043[b]	[67]
MT1-MMP	GDK E	(Gly-Pro-Hyp)$_4$-Gly-Asp-Lys-Gly-Pro-Lys(Mca)-Gly-Pro-Gln-Gly~Leu-Arg-Gly-Gln-Lys(Dnp)-Gly-Val-Arg-Gly-Asp-Lys-(Gly-Pro-Hyp)$_4$-NH$_2$	1.291	12.0 ± 0.97	107,577[b]	[67]
MT1-MMP	KGD A	Lys-Gly-Asp-Hyp-(Gly-Pro-Hyp)$_4$-Gly-Pro-Lys(Mca)-Gly-Pro-Gln-Gly~Leu-Arg-Gly-Gln-Lys(Dnp)-Gly-Val-Arg-(Gly-Pro-Hyp)$_3$-Gly-Pro-Lys-Gly-Asp-Hyp-NH$_2$	0.3	14.1 ± 0.26	24,077[b]	[67]
MT1-MMP	KGD B	Gly-Pro-Lys-Gly-Asp-Hyp-(Gly-Pro-Hyp)$_3$-Gly-Pro-Lys(Mca)-Gly-Pro-Gln-Gly~Leu-Arg-Gly-Gln-Lys(Dnp)-Gly-Val-Arg-(Gly-Pro-Hyp)$_2$-Gly-Pro-Lys-Gly-Asp-Hyp-Gly-Pro-Hyp-NH$_2$	0.4	17.1 ± 3.6	23,486[b]	[67]
MT1-MMP	KGD C	Gly-Pro-Hyp-Gly-Pro-Lys-Gly-Asp-Hyp-(Gly-Pro-Hyp)$_2$-Gly-Pro-Lys(Mca)-Gly-Pro-Gln-Gly~Leu-Arg-Gly-Gln-Lys(Dnp)-Gly-Val-Arg-Gly-Pro-Hyp-Gly-Pro-Lys-Gly-Asp-Hyp-(Gly-Pro-Hyp)$_2$-NH$_2$	0.4	15.5 ± 4.1	23,012[b]	[67]

(continued)

Table 1 (continued)

Enzyme	Substrate abbreviation	Substrate sequence	k_{cat} (s^{-1})	K_M (μM)	k_{cat}/K_M[a] (s^{-1} M^{-1})	Reference
MT1-MMP	KGD D	$(Gly-Pro-Hyp)_2$-Gly-Pro-Lys-Gly-Asp-Hyp-Gly-Pro-Hyp-Gly-Pro-Lys(Mca)-Gly-Pro-Gln-Gly~Leu-Arg-Gly-Gln-Lys(Dnp)-Gly-Val-Arg-Gly-Pro-Lys-Gly-Asp-Hyp-$(Gly-Pro-Hyp)_3$-NH_2	0.5	21.2 ± 4.5	24,300[b]	[67]
MT1-MMP	KGD E	$(Gly-Pro-Hyp)_3$-Gly-Pro-Lys-Gly-Asp-Hyp-Gly-Pro-Lys(Mca)-Gly-Pro-Gln-Gly~Leu-Arg-Gly-Gln-Lys(Dnp)-Gly-Val-Lys-Gly-Asp-Hyp-$(Gly-Pro-Hyp)_4$-NH_2	0.6	25.1 ± 13.1	23,260[b]	[67]
MMP-1	fTHP-15(5-Fam/Dabcyl)	$(Gly-Pro-Hyp)_5$-Gly-Pro-Lys(5-Fam)-Gly-Pro-Gln-Gly~Leu-Arg-Gly-Gln-Lys(Dabcyl)-Gly-Val-Arg-$(Gly-Pro-Hyp)_5$-NH_2	0.0217	7.3	2968[b]	[55]
MMP-1	fTHP-15	$(Gly-Pro-Hyp)_5$-Gly-Pro-Lys(Mca)-Gly-Pro-Gln-Gly~Leu-Arg-Gly-Gln-Lys(Dnp)-Gly-Val-Arg-$(Gly-Pro-Hyp)_5$-NH_2	0.0042	2.31	1833[b]	[55]
MMP-13	fTHP-15(5-Fam/Dabcyl)	$(Gly-Pro-Hyp)_5$-Gly-Pro-Lys(5-Fam)-Gly-Pro-Gln-Gly~Leu-Arg-Gly-Gln-Lys(Dabcyl)-Gly-Val-Arg-$(Gly-Pro-Hyp)_5$-NH_2	1.1	16.76	65,964[b]	[55]
MMP-13	fTHP-15	$(Gly-Pro-Hyp)_5$-Gly-Pro-Lys(Mca)-Gly-Pro-Gln-Gly~Leu-Arg-Gly-Gln-Lys(Dnp)-Gly-Val-Arg-$(Gly-Pro-Hyp)_5$-NH_2	0.948	20.25	46,815[b]	[55]
MT1-MMP	fTHP-15(5-Fam/Dabcyl)	$(Gly-Pro-Hyp)_5$-Gly-Pro-Lys(5-Fam)-Gly-Pro-Gln-Gly~Leu-Arg-Gly-Gln-Lys(Dabcyl)-Gly-Val-Arg-$(Gly-Pro-Hyp)_5$-NH_2	0.0055	4.37	1249[b]	[55]

(continued)

Table 1
(continued)

Enzyme	Substrate abbreviation	Substrate sequence	k_{cat} (s^{-1})	K_M (μM)	k_{cat}/K_M[a] (s^{-1} M^{-1})	Reference
MT1-MMP	fTHP-15	$(Gly-Pro-Hyp)_5$-Gly-Pro-Lys(Mca)-Gly-Pro-Gln-Gly~Leu-Arg-Gly-Gln-Lys(Dnp)-Gly-Val-Arg-$(Gly-Pro-Hyp)_5$-NH_2	0.0259	13.1	1981[b]	[55]
MT6-MMP/ MMP-25	fTHP-15(5-Fam/ Dabcyl)	$(Gly-Pro-Hyp)_5$-Gly-Pro-Lys(5-Fam)-Gly-Pro-Gln-Gly~Leu-Arg-Gly-Gln-Lys(Dabcyl)-Gly-Val-Arg-(Gly-Pro-$Hyp)_5$-NH_2	ND	ND	ND	[84]
MMP-2	α1(V)436–447 fTHP	$(Gly-Pro-Hyp)_5$-Gly-Pro-Lys(Mca)-Gly-Pro-Pro-Gly~Val-Val-Gly-Glu-Lys(Dnp)-Gly-Glu-Gln-$(Gly-Pro-Hyp)_5$-NH_2	0.062	4.4	14,000[c]	[47]
MMP-9	α1(V)436–447 fTHP	$(Gly-Pro-Hyp)_5$-Gly-Pro-Lys(Mca)-Gly-Pro-Pro-Gly~Val-Val-Gly-Glu-Lys(Dnp)-Gly-Glu-Gln-$(Gly-Pro-Hyp)_5$-NH_2	0.044	8.1	5500[c]	[47]
MMP-9	α1(V)436–447 fTHP	$(Gly-Pro-Hyp)_5$-Gly-Pro-Lys(Mca)-Gly-Pro-Pro-Gly~Val-Val-Gly-Glu-Lys(Dnp)-Gly-Glu-Gln-$(Gly-Pro-Hyp)_5$-NH_2	3.99 ± 0.13	77.8 ± 5.8	51,360 ± 1676	[52, 64]
MMP-9 CAT	α1(V)436–447 fTHP	$(Gly-Pro-Hyp)_5$-Gly-Pro-Lys(Mca)-Gly-Pro-Pro-Gly~Val-Val-Gly-Glu-Lys(Dnp)-Gly-Glu-Gln-$(Gly-Pro-Hyp)_5$-NH_2	1.17 ± 0.21	22.5 ± 1.7	52,249 ± 9192	[52, 64]
MMP-12	α1(V)436–447 fTHP	$(Gly-Pro-Hyp)_5$-Gly-Pro-Lys(Mca)-Gly-Pro-Pro-Gly~Val-Val-Gly-Glu-Lys(Dnp)-Gly-Glu-Gln-$(Gly-Pro-Hyp)_5$-NH_2	2.35	24.8	97,600[c]	[65]
MMP-13	pfTHP-1	$(Gly-Pro-Hyp)_4$-Gly-Pro-Lys(Mca)-Gly-Sar-Ala-Gly~Gln-Arg-Gly-Val-Lys(Dnp)-Gly-*N*Leu-Hyp-(Gly-Pro-$Hyp)_4$-NH_2	0.9 ± 0.05	21.4 ± 1.1	42,000[c]	[68]

(continued)

Table 1 (continued)

Enzyme	Substrate abbreviation	Substrate sequence	k_{cat} (s^{-1})	K_M (μM)	k_{cat}/K_M[a] (s^{-1} M^{-1})	Reference
MMP-13	Ac-pfTHP-1	Ac-(Gly-Pro-Hyp)$_4$-Gly-Pro-Lys(Mca)-Gly-Sar-Ala-Gly~Gln-Arg-Gly-Val-Lys(Dnp)-Gly-*N*Leu-Hyp-(Gly-Pro-Hyp)$_4$-NH_2	0.6 ± 0.05	83.3 ± 4.3	7200[e]	[68]

[a]Assays were performed at 37 °C, pH 7.5 except where noted. *ND* not determined
[b]Assay performed at 25 °C
[c]Assay performed at 27 °C
[d]fTHP-4 and fTHP-15 have the same sequence
[e]Assay performed at 30 °C
[f]Substitutions relative to fTHP-4 are indicated in bold

possible and optimal substrate design represents a compromise of the aforementioned factors (Fig. 1).

Additional fTHPs have been constructed incorporating L- or D-2-amino-3-(7-methoxy-4-coumaryl)propionic acid (Amp) or L- or D-2-amino-3-(6,7-dimethoxy-4-coumaryl)propionic acid (Adp) as the fluorophore and Dnp as the quencher [48]. The desired sequences were C_6-(Gly-Pro-Hyp)$_5$-Gly-Pro-[Amp/Adp]-Gly-Pro-Gln-Gly~Leu-Arg-Gly-Gln-Lys(Dnp)-Gly-Val-Arg-(Gly-Pro-Hyp)$_5$-NH_2. All four fTHPs formed stable triple-helices. MMP-2 rates of hydrolysis for all fTHPs were considerably more rapid than MMP-1. Evaluation of individual kinetic parameters indicated that MMP-2 bound to the fTHPs more efficiently than MMP-1. Comparison to a triple-helical substrate incorporating the same sequence but with a different fluorophore [Lys(Mca)] demonstrated that the shorter side-chain of Amp or Adp was better tolerated by MMP-1 and MMP-2. Adp may well be the fluorophore of choice for fTHPs, as (a) fTHPs incorporating Adp were obtained in significantly higher yields than the Amp-containing fTHPs, (b) Adp has a larger Stokes shift than either Amp or Lys(Mca), and thus has less chance of self-quenching, (c) Adp has a relatively high quantum yield, (d) the Adp/Dnp pair is compatible with multiwell plate reader formats, and (e) MMPs better tolerate Adp compared with Lys(Mca).

We have also examined the preparation of fTHP substrates using 5-Fam as the fluorophore and Dabcyl as the quencher [55]. The N^{α}-(9-fluorenylmethoxycarbonyl)-N^{ε}-(5-carboxyfluorescein)-L-lysine [Fmoc-Lys(5-Fam)] building block was synthesized utilizing two distinct synthetic routes. The first involved copper complexation of Lys while the second utilized Fmoc-Lys with microwave irradiation. Both approaches allowed convenient production of a very pure final product at a reasonable cost. Fmoc-Lys(5-Fam) and Fmoc-Lys(Dabcyl) were incorporated into the sequence of a THP substrate utilizing automated solid-phase peptide synthesis protocols. A second substrate was assembled where Mca was the fluorophore and Dnp was the quencher. Circular dichroism (CD) spectroscopy was used to determine the influence of the fluorophore/quencher pair on the stability of the triple-helix. The activity of the two substrates was examined with MMP-1, MMP-13, and MT1-MMP. The combination of 5-Fam as fluorophore and Dabcyl as quencher resulted in a triple-helical substrate that, compared with the fluorophore/quencher pair of Mca/Dnp, had a slightly destabilized triple-helix but was hydrolyzed more rapidly by MMP-1 and MMP-13 and had greater sensitivity.

Due to the differing sequences of the fTHPs, their relative triple-helical thermal stabilities can vary [50, 51]. In order to obtain fTHPs that are stable under near-physiological conditions, as well as to examine the effects of substrate thermal stability on MMP activity, the "peptide-amphiphile" approach has been utilized [56–59]. Peptide-amphiphiles (PAs) mimic defined topological structures by incorporating an amino acid sequence with the propensity to form a triple-helix as the polar head group and a dialkyl or monoalkyl hydrocarbon chain as the nonpolar tail [56–59]. Desirable peptide head group melting temperature (T_m) values can be achieved for in vivo use, as triple-helical PAs have been constructed with T_m values ranging from 30 to 70 °C [40, 41, 46, 57, 58, 60–62]. fTHP PAs have been created that, based on their T_m values, are suitable for studying MMP activity under in vivo conditions.

1.4 Dissecting MMP Collagenolytic Activities

A series of fTHPs has been assembled that incorporate sequences based on a consensus types I-III collagen 769–783 region (Table 1). The fTHP substrates have been utilized to determine individual kinetic parameters (Table 1) and activation energies for hydrolysis of triple-helices [46, 47, 49–53, 63–69]. MMP-1, MMP-2, MMP-8, MMP-9, MMP-13, and MT1-MMP hydrolysis of the consensus types I-III collagen sequence (designated fTHP-15) occurred at the Gly~Leu bond, the analogous bond cleaved by these MMPs in the native collagen chains [70]. Thus, the incorporation of the fluorophore/quenching pair of Mca and Dnp did not affect the ability of the enzyme to recognize or cleave the substrate. A modified version of the consensus types I-III collagen fTHP, in which the P_1' subsite incorporated a Cys(4-methoxybenzyl) group (fTHP-9), was hydrolyzed efficiently by MT1-MMP, MMP-8, and MMP-13, but poorly by other MMPs [49, 50].

Analysis of MMP-1/THP interactions by hydrogen/deuterium exchange mass spectrometry (HDX MS) followed by evaluation of wild-type and mutant MMP-1 kinetics using fTHPs led to the identification of three noncatalytic regions in MMP-1 (residues 285–295, 302–316, and 437–457) and two specific residues (Ile290 and Arg291) that participate in collagenolysis [52]. Ile290 and Arg291 contributed to recognition of triple-helical structure, and facilitated both the binding and catalysis of the triple-helix. Evidence from this and prior studies indicated that the MMP-1 CAT and HPX domains collaborate in collagen catabolism by properly aligning the triple-helix and coupling conformational states to facilitate hydrolysis. This study was the first to document the roles of specific residues within the MMP-1 HPX domain in substrate binding and turnover.

To evaluate the role of charged residue clusters on the regulation of MMP collagenolysis, a series of ten fTHPs were constructed in which either Lys-Gly-Asp or Gly-Asp-Lys motifs replaced Gly-Pro-Hyp repeats [67]. The stabilities of fTHPs containing the two different motifs were analyzed, and kinetic parameters for substrate hydrolysis by six MMPs determined. A general trend for virtually all enzymes was that, as Gly-Asp-Lys motifs were moved from the extreme *N*- and *C*-termini to the interior next to the cleavage site sequence, k_{cat}/K_M values increased. Additionally, all Gly-Asp-Lys fTHPs were as good or better substrates than the parent fTHP, in which Gly-Asp-Lys was not present. In turn, the Lys-Gly-Asp fTHPs were also always better substrates than the parent THP, but the magnitude of the difference was considerably less compared with the Gly-Asp-Lys series. Of the MMPs tested, MMP-2 and MMP-9 most greatly favored the presence of charged residues, with preference for the Gly-Asp-Lys series. Lys-Gly-[Asp/Glu] motifs are more commonly found near potential MMP cleavage sites than Gly-[Asp/Glu]-Lys motifs. As Lys-Gly-Asp is not as favored by MMPs as Gly-Asp-Lys, the Lys-Gly-Asp motif appears advantageous over the Gly-Asp-Lys motif by preventing unwanted MMP hydrolysis. More specifically, the lack of Gly-Asp-Lys clusters may diminish potential MMP-2 and MMP-9 collagenolytic activity. This study indicated that MMPs have interactions spanning the P_{23}–P_{23}' subsites of collagenous substrates.

Although collagenolytic MMPs possess common domain organizations, there are subtle differences in their processing of collagenous triple-helical substrates. We incorporated peptoid residues in THPs and examined MMP activities toward these peptomeric chimeras [68].

Several different peptoid residues were incorporated into triple-helical substrates in subsites P_3, P_1, P_1', and P_{10}' individually or in combination, and the effects of the peptoid residues were evaluated on the activities of full-length MMP-1, MMP-8, MMP-13, and MT1-MMP. Most peptomers showed little discrimination between MMPs. However, a peptomer containing *N*-methyl Gly (sarcosine) in the P_1' subsite and *N*-isobutyl Gly (*N*Leu) in the P_{10}' subsite was hydrolyzed efficiently only by MMP-13 (nomenclature relative to α1(I)772–786 sequence). Cleavage site analysis showed hydrolysis at the Gly-Gln bond, indicating a shifted binding of the triple-helix compared to the parent sequence. Favorable hydrolysis by MMP-13 was not due to sequence specificity or instability of the substrate triple-helix, but rather was based on the specific interactions of the P_7' peptoid residue with the MMP-13 HPX domain. A FRET triple-helical peptomer was constructed and found to be readily processed by MMP-13, not cleaved by MMP-1 and MMP-8, and weakly hydrolyzed by MT1-MMP. The influence of the triple-helical structure containing peptoid residues on the interaction between MMP subsites and individual substrate residues may provide additional information for the mechanism of collagenolysis and the understanding of collagen specificity.

MMP selectivity has been observed with a type V collagen derived fTHP [47]. The α1(V)436–447 fTHP was hydrolyzed efficiently by MMP-2, MMP-9, and MMP-12 (Table 1), but was not hydrolyzed by MMP-1, MMP-3, MMP-13, or MT1-MMP [47, 65]. The type V collagen sequence Gly-Pro-Pro-Gly~Val-Val-Gly-Glu-Lys-Gly-Glu-Gln, as a single-stranded peptide, was hydrolyzed extremely slowly by either MMP-2 or MMP-9 [47]. Therefore, it is the triple-helical structure, along with the sequence, that imparts substrate specificity among various MMPs [47].

fTHPs have provided insights into (a) the relative triple-helical peptidase activities of the various collagenolytic MMPs (Fig. 2), (b) the collagen preferences of these MMPs, and (c) the relative roles of MMP domains and specific residues in efficient collagenolysis [40, 41, 46–53, 63–69, 71, 72]. fTHPs have been noted as being particularly advantageous for dissecting the mechanism of collagenolysis [1].

2 Materials

2.1 Manual Peptide Synthesis

Standard peptide reaction vessels (Chemglass Life Sciences or Peptides International) are used for manual peptide synthesis.

2.2 Automated Peptide Synthesis

A conventional peptide synthesizer such as the PS3 (Protein Technologies Inc.) or a microwave assisted peptide synthesizer such as the Liberty Blue (CEM Corp.).

2.3 Solid Phase for Peptide Synthesis

Resins for solid-phase synthesis, including Fmoc-Rink amide 4-methylbenzhydrylamine (MBHA) (Rapp-Polymere, EMD Millipore or Advanced ChemTech).

2.4 Amino Acids and Other Peptide/Peptoid Synthesis Reagents

1. Fmoc-protected amino acids (Sigma-Aldrich).
2. O-(6-chlorobenzotriazol-1-yl)-*N*,*N*,N′,*N*-tetramethyluronium hexafluorophosphate (HCTU).
3. 1-hydroxybenzotriazole (HOBt).
4. *N*,*N*′-diisopropylcarbodiimide (DIC).

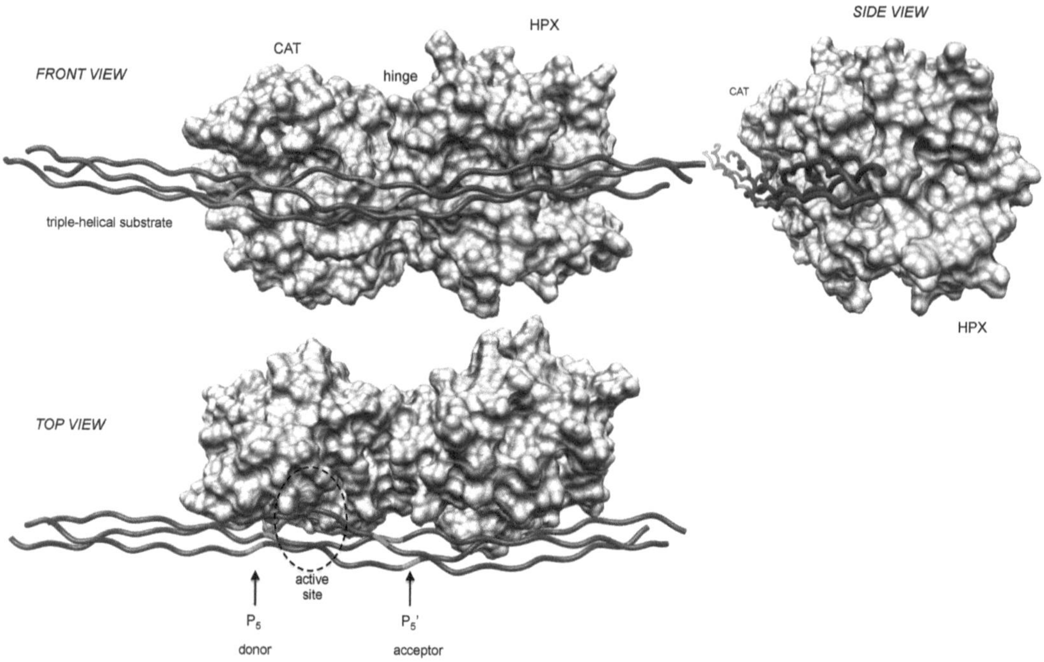

Fig. 1 The design of MMP triple-helical substrates. Illustration of a MMP/peptide complex showing that multiple contacts are formed between a triple-helical collagen-model peptide and the MMP CAT and HPX domains. The optimal positions for donor/acceptor moieties when designing synthetic fluorescent peptides were found to be at the P_5 subsite (behind the active site) and the P_5' subsite (within the hinge domain). The active site is marked with a *dashed line*

5. *N*,*N*-diisopropylethylamine (DIEA).
6. 1,8-diazabicyclo[5.4.0]undec-7-ene (DBU).
7. Hydrazine hydrate.
8. [dimethylamino(triazolo[4,5-b]pyridin-3-yloxy)methylidene]-dimethylazanium hexafluorophosphate (HATU).
9. 4-dimethylaminopyridine (DMAP).
10. Trifluoroacetic acid (TFA, peptide synthesis grade).
11. Piperidine.
12. Ammonia solution.
13. Bromoacetic acid.
14. Acetic acid.
15. Ninhydrin test kit (Anaspec).
16. Amines of choice (commercial sources for these chemical can be found using an online chemical directory search, such as ChemExper Chemical Directory, www.chemexper.com).

2.5 Fluorophore/Quencher Containing Synthetic Building Blocks

The syntheses of Fmoc-chemistry compatible fluorophore and quencher building blocks are described in detail in the literature (Table 2) and are commercially available.

2.6 Organic Solvents

1. Dimethylformamide (DMF).
2. N-methylpyrrolidone (NMP).
3. Dimethylsulfoxide (DMSO).
4. Dichloromethane (DCM).
5. Methanol (MeOH).
6. Acetonitrile.

2.6.1 Fmoc Deprotection Solution

Mix 20 mL DBU, 100 mL of piperidine, and 880 mL of NMP.

2.6.2 Dde Deprotection Solution

The 1-(4,4-dimethyl-2,6-dioxocyclohexylidene)ethyl (Dde) protecting groups are removed using 2% hydrazine in DMF. This solution is prepared by mixing 2 mL hydrazine hydrate and 98 mL DMF.

2.6.3 Peptide Cleavage Cocktail

To prepare 5 mL of such a solution mix 250 μL of H_2O and 4.75 mL of TFA. For every 100 mg of resin use 1.5 mL cleavage cocktail (*see* **Note 1**).

2.7 HPLC Characterization and Purification

1. Agilent 1200 Infinity series HPLC (Agilent Technologies).
2. Analytical HPLC: Vydac C18 column (5 μm, 300 Å, 150 mm × 4.6 mm) (Grace Davison).
3. Preparative HPLC: Vydac C18 column (15–20 μm, 300 Å, 250 mm × 22 mm) (Grace Davison).
4. Mobile phase: Solvent A—0.1% TFA in H_2O. Solvent B—0.1% TFA in acetonitrile.
5. Analytical gradient—2 to 100% B over 20 min at a flowrate of 1 mL/min.
6. Preparative gradient—5 to 50% B in 60 min at a flowrate of 10 mL/min.
7. Detection is carried out at λ = 220 and 280 nm.

2.8 Mass Spectrometry

1. Voyager DE-PRO Biospectrometry workstation (Applied Biosystems) or Bruker Microflex LF mass spectrometer (Bruker Daltonics).
2. MALDI matrix - α-cyano-4-hydroxycinnamic acid (HCCA) or 2,5-dihydroxybenzoic acid (DHB) (Sigma-Aldrich or Fisher Scientific).

2.9 Circular Dichroism (CD) Spectroscopy

1. Jasco J-810 spectropolarimeter (Jasco).
2. Record over a wavelength (λ) range of 190–250 nm with a sample concentration of 0.1 mg/mL in 0.5% acetic acid solution (v/v).
3. Thermal transition curves are obtained by recording the molar ellipticity ([θ]) at λ = 225 nm with an increase in temperature of 20 °C/h over a range of 5–80 °C. The temperature is controlled by a JASCO PTC-348WI (Jasco) temperature control unit.

2.10 Peptide Lyophilization

Labconco FreeZone lyophilizer (freeze dryer) system (Labconco Corp).

2.11 Absorption Spectrophotometry

1. NanoDrop 2000c UV-Vis spectrophotometer (Thermo Fisher Scientific).
2. Fluorescent dye-conjugated peptide is measured using the specific wavelength (λ) and molar absorption coefficient (ε) characteristic for the fluorophore. Triple-helical peptide-conjugate concentration is calculated according to the Beer-Lambert law, using the following formula:

$$C = A / (l \times 3 \times \varepsilon).$$

where C = peptide concentration (M), A = measured absorbance, and l = cell path length (cm). Please note that in case of triple-helical peptides the molar absorption coefficient ε has to be multiplied by the number of fluorophore groups present in a molecule.

2.12 Enzymatic Assay Buffers, Stock Solutions, and Reagents

1. Chymotrypsin stock: 1 mg/ml Chymotrypsin in 1 mM HCl.
2. PMSF stock: 0.2 M phenylmethyl sulfonyl fluoride (PMSF) in ethanol.
3. APMA stock: Prepare 100 mM 4-aminophenylmercuric acetate (APMA) (Sigma) in DMSO and store at 4 °C (stable for several months).
4. Dimethylsulfoxide (DMSO).
5. 0.1 M $ZnCl_2$.
6. Trypsin-3 (R&D Systems).
7. AEBSF.
8. Furin.
9. SSM 3 trifluoroacetate (*rel*-*N*,*N*″-[[(1*R*,3*S*,4*S*,6*R*)-4,6-bis[(aminoiminomethyl)amino]-1,3-cyclohexanedyil]bis(oxy-4,1-phenylene)]bisguanidine tetratrifluoroacetate salt) (R&D Systems).
10. ProMMP-1, proMMP-2, proMMP-3, proMMP-8, proMMP-9, proMMP-12, proMMP-13, proMT1-MMP, and proMT2-MMP (R&D Systems).
11. Knight single-stranded peptide (SSP) [Mca-Lys-Pro-Leu-Gly--Leu-Lys(Dnp)-Ala-Arg-NH_2] was synthesized using methods described previously [17, 73]. Knight SSP can be prepared as a 20 mM stock solution in water and stored at −20 °C [73].
12. Enzyme assay buffer (EAB): 50 mM Tris–HCl, pH 7.5, 100 mM NaCl, 10 mM $CaCl_2$, 0.05% Brij 35.
13. TS buffer: 50 mM Tris–HCl pH 7.5, 150 mM NaCl, 0.01% Brij-35, 10 mM $CaCl_2$, 1 μM $ZnCl_2$.

14. fTHP-15 and α1(V)436–447 fTHP peptide stock solutions: Dissolve fTHP-15 or α1(V)436–447 fTHP in DMSO and store at 4 °C for up to 1 year.

3 Methods

3.1 Peptide Synthesis Protocols

For the incorporation of individual amino acids by Fmoc-solid-phase methodology HCTU/HOBt activation is recommended [74]. Reagent amounts including Fmoc amino acids (Fmoc-AA), HCTU, HOBt, and DIEA are calculated in relation to peptide synthesis scale (typically expressed in mmol).

3.1.1 General Amino Acid Coupling Cycles

For Fmoc-amino acid couplings the following molar excess is recommended: Fmoc-AA = 3 equiv., HCTU = 2.5 equiv., and DIEA = 5.5 equiv. Recommended coupling time using a conventional automated peptide synthesizer is 1 h.

3.1.2 General Fmoc Deprotection Cycle

The Fmoc protecting group is removed by gentle treatment of peptidyl-resin using standard Fmoc deprotection solution. This step can be performed on an automated peptide synthesizer using an appropriate deprotection program or manually as follows:

1. Place the resin in a peptide synthesis reaction vessel.
2. Add the appropriate amount of Fmoc deprotection solution (three to five times the resin volume).
3. Agitate the resin for 5 min. Decant the solution.
4. Repeat **step 2**.
5. Agitate the resin for another 15 min. Decant the solution.
6. Wash resin (three times) using 5 mL of DMF.

3.1.3 Removal of the Dde Protecting Group

The Dde protecting group is removed from Lys residues according to the following steps:

1. Place the resin in a peptide synthesis reaction vessel.
2. Add the appropriate amount of Dde deprotection solution (four times the resin volume).
3. Agitate the resin for 5 min and then decant the solution.
4. Repeat the treatment with Dde deprotection solution two more times before decanting off the solution.
5. Wash the resin three times using 5 mL of DMF.

3.1.4 "On-Resin" Approach for Synthesis of THPs Containing 5-FAM

This protocol was adapted from [75].

1. The peptide is synthesized using an automated peptide synthesizer and general coupling conditions for the incorporation of individual amino acids (see above). The Fmoc group is removed

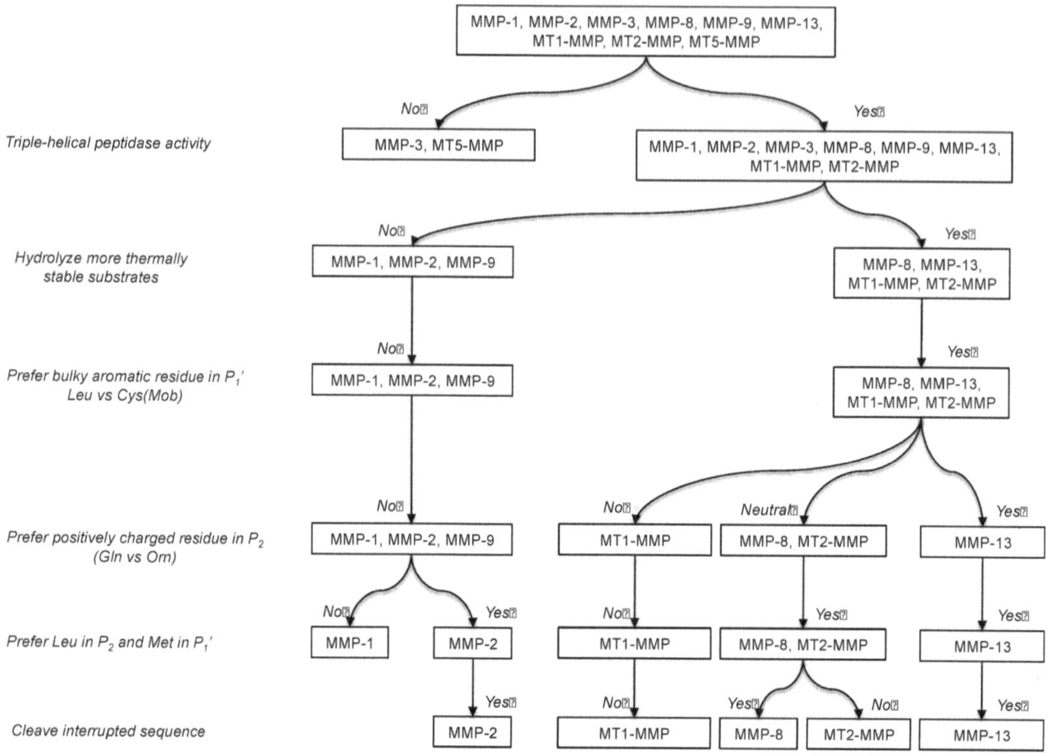

Fig. 2 Flowchart depicting MMP triple-helical peptidase activities. The nine MMPs examined in this study were initially subdivided based upon their ability to cleave a triple-helical substrate. Triple-helical peptidase MMPs were further subdivided upon their ability to process more thermally stable substrates and finally, based on their specificity for residues within the P_1' and P_2 subsites. Ultimately, the flowchart illustrates that "collagenolytic" MMPs display distinct proteolytic behaviors. Figure modified from [51]

as outlined (see above) (*see* **Note 1**) and the 5-FAM moiety is incorporated at the end of the peptide synthesis.

2. The side chain Dde protecting groups are removed using the Dde deprotection solution and conditions discussed (see above).
3. After Dde deprotection, wash the resin in the peptide synthesis reaction vessel sequentially using 5 mL DMF, 10 mL of 10% aqueous DMF, 10 mL DCM, and 15 mL MeOH.
4. Place the reaction vessel (no cap) in a desiccator connected to a vacuum pump and dry the peptide-resin in vacuo for 4–6 h.
5. Conjugate the 5-FAM fluorescent dye to the ε-amino groups of Lys using the following conditions (for a typical 0.1 mmol scale synthesis):
   (a) Place the resin in a dry reaction vessel.
   (b) Prepare a solution containing 225 mg of 5-FAM dye, 81 mg HOBt (monohydrate), 185 μL of DIC, and 73 mg of DMAP in 5 mL of DMF.

(c) Add this mixture to the resin and agitate for 16 h. Filter the resin and sequentially wash with 5 mL of DMF, DCM, and MeOH.

(d) Perform a ninhydrin test. A negative result indicates complete coupling of 5-FAM.

(e) Cleave the peptide from the resin using a solution containing 5% thioanisole and 90% TFA in water (use 1.5 mL of cleavage cocktail per 100 mg of dry resin). Incubate at room temperature for 3 h to ensure complete cleavage and side-chain deprotection of the peptide. Perform this step under the fume hood (*see* **Note 2**).

(f) Collect the filtrate and add 10–15 mL of water. Perform this step under the fume hood.

(g) Freeze and then lyophilize.

(h) Purify the peptide by preparative RP-HPLC. Then analyze by analytical HPLC and MALDI-TOF mass spectrometry to check the purity and mass of the peptide.

(i) The concentration of the 5-FAM-containing THP peptide can be determined by absorption spectroscopy (λ = 495 nm, ε = 75,000 cm^{-1} M^{-1}) on a NanoDrop spectrophotometer.

3.1.5 "On-Resin" Approach for Synthesis of THPs Containing DABCYL

This protocol is adapted from [76]. DABCYL is fairly soluble in DMF, but has much greater solubility in NMP.

1. Remove side chain Dde protecting groups using the Dde deprotection solution and conditions (see above).

Table 2

Fmoc fluorophore and quencher building block synthesis and commercial availability

Building block	Literature citation for synthesis of building block	Commercial source[a]
Fluorophore		
Fmoc-Lys(5-Fam)-OH	[55]	D, F
Fmoc-Lys(Mca)-OH	[85, 86]	A, B, C, D, E, F, G
Quencher		
Fmoc-Lys(Dnp)-OH	[87]	A, B, C, D, E, F, G
Fmoc-Lys(DABCYL)-OH	[88]	B, C, E, F

[a]A—Sigma-Aldrich (St. Louis, USA); B—EMD Millipore (Darmstadt, Germany); C—Fisher Scientific (Atlanta, USA); D—Anaspec (Fremont, USA); E—Chemimpex (Wood Dale, USA); F—Chempep, Inc. (Wellington, USA); G—Bachem (Torrance, USA)

2. Following Dde deprotection, wash the resin in the peptide synthesis reaction vessel sequentially with 5 mL DMF, 10 mL of 10% aqueous DMF, 10 mL DCM, and 15 mL MeOH.
3. In a separate flask dissolve 4 equiv. (in relation to synthesis scale) of DABCYL in 10 mL NMP. Then activate the DABCYL carboxyl group by adding 4.4 equiv. of DIC and 8 equiv. of HOBt along with 8 equiv. of DIPEA.
4. Add this solution to the peptide resin and allow the acylation reaction to continue overnight. (*Note*: Coupling of the DABCYL derivative is fairly slow.)
5. After the DABCYL coupling, wash the resin in the peptide synthesis reaction vessel sequentially with 5 mL DMF, 10 mL DCM, and 15 mL MeOH. Repeat this washing step until the wash solution is transparent.
6. Perform a ninhydrin test. A negative result indicates complete coupling of DABCYL. Otherwise repeat **steps 3–6**.

3.1.6 Synthesis of THP Substrates Containing Peptoid Residues

This protocol (Fig. 3) is adapted from [68]. Peptoid residues can be conveniently introduced during a standard peptide synthesis using the submonomeric approach. The incorporation of each peptoid unit requires two steps: acylation using chloro- or bromo-acetic acid followed by substitution with the appropriate amine [77] as outlined below:

1. Prepare all solvents and reagents in a fume hood.
2. Place the resin in a dry reaction vessel.
3. If necessary, remove Fmoc protecting groups by agitating the resin with 20% piperidine in DMF (3 × 6 min) using four to five times the resin volume.
4. Wash the resin (3 min each) three times with 10 mL of DMF followed by three times with 10 mL of DCM.
5. Dissolve bromoacetic acid (10 equiv. relative to resin loading) and DIC (10 equiv) in an amount of DMF which equals 2–3 volumes of resin and add this solution to the resin.
6. Allow the resin to agitate at room temperature for 30 min.
7. Wash the resin three times with 10 mL of DMF for 3 min each.
8. Dissolve the amine (25 equiv) in DMSO, NMP, or DMF.
9. Upon addition of the amine solution to the resin, allow the resin to agitate at room temperature for 2 h. In some cases, this displacement step may take up to 12 h (overnight).
10. Wash the resin (3 min each) three times with 10 mL of DMF followed by three times with 10 mL of DCM.
11. Perform the ninhydrin test when necessary.

12. Continue with peptide synthesis until the desired sequence is obtained.

3.2 Characterization of fTHPs

The peptide molecular mass is confirmed by MALDI-TOF mass spectrometry using HCCA or SA MALDI matrices. While CD spectroscopy [40, 41, 46, 57, 58, 61, 78–80] is used to determine the degree of triple-helicity of each peptide. The general protocol is as follows:

1. Dissolve the peptide in an appropriate buffer to a final concentration of 0.1 mg/mL. Incubate the peptide solution for 24–48 h at 4 °C to allow triple-helix formation. *See* **Note 3**.
2. Record a series of spectra (n = 5–10) over the range of λ = 190–250 nm. The molar ellipticity for a typical spectrum of a triple-helical peptide displays a positive peak at λ =225 nm and a minimum at λ = 205 nm.
3. Determine the melting temperature (T_m) by monitoring the change in molar ellipticity at λ = 225 nm as the temperature is increased at a rate of 20 °C/h from 5 to 80 °C. Data are plotted using GraphPad Prism software (GraphPad) and, where possible, the T_m value is defined as the inflection point of the first derivative of the sigmoidal melting curve. Alternatively, T_m is evaluated from the midpoint of the transition curve.

3.3 Enzyme Activation

3.3.1 Activation of proMMP-1

1. Dilute proMMP-1 to 100 μg/mL in TS buffer (*see* **Note 4**).
2. Activate proMMP-1 by adding APMA to a final concentration of 1 mM.
3. Incubate at 37 °C for 2–3 h.
4. Immediately after incubation dilute MMP-1 to 20 μg/mL and store at −80 °C until ready to use.
5. Activity is initially measured with the Knight substrate (MMP-1 exhibits a weak activity toward fTHP-15).

3.3.2 Activation of proMMP-2

1. Dilute proMMP-2 to 100 μg/mL in TS buffer (*see* **Note 4**).
2. Activate proMMP-2 by adding APMA to a final concentration of 1 mM.
3. Incubate at 37 °C for 24 h.
4. Immediately after incubation dilute MMP-2 to 20 μg/mL and store at −80 °C until ready to use.
5. MMP-2 can also be obtained in active form from EMD Millipore (cat # PF023).

3.3.3 Activation of proMMP-3

1. Activate proMMP-3 at 20 μg/mL in EAB containing 5 μg/mL chymotrypsin.
2. Incubate at 37 °C for 30 min.

3. Terminate activation by adding PMSF to a final concentration of 2 mM. Pre-warm the PMSF to 37 °C just prior to adding to the sample.
4. Store at −80 °C for future use.
5. Measure activity using the Knight substrate at 10 μM final substrate concentration.

3.3.4 Activation of proMMP-8

1. Dilute proMMP-8 to 100 μg/mL in TS buffer (*see* **Note 4**).
2. Activate proMMP-8 by adding APMA to a final concentration of 1 mM.
3. Incubate at 37 °C for 1 h.
4. Immediately after incubation dilute MMP-8 to 20 μg/mL and store at −80 °C until ready to use.

3.3.5 Activation of proMMP-9

1. Dilute proMMP-9 to 100 μg/mL in TS buffer (*see* **Note 4**).
2. Activate proMMP-9 by adding APMA to a final concentration of 1 mM.
3. Incubate at 37 °C for 24 h.
4. Immediately after incubation dilute MMP-9 to 20 μg/mL and store at −80 °C until ready to use.
5. MMP-9 can also be obtained in active form from EMD Millipore (cat # PF140).

3.3.6 Activation of proMMP-12

1. Dilute proMMP-12 to 100 μg/mL in TS buffer (*see* **Note 4**).
2. Activate proMMP-12 by adding APMA to a final concentration of 1 mM.
3. Incubate at 37 °C for 24 h.
4. Immediately after incubation dilute MMP-12 to 20 μg/mL and store at −80 °C until ready to use.

3.3.7 Activation of proMMP-13

1. Dilute proMMP-13 to 100 μg/mL in TS buffer (*see* **Note 4**).
2. Activate proMMP-13 by adding APMA to a final concentration of 1 mM.
3. Incubate at 37 °C for 2 h.
4. Immediately after incubation dilute MMP-13 to 20 μg/mL and store at −80 °C until ready to use.

3.3.8 Activation of proMT1-MMP

Method 1

1. Prepare activation buffer containing 50 mM Tris pH 9.0, 1 mM $CaCl_2$ and 0.5% (v/v) Brij-35.
2. Activate proMT1-MMP at 40 μg/mL with 0.86 μg/mL recombinant furin in activation buffer.
3. Incubate at 37 °C for 1.5 h.

4. Stop the reaction by adding SSM 3 trifluoroacetate to a final concentration of 1 μM.

Method 2

1. Prepare a stock solution of 100 mM 4-(2-aminoethyl-benzensulfonyl fluoride hydrochloride) (AEBSF) (R&D Systems, cat # EI001) in deionized water (*see* **Note 5**).
2. Activate 100 μg/mL proMT1-MMP by adding recombinant human active trypsin-3 at a final concentration of 5 μg/mL in TS buffer.
3. Incubate at 37 °C for 2 h.
4. Stop the reaction by adding AEBSF to a final concentration of 1 mM and incubate at room temperature for 15 min.
5. Store the activated enzyme at −80 °C for future use.

3.3.9 Activation of proMT2-MMP (MMP-15)

1. Prepare AEBSF 100 mM stock in deionized water.
2. Activate proMT2-MMP at 100 μg/mL with 0.1 μg/mL trypsin-3 in TS buffer.
3. Incubate at 37 °C for 2 h.
4. Add AEBSF to a final concentration of 1 mM and incubate at room temperature for 15 min to stop the reaction.
5. Store the activated enzyme at −80 °C for future use.

3.4 Enzymatic Assays Using fTHPs

Manual assays are performed in black low volume 384-well plates (Greiner Bio-One, Monroe, NC) using a microplate reader such as BioTek Synergy 4 running Gen5 1.11, Gen5 2.00, and Gen5 2.01 software as described previously [64, 81]. MMP-3 should not cleave fTHP substrates and can be used as a negative control. Its activity can be tested with the Knight SSP. Enzymatic assays are performed as follows:

1. Prepare 100 μM fTHP-15 and α1(V)436–447 fTHP stock solutions in TS buffer and incubate overnight at 4 °C to allow for the proper folding of the peptide triple-helix. The working stock is stable for 1 week.
2. Prepare a 12-point serial dilution of fTHP substrate with the highest concentration being 50 μM.
3. Add 5 μL of each peptide solution to 12 wells of the 384-well plate.
4. Measure the initial fluorescence in relative fluorescence units (RFUs).
5. Add 5 μL of a twofold enzyme stock solution to each well.
6. Monitor fluorescence for 15 min at 25 °C to determine the initial reaction rate.

7. Store the plates at ambient temperature for 24 h before taking a final reading. All measurements are performed in triplicate.
8. Plot fluorescence (RFU) versus time (s). The rate of hydrolysis (Corr *Vi*; μM/s) is determined by dividing the slope of the plot by the fluorescence change after 24 h, and then multiplying by the substrate concentration (μM) {Corr *Vi* = (Slope/Change in RFU) × [peptide]} (*see* **Note 6**).
9. Kinetic parameters are calculated from Lineweaver-Burke, Hanes-Woolf, and Eadee-Hofstee linear plots (substrate concentration versus reaction velocity) as well as nonlinear Michaelis-Menton regression using GraphPad Prism.

4 Notes

1. During peptide synthesis the last, *N*-terminal amino acid to be incorporated must be coupled as a Boc and not Fmoc-derivative to be compatible with Dde group removal.
2. If the yield or purity of crude peptide is low a different cleavage cocktail should be considered [82].
3. Due to aggregation of triple-helical peptides, use a concentration of 2–500 μM.
4. ProMMP-1, proMMP-3, proMMP-8, or proMMP-12 (20 ng/μL) can be activated with a chymotrypsin/trypsin mixture (final concentration of 5 ng/μL for each) for 30 min at 37 °C. In our experience, this mode of activation results in greater enzymatic activity compared with the manufacturer's recom-

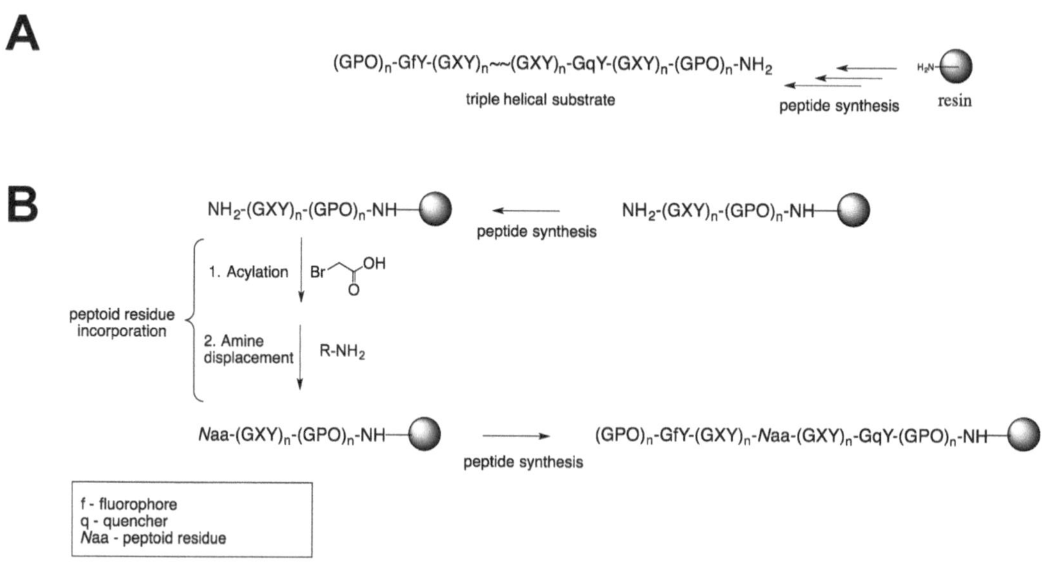

Fig. 3 Synthesis of fTHP substrates. (**a**) Peptidic fTHPs and (**b**) fTHP containing a peptoid residue

mended activation protocol. The reaction is stopped by the addition of 2 mM PMSF. Enzyme activity is tested with 5 μM Knight SSP in TS buffer.

5. AEBSF (20 or 100 mM) stock solution prepared in water is slightly acidic. If a final pH greater than 7 is required, the pH should be adjusted just prior to use. Stock solutions are stable for up to 2 months at −20 °C or 1 week at 4 °C. AEBSF is inactivated by 50% by incubating at 37 °C after 5 h.
6. The inner filter effect (IFE) is often observed when using fluorogenic substrates at high concentrations. Decreases in fluorescence due to IFE exceed 10% once the sum of the absorbances at excitation and emission wavelengths is greater than 0.08 [81]. Correction factors can be applied to take into account IFEs [83]. Alternatively, fitting multiple progress curves can be used to obtain accurate reaction kinetics even in the presence of IFEs [81].

Acknowledgments

We gratefully acknowledge the National Institutes of Health (EB000289, AR063795, and CA098799), the Multiple Sclerosis National Research Institute, and the Texas Higher Education STAR Award Program for support of our laboratory's research on MMPs.

References

1. Lombard C, Saulnier J, Wallach J (2005) Assays of matrix metalloproteinases (MMPs) activities: a review. Biochimie 87:265–272
2. Nagase H, Fields GB (1996) Human matrix metalloproteinase specificity studies using collagen sequence-based synthetic peptides. Biopolymers 40:399–416
3. Fields GB (2001) Using fluorogenic peptide substrates to assay matrix metalloproteinases. In: Clark IM (ed) Methods in Molecular Biology 151: Matrix Metalloproteinase Protocols. Humana, Totowa, NJ, pp 495–518
4. Ohkubo S, Miyadera K, Sugimoto Y, K-i M, Wierzba K, Yamada Y (1999) Identification of substrate sequences for membrane type-1 matrix metalloproteinase using bacteriophage peptide display library. Biochem Biophys Res Commun 266:308–313
5. Deng S-J, Bickett DM, Mitchell JL, Lambert MH, Blackburn RK, Carter HL III, Neugebauer J, Pahel G, Weiner MP, Moss ML (2000) Substrate specificity of human collagenase 3 assessed using a phage-displayed peptide library. J Biol Chem 275:31422–31427
6. Turk BE, Huang LL, Piro ET, Cantley LC (2001) Determination of protease cleavage site motifs using mixture-based oriented peptide libraries. Nat Biotechnol 19:661–667
7. Kridel SJ, Sawai H, Ratnikov BI, Chen EI, Li W, Godzik A, Strongin AY, Smith JW (2002) A unique substrate binding mode discriminates membrane type 1-matrix metalloproteinase (MT1-MMP) from other matrix metalloproteinases. J Biol Chem 277:23788–23793
8. Chen EI, Kridel SJ, Howard EW, Li W, Godzik A, Smith JW (2002) A unique substrate recognition profile for matrix metalloproteinase-2. J Biol Chem 277:4485–4491
9. Park HI, Turk BE, Gerkema FE, Cantley LC, Sang Q-XA (2002) Peptide substrate specificities and protein cleavage sits of human endometase/matrilysin-2/matrix metalloproteinase-26. J Biol Chem 277:35168–35175

10. Pan W, Arnone M, Kendall M, Grafstrom RH, Seitz SP, Wasserman ZR, Albright CF (2003) Identification of peptide substrates for human MMP-11 (stromelysin-3) using phage display. J Biol Chem 278:27820–27827
11. Knight CG (1995) Fluorometric assays of proteolytic enzymes. Methods Enzymol 248:18–34
12. Gershkovich AA, Kholodovych VV (1996) Fluorogenic substrates for proteases based on intramolecular fluorescence energy transfer (IFETS). J Biochem Biophys Methods 33:135–162
13. Stack MS, Gray RD (1989) Comparison of vertebrate collagenase and gelatinase using a new fluorogenic substrate peptide. J Biol Chem 264(8):4277–4281
14. Niedzwiecki L, Teahan J, Harrison RK, Stein RL (1992) Substrate specificity of the human matrix metalloproteinase stromelysin and the development of continuous fluorometric assays. Biochemistry 31:12618–12623
15. Knight CG, Willenbrock F, Murphy G (1992) A novel coumarin-labelled peptide for sensitive continuous assays of the matrix metalloproteinases. FEBS Lett 296(3):263–266
16. Bouvier J, Schneider P, Malcolm B (1993) A fluorescent peptide substrate for the surface metalloprotease of *Leishmania*. Exp Parasitol 76:146–155
17. Nagase H, Fields CG, Fields GB (1994) Design and characterization of a fluorogenic substrate selectively hydrolyzed by stromelysin 1 (matrix metalloproteinase-3). J Biol Chem 269:20952–20957
18. Bickett DM, Green MD, Berman J, Dezube M, Howe AS, Brown PJ, Roth JT, McGeehan GM (1993) A high throughput fluorogenic substrate for interstitial collagenase (MMP-1) and gelatinase (MMP-9). Anal Biochem 212:58–64
19. Beekman B, Wouter J, Bloemhoff W, Ronday K, Tak PP, te Koppele JM (1996) Convenient fluorometric assay for matrix metalloproteinase activity and its application in biological media. FEBS Lett 390:221–225
20. Geoghegan KF, Emery MJ, Martin WH, McColl AS, Daumy GO (1993) Site-directed double fluorescent tagging of human renin and collagenase (MMP-1) substrate peptides using the periodate oxidation of *N*-terminal serine. An apparently general strategy for provision of energy-transfer substrates for proteases. Bioconjug Chem 4:537–544
21. Geoghegan KF (1996) Improved method for converting an unmodified peptide to an energy transfer substrate for proteinase. Bioconjug Chem 7:385–391
22. Rakhmanova V, Meyer R, Tong X (2007) New substrates for FRET-based assays: Use of a long-wavelength fluorophore to detect the activity of MMPs. Gen Eng Biotechnol News 15:40
23. Moss ML, Rasmussen FH (2007) Fluorescent substrates for the proteinases ADAM17, ADAM10, ADAM8, and ADAM12 useful for high-throughut inhibitor screening. Anal Biochem 366:144–148
24. Buchardt J, Ferreras M, Krog-Jensen C, Delaisse J-M, Foged NT, Meldal M (1999) Phosphinic peptide matrix metalloproteinase-9 inhibitors by solid-phase synthesis using a building block approach. Chem A Eur J 5:2877–2884
25. Buchardt J, Schiodt CB, Krog-Jensen C, Delaissé J-M, Foged NT, Meldal M (2000) Solid phase combinatorial library of phosphinic peptides for discovery of matrix metalloproteinase inhibitors. J Comb Chem 2:624–638
26. Ito AS, Turchiello RDF, Hirata IY, Cezari MHS, Meldal M, Juliano L (1999) Fluorescent properties of amino acids labeled with ortho-aminobenzoic acid. Biospectroscopy 4:395–402
27. de Souzaa ES, Hiratab IY, Julianob L, Itoa AS (2000) End-to-end distance distribution in bradykinin observed by Förster resonance energy transfer. Biochim Biophys Acta 1474:251–261
28. Takara M, Ito AS (2005) General and specific solvent effects in optical spectra of ortho-aminobenzoic acid. J Fluoresc 15:171–177
29. George J, Teear ML, Norey CG, Burns DD (2003) Evaluation of an imaging platform during the development of a FRET protease assay. J Biomol Screen 8:72–80
30. Knight CG (1991) A quenched fluorescent substrate for thimet peptidase containing a new fluorescent amino acid, DL-2-amino-3-(7-methoxy-4-coumaryl)propionic acid. Biochem J 274:45–48
31. Woodhead-Galloway J (1980) In: Collagen: The anatomy of a protein. Edward Arnold Limited, London, pp. 10–19
32. Shoulders MD, Raines RT (2009) Collagen structure and stability. Annu Rev Biochem 78:929–958
33. Fields GB, Prockop DJ (1996) Perspectives on the synthesis and application of triple-helical, collagen-model peptides. Biopolymers (Pept Sci) 40:345–357
34. Fields GB (2010) Synthesis and biological applications of collagen-model triple-helical peptides. Org Biomol Chem 8:1237–1258

35. Jenkins CL, Raines RT (2002) Insights on the conformational stability of collagen. Nat Prod Rep 19:49–59
36. Brodsky B, Shah NK (1995) The triple-helix motif in proteins. FASEB J 9:1537–1546
37. Koide T (2005) Triple helical collagen-like peptides: engineering and applications in matrix biology. Connect Tissue Res 46:131–141
38. Koide T (2007) Designed triple-helical peptides as tools for collagen biochemistry and matrix engineering. Phil Trans R Soc B 362:1281–1291
39. Brodsky B, Thiagarajan G, Madhan B, Kar K (2008) Triple-helical peptides: an approach to collagen conformation, stability, and self-association. Biopolymers 89:345–353
40. Lauer-Fields JL, Tuzinski KA, Shimokawa K, Nagase H, Fields GB (2000) Hydrolysis of triple-helical collagen peptide models by matrix metalloproteinases. J Biol Chem 275:13282–13290
41. Lauer-Fields JL, Nagase H, Fields GB (2000) Use of Edman degradation sequence analysis and matrix-assisted laser desorption/ionization mass spectrometry in designing substrates for matrix metalloproteinases. J Chromatogr A 890:117–125
42. Ottl J, Battistuta R, Pieper M, Tschesche H, Bode W, Kühn K, Moroder L (1996) Design and synthesis of heterotrimeric collagen peptides with a built-in cystine-knot. FEBS Lett 398:31–36
43. Ottl J, Gabriel D, Murphy G, Knäuper V, Tominaga Y, Nagase H, Kröger M, Tschesche H, Bode W, Moroder L (2000) Recognition and catabolism of synthetic heterotrimeric collagen peptides by matrix metalloproteinases. Chem Biol 7:119–132
44. Ottl J, Moroder L (1999) Disulfide-bridged heterotrimeric collagen peptides containing the collagenase cleavage site of collagen type I: synthesis and conformational properties. J Am Chem Soc 121:653–661
45. Bertini I, Fragai F, Luchinat C, Melikian M, Toccafondi M, Lauer JL, Fields GB (2012) Structural basis for matrix metalloproteinase 1 catalyzed collagenolysis. J Am Chem Soc 134:2100–2110
46. Lauer-Fields JL, Broder T, Sritharan T, Nagase H, Fields GB (2001) Kinetic analysis of matrix metalloproteinase triple-helicase activity using fluorogenic substrates. Biochemistry 40:5795–5803
47. Lauer-Fields JL, Sritharan T, Stack MS, Nagase H, Fields GB (2003) Selective hydrolysis of triple-helical substrates by matrix metalloproteinase-2 and -9. J Biol Chem 278:18140–18145
48. Lauer-Fields JL, Kele P, Sui G, Nagase H, Leblanc RM, Fields GB (2003) Analysis of matrix metalloproteinase activity using triple-helical substrates incorporating fluorogenic L- or D-amino acids. Anal Biochem 321:105–115
49. Minond D, Lauer-Fields JL, Nagase H, Fields GB (2004) Matrix metalloproteinase triple-helical peptidase activities are differentially regulated by substrate stability. Biochemistry 43:11474–11481
50. Minond D, Lauer-Fields JL, Cudic M, Overall CM, Pei D, Brew K, Visse R, Nagase H, Fields GB (2006) The roles of substrate thermal stability and P2 and P1′ subsite identity on matrix metalloproteinase triple-helical peptidase activity and collagen specificity. J Biol Chem 281:38302–38313
51. Minond D, Lauer-Fields JL, Cudic M, Overall CM, Pei D, Brew K, Moss ML, Fields GB (2007) Differentiation of secreted and membrane-type matrix metalloproteinase activities based on substitutions and interruptions of triple-helical sequences. Biochemistry 46:3724–3733
52. Lauer-Fields JL, Chalmers MJ, Busby SA, Minond D, Griffin PR, Fields GB (2009) Identification of specific hemopexin-like domain residues that facilitate matrix metalloproteinase collagenolytic activity. J Biol Chem 284:24017–24024
53. Lauer-Fields JL, Fields GB (2002) Triple-helical peptide analysis of collagenolytic protease activity. Biol Chem 383:1095–1105
54. Lauer-Fields JL, Juska D, Fields GB (2002) Matrix metalloproteinases and collagen catabolism. Biopolymers (Pept Sci) 66:19–32
55. Tokmina-Roszyk M, Tokmina-Roszyk D, Fields GB (2013) The synthesis and application of Fmoc-Lys(FAM) building blocks. Biopolymers (Pept Sci) 100:347–355
56. Berndt P, Fields GB, Tirrell M (1995) Synthetic lipidation of peptides and amino acids: monolayer structure and properties. J Am Chem Soc 117:9515–9522
57. Yu Y-C, Berndt P, Tirrell M, Fields GB (1996) Self-assembling amphiphiles for construction of protein molecular architecture. J Am Chem Soc 118:12515–12520
58. Yu Y-C, Tirrell M, Fields GB (1998) Minimal lipidation stabilizes protein-like molecular architecture. J Am Chem Soc 120:9979–9987
59. Yu Y-C, Roontga V, Daragan VA, Mayo KH, Tirrell M, Fields GB (1999) Structure and dynamics of peptide-amphiphiles incorporating

triple-helical proteinlike molecular architecture. Biochemistry 38:1659–1668

60. Fields GB, Lauer JL, Dori Y, Forns P, Yu Y-C, Tirrell M (1998) Protein-like molecular architecture: biomaterial applications for inducing cellular receptor binding and signal transduction. Biopolymers 47:143–151
61. Malkar NB, Lauer-Fields JL, Borgia JA, Fields GB (2002) Modulation of triple-helical stability and subsequent melanoma cellular responses by single-site substitution of fluoroproline derivatives. Biochemistry 41:6054–6064
62. Malkar NB, Lauer-Fields JL, Juska D, Fields GB (2003) Characterization of peptide-amphiphiles possessing cellular activation sequences. Biomacromolecules 4:518–528
63. Lauer-Fields JL, Sritharan T, Kashiwagi M, Nagase H, Fields GB (2007) Substrate conformation modulates aggrecanase (ADAMTS-4) affinity and sequence specificity: suggestion of a common topological specificity of functionally diverse proteases. J Biol Chem 282:142–150
64. Lauer-Fields JL, Whitehead JK, Li S, Hammer RP, Brew K, Fields GB (2008) Selective modulation of matrix metalloproteinase 9 (MMP-9) functions via exosite inhibition. J Biol Chem 283:20087–20095
65. Bhaskaran R, Palmier MO, Lauer-Fields JL, Fields GB, Van Doren SR (2008) MMP-12 catalytic domain recognizes triple-helical peptide models of collagen V with exosites and high activity. J Biol Chem 283:21779–21788
66. Robichaud TK, Steffensen B, Fields GB (2011) Exosite interactions impact matrix metalloproteinase collagen specificities. J Biol Chem 286:37535–37542
67. Lauer JL, Bhowmick M, Tokmina-Roszyk D, Lin Y, Van Doren SR, Fields GB (2014) The role of collagen charge clusters in the regulation of matrix metalloproteinase activity. J Biol Chem 289:1981–1992
68. Stawikowski MJ, Stawikowska R, Fields GB (2015) Collagenolytic matrix metalloproteinase activities towards peptomeric triple-helical substrates. Biochemistry 54:3110–3121
69. Zhao Y, Marcink T, Gari RRS, Marsh BP, King GM, Stawikowska R, Fields GB, Van Doren SR (2015) Transient collagen triple helix binding to a key metalloproteinase in invasion and development. Structure 23:257–269
70. Woessner JF, Nagase H (2000) Matrix metalloproteinases and TIMPs. Oxford University Press, Oxford
71. Lauer-Fields JL, Nagase H, Fields GB (2004) Development of a solid-phase assay for analysis of matrix metalloproteinase activity. J Biomol Tech 15:305–316
72. Hurst DR, Schwartz MA, Ghaffari MA, Jin Y, Tschesche H, Fields GB, Sang Q-XA (2004) Catalytic- and ecto-domains of membrane type 1-matrix metalloproteinase have similar inhibition profiles but distinct endopeptidase activities. Biochem J 377:775–779
73. Neumann U, Kubota H, Frei K, Ganu V, Leppert D (2004) Characterization of Mca-Lys-Pro-Leu- Gly-Leu-Dpa-Ala-Arg-NH2, a fluorogenic substrate with increased specificity constants for collagenases and tumor necrosis factor converting enzyme. Anal Biochem 328:166–173
74. Fields GB, Tian Z, Barany G (1993) Principles and practice of solid-phase peptide synthesis. In: Grant GA (ed) Synthetic peptides: A user's guide. W.H. Freeman & Co., New York, NY, pp 77–183
75. Zhang X, Bresee J, Cheney PP, Xu B, Bhowmick M, Cudic M, Fields GB, Edwards WB (2014) Evaluation of a triple-helical peptide with quenched fluorophores for optical imaging of MMP-2 and MMP-9 proteolytic activity. Molecules 19:8571–8588
76. Pennington MW (1994) Site specific chemical modification procedures. In: Pennington MW, Dunn BM (eds) Peptide synthesis protocols: methods in molecular biology 35. Humana Press, Totowa, NJ, pp 171–186
77. Zuckermann RN, Kerr JM, Kent SBH, Moos WH (1992) Efficient method for the preparation of peptoids [oligo(N-substituted glycines)] by submonomer solid phase synthesis. J Am Chem Soc 114:10646–10647
78. Fields CG, Lovdahl CM, Miles AJ, Matthias-Hagen VL, Fields GB (1993) Solid-phase synthesis and stability of triple-helical peptides incorporating native collagen sequences. Biopolymers 33:1695–1707
79. Fields CG, Mickelson DJ, Drake SL, McCarthy JB, Fields GB (1993) Melanoma cell adhesion and spreading activities of a synthetic 124-residue triple-helical "mini-collagen". J Biol Chem 268:14153–14160
80. Grab B, Miles AJ, Furcht LT, Fields GB (1996) Promotion of fibroblast adhesion by triple-helical peptide models of type I collagen-derived sequences. J Biol Chem 271:12234–12240
81. Palmier MO, Van Doren SR (2007) Rapid determination of enzyme kinetics from fluorescence: overcoming the inner filter effect. Anal Biochem 371:43–51
82. Guy CA, Fields GB (1997) Trifluoroacetic acid cleavage and deprotection of resin-bound pep-

tides following synthesis by fmoc chemistry. Methods Enzymol 289:67–83

83. Liu Y, Kati W, Chen C-M, Tripathi R, Molla A, Kohlbrenner W (1999) Use of a fluorescence plate reader for measuring kinetic parameters with inner filter effect correction. Anal Biochem 267:331–335
84. Amar S, Fields GB (2012) Production and characterization of matrix metalloproteinases implicated in multiple sclerosis. In: Kokotos G, Constantinou-Kokotou V, Matsoukas J (eds) Peptides 2012 (proceedings of the thirty-second european peptide symposium). European Peptide Society, Athens, pp 102–103
85. Katritzky AR, Yoshioka M, Narindoshvili T, Chung A, Johnson JV (2008) Fluorescent labeling of peptides on solid phase. Org Biomol Chem 6:4582–4586
86. Malkar NB, Fields GB (2001) Synthesis of *Nα*-(fluoren-9-ylmethoxycarbonyl)-*Nε*-[(7-methoxycoumarin-4-yl)acetyl]-L-lysine for use in solid-phase synthesis of fluorogenic substrates. Lett Pept Sci 7:263–267
87. Anastasi A, Knight CG, Barrett AJ (1993) Characterization of the bacterial metalloendopeptidase pitrilysin by use of a continuous fluorescence assay. Biochem J 290:601–607
88. Wang GT, Krafft GA (1992) Automated synthesis of fluorogenic protease substrates: design of probes for Alzheimer's disease-associated proteases. Bioorg Med Chem Lett 2:1665–1668

Chapter 9

Time-Resolved Analysis of Matrix Metalloproteinase Substrates in Complex Samples

Pascal Schlage, Fabian E. Egli, and Ulrich auf dem Keller

Abstract

Identification of physiological substrates is the key to understanding the pleiotropic functions of matrix metalloproteinases (MMPs) in health and disease. Quantitative mass spectrometry-based proteomics has revolutionized current approaches in protease substrate discovery and helped to unravel many new MMP activities in complex biological systems. Multiplexing further extended the capabilities of these techniques and facilitated more complicated experimental designs that include multiple proteases or monitoring the activity of a single protease at more than one concentration or at multiple time points with a complex test proteome. In this chapter, we provide a protocol for time-resolved iTRAQ-based Terminal Amine Isotopic Labeling of Substrates (TAILS), with the focus on MMP substrate identification and characterization in cell culture supernatants and introduce an automated procedure for the interpretation of time-resolved iTRAQ-TAILS datasets.

Key words Protease, Substrate discovery, Proteomics, iTRAQ-TAILS, Time-resolved degradomics

1 Introduction

In recent years, a variety of proteomics-based high throughput techniques have been developed that allow for the identification of matrix metalloproteinase (MMP) substrates and their corresponding cleavage sites [1–3]. These substrate discovery approaches often use an experimental setup that compares two conditions (control vs. protease-treated) and apply rather arbitrary and long incubation times [4]. Although successful, these 2plex experiments often suffer from high numbers of false-positive identifications and may lead to long protein lists containing hundreds of potential substrate candidates, aggravating the selection of high-confidence hits for follow-up studies. In order to better discriminate between bystander substrates, i.e., unphysiological cleavages caused by unnaturally high concentrations of the test protease or too long incubation times, and physiologically relevant endogenous cleavages, more sophisticated approaches have been introduced that are

Charles A. Galea (ed.), *Matrix Metalloproteases: Methods and Protocols*, Methods in Molecular Biology, vol. 1579,
DOI 10.1007/978-1-4939-6863-3_9, © Springer Science+Business Media LLC 2017

able to quantify cleavages in more than two conditions, e.g., time-series experiments [5, 6]. One of these, iTRAQ-based Terminal Amine Isotopic Labeling of Substrates (TAILS) [7], is a substrate discovery workflow that allows for the monitoring of cleavages in up to eight experimental conditions and therefore is ideally suited to test different test protease concentrations or incubation times in a single experiment [5, 8]. These methods enable the determination of cleavage efficiencies for different substrates and thus significantly improve the selection of candidate proteins for functional characterization.

Time-resolved iTRAQ-TAILS monitors cleavages of a test protease at multiple time points of incubation and has been successfully applied to identify high-confidence substrates of MMP10 in fibroblast and keratinocyte culture supernatants [5, 8]. Importantly, classification of cleavage events using time-dependent abundance clusters of neo-N-terminal peptides, rather than based on single time point protease/control ratios, significantly enhanced the specificity and sensitivity of substrate discovery. Moreover, time-resolved iTRAQ-TAILS categorized substrates by cleavage site specificities and correlated cleavage efficiencies with secondary and tertiary structures of substrate proteins. Finally, this extended multiplexed substrate discovery workflow narrowed in on highly relevant MMP10 substrate candidates with known physiological functions, generating valid testable hypotheses for mechanistic dissection.

In time-resolved iTRAQ-TAILS experiments, a complex proteome is incubated with a test protease, sampled at multiple time points after incubation, differentially labeled at the proteome level with iTRAQ reagents and then analyzed by LC-MS/MS both prior (preTAILS) and after (TAILS) enrichment of protein N termini (Fig. 1). Including the preTAILS analyses is critical since it allows for the quantification of internal tryptic peptides via iTRAQ-labeling of lysine residues. By default, these do not change in abundance over time and serve, together with natural protein N termini, as a negative dataset for high confidence filtering of cleavage events using fuzzy clustering algorithms. In addition to assessing the progressive action of the test protease at multiple time points, proteome samples not treated with protease are analyzed at the beginning and at the end of the incubation period to control for basal activity of endogenous proteases in the sample. Finally, identified cleavage events are assigned with the help of their time-dependent abundance profiles to groups of substrates that are processed by the test protease with different efficiencies.

Here, we present a detailed procedure for time-resolved protease substrate degradomics using an 8plex-iTRAQ-TAILS protocol. Although in principle the workflow could also be applied to identify and characterize substrate cleavages of test proteases that act on intracellular proteins, we focus our discussion on proteomes

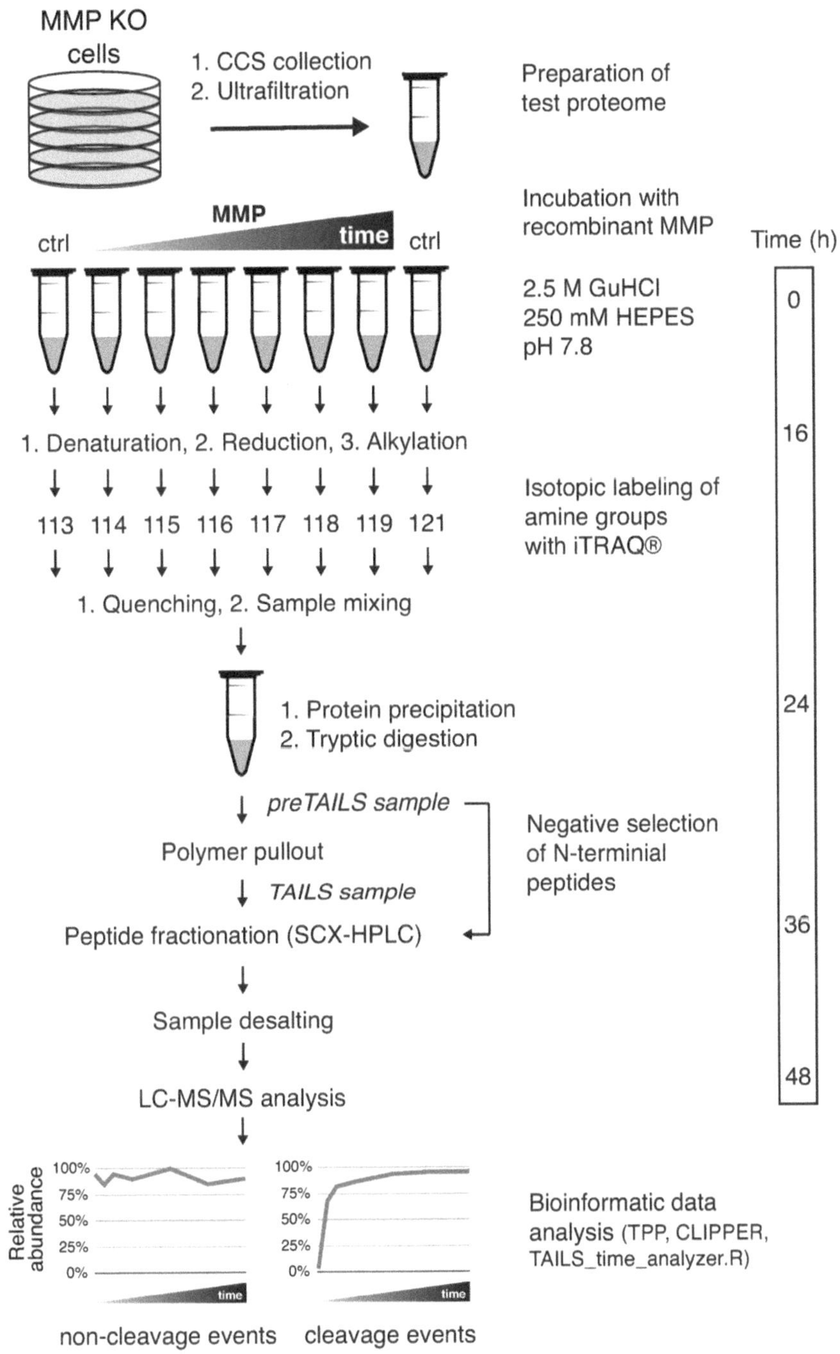

Fig. 1 Overview of the time-resolved iTRAQ-TAILS workflow. Cell culture supernatant (CCS) is collected from MMP-deficient cells (MMP KO) and incubated with the test MMP. Aliquots are removed at given time points and subjected to the 8plex-iTRAQ-TAILS protocol. Including control samples at time zero and at the end of the incubation time controls for basal proteolysis in the sample proteome. Peptides are analyzed by LC-MS/MS prior to (preTAILS sample) and after (TAILS sample) N-terminal enrichment to enable stringent categorization of true protease cleavages and non-cleavage events with the help of the automated analysis pipeline

prepared from cell culture supernatants (i.e., the secretome), which are the prime targets of MMPs. We provide detailed instructions on how to obtain a suitable secretome from adherent cells; however, this may also be performed with proteomes from alternative sources, which have been prepared in buffer conditions compatible with recombinant protease treatment. Data were analyzed using fuzzy clustering for stringent filtering of true cleavage events and subclustering of cleavages by efficiencies. Fuzzy clustering is discussed in detail and a novel R script is described that takes output from an updated CLIPPER [9] version and automates interpretation of time-resolved iTRAQ-TAILS datasets.

2 Materials

2.1 Secretome Collection and Concentration

1. Cell culture medium (free of protein, serum, phenol red, and antibiotics).
2. Phosphate-buffered saline (PBS).
3. Filter with 0.22 μm pore size.
4. Phenylmethylsulfonylfluoride (PMSF).
5. Amicon Ultra-15 Centrifugal Filter Units (3 kDa MWCO, Millipore).
6. Amine-free buffer with optimized pH for MMP activity, e.g., 50 mM HEPES, pH 7.8.
7. Bradford protein assay (Bio-Rad).

2.2 Test Protease Treatment

1. (Optional) 10 mM APMA (4-aminophenylmercuric acetate) (*see* **Note 1**).
2. Recombinant MMP (R&D Systems).
3. 1 M calcium chloride ($CaCl_2$).
4. 1 M sodium chloride (NaCl).

2.3 8plex-iTRAQ-TAILS Protocol

1. 8 M Guanidine hydrochloride (GuHCl).
2. 1 M 4-(2-hydroxyethyl)-1-piperazineethanesulfonic acid (HEPES, pH 7.8).
3. 350 mM Tris(2-carboxyethyl)phosphine hydrochloride (TCEP).
4. 250 mM iodoacetamide (freshly prepared).
5. 8plex iTRAQ® Reagents (AB Sciex).
6. Dimethyl sulfoxide (DMSO).
7. 1 M Ammonium bicarbonate (NH_4HCO_3).
8. Acetone (−20 °C).
9. Methanol (MeOH; −20 °C).

10. 100 mM Sodium hydroxide (NaOH).
11. 1 M Hydrochloric acid (HCl).
12. 1 μg/μl Trypsin in 50 mM acetic acid (Trypsin Gold, Promega)
13. Hyperbranched polyglycerol-aldehydes (HPG-ALD) polymer. Store polymer aliquots under inert gas at −80 °C. Polymer is available without commercial or company restriction from Flintbox Innovation Network, The Global Intellectual Exchange and Innovation Network (http://www.flintbox.com/public/project/1948/).
14. 5 M sodium cyanoborohydride ($NaBH_3CN$) solution in 1 M NaOH (Sigma).
15. Amicon® Ultra-0.5 Centrifugal Filter Units (30 kDa MWCO, Millipore).

2.4 Strong Cation Exchange (SCX)—High Performance Liquid Chromatography (HPLC)

1. Agilent Technologies 1100/1200 series HPLC system.
2. PolySULFOETHYL A column, 200 × 2.1 mm, 5 μm 300-Å (PolyLC Inc.).
3. 50% Phosphoric acid (H_3PO_4).
4. SCX buffer A: 10 mM KH_2PO_4 pH 2.7 in 25% acetonitrile.
5. SCX buffer B: 10 mM KH_2PO_4 pH 2.7, 0.5 M KCl in 25% acetonitrile.

2.5 Desalting, Pooling and Clean-Up of Fractionated Peptides

1. SpeedVac concentrator.
2. OMIX C18 pipette tips, 10–100 μl (Agilent Technologies).
3. Equilibration solution: 3% acetonitrile, 0.1% trifluoroacetic acid (TFA).
4. 50% Acetonitrile.
5. Elution solution: 80% acetonitrile, 0.1% trifluoroacetic acid (TFA).
6. 3% Acetonitrile, 0.1% formic acid (FA).

2.6 LC-MS/MS Analysis

1. Hybrid mass spectrometer (e.g., Thermo Scientific: Orbitrap XL, Orbitrap Velos or Q Exactive) coupled to a capillary liquid chromatography system with a tip column packed with C18 material (e.g., 75 μm × 150 mm; AQ, 3 μm 200 Å, Bischoff GmbH).

2.7 Data Analysis

1. Search engine for spectrum to peptide assignment (e.g., Mascot, X! Tandem).
2. Trans-Proteomic Pipeline (TPP) (v7.1 POLAR VORTEX rev 1) [10] (http://sourceforge.net/projects/sashimi/files/Trans-Proteomic%20Pipeline%20%28TPP%29/TPP%20v4.7%20%28polar%20vortex%29%20rev%201/).

3. CLIPPER [9] (version for TPP_v_4_7_1) (http://clipserve.clip.ubc.ca/tails/).
4. Functional R installation (https://www.r-project.org) and the Mfuzz Package [11].
5. R script 'TAILS_time_analyzer.R' (http://www.mhs.biol.ethz.ch/research/werner/group-members-werner/aufdem-keller.html).

3 Methods

3.1 Secretome Collection and Concentration

1. Wash 70% confluent MMP knockout (KO) cells (*see* **Note 2**) three times with PBS.
2. Incubate MMP KO cells with cell culture media (free of protein, serum, phenol red, and antibiotics).
3. After 24–48 h (*see* **Note 3**) collect the culture media and centrifuge for 5 min at 400 × *g* at room temperature to remove dead cells.
4. Collect supernatant and add serine protease inhibitor (0.5 mM PMSF final) on ice.
5. Centrifuge supernatant for 30 min at 3,000 × *g* at 4 °C to remove cell debris.
6. Collect supernatant and filter it through a 0.22 μm filter.
7. Freeze supernatants at −20 °C for storage or continue with **step 8**.
8. Add water to Amicon Ultra-15 Centrifugal Filter Units and wash filter to remove glycerol by centrifugation for 35 min at 4 °C (for optimal speed see manufacturer's instructions).
9. Concentrate supernatants by multiple centrifugation steps at 4 °C.
10. Exchange buffer by diluting the concentrated supernatants with 10 volumes of 50 mM HEPES, pH 7.8 and centrifugation.
11. Repeat **step 10** two times.
12. Reduce the volume of the concentrated supernatant to 1 ml by further centrifugation.
13. If precipitates occur, try to dissolve them or remove them by centrifugation.
14. Take an aliquot and determine protein concentration using the Bradford protein assay.
15. Freeze concentrated secretome at −80 °C or continue with test protease treatment.

3.2 Test Protease Treatment

1. Activate recombinant MMP by auto-activation or with APMA according to manufacturer's instructions (*see* **Note 4**).
2. Adjust protein concentration to 2 μg/μl with 50 mM HEPES, pH 7.8 and add 10 mM $CaCl_2$ and 100 mM NaCl.
3. Add activated MMP (protease-treated sample), or 50 mM HEPES, pH 7.8 without MMP (control sample).
4. Take aliquots (0.25 mg protein for protease-treated and control samples) at different time points, e.g., 0 and 16 h for control and 1, 2, 4, 8, 12, 16 h for protease-treated samples (*see* **Note 5**).
5. Stop the reaction by freezing at −20 °C or directly continue with 8plex-iTRAQ-TAILS protocol.

3.3 8plex-iTRAQ-TAILS Protocol

1. Denature aliquots (0.25 mg protein at 2 μg/μl) by adding 75 μl of 8 M GuHCl and 55 μl of 1 M HEPES buffer (pH 7.8) (final concentrations: 1 μg/μl protein, 2.5 M GuHCl, 250 mM HEPES, pH 7.8; total volume: 255 μl) and incubation at 65 °C for 15 min.
2. Reduce cysteine residues by adding 1 mM TCEP and incubate at 65 °C for 45 min.
3. Add 5 mM iodoacetamide and incubate at 65 °C for 30 min to alkylate cysteine residues.
4. Add 260 μl of DMSO to each vial of iTRAQ reagents (1 mg each) and mix by pipetting.
5. Differentially label proteins by adding diluted iTRAQ reagents (protein:iTRAQ weight ratio of 1:4) resulting in a final DMSO concentration of 50%.
6. Incubate at room temperature for 30 min.
7. Quench labeling reaction by adding 0.1 M NH_4HCO_3 and incubate for 15 min at room temperature.
8. Combine all samples.
9. Precipitate proteins by adding eight sample volumes of ice cold acetone and one sample volume of ice cold MeOH followed by incubation at −80 °C for 2 h.
10. Centrifuge sample at 10,000 × *g* for 30 min at 4 °C.
11. Wash pellet with 15 ml cold MeOH and repeat **step 10**.
12. Air dry pellet.
13. Resuspend dried pellet in 200 μl 0.1 M NaOH.
14. Adjust the protein concentration to 1 μg/ml and 100 mM HEPES, pH 7.8 by adding the appropriate amount of 1 M HEPES, pH 7.8 and water.
15. Add Trypsin at a final protease:protein ratio of 1:100 (w/w) and incubate overnight at 37 °C.

16. (Optional) Test completeness of digest by running 3 μg of protein before and after tryptic digest by SDS-PAGE followed by silver staining (*see* **Note 6**).
17. Take 10% of peptide solution and store at −20 °C (preTAILS sample).
18. Adjust the pH of the remaining solution to pH 6–7 with 1 M HCl.
19. Add a threefold excess (w/w) of HPG-ALD polymer (*see* **Note 7**) and 50 mM $NaBH_3CN$ solution and incubate overnight at 37 °C.
20. Collect unbound peptides in the flow-through by filtrating the polymer solution in multiple steps on the same Amicon Ultra-0.5 ml Centrifugal Filter Units (30 kDa cut-off) at 13,000 × *g* for 10 at room temperature. Combine filtrates in a single tube.
21. Wash the polymer by adding 200 μl of 0.1 M NH_4HCO_3 to filtration device and centrifuge again.
22. Combine the flow-throughs of **steps 20** and **21** and store at −20 °C (TAILS sample).

3.4 Peptide Fractionation by Strong Cation Exchange (SCX)—High Performance Liquid Chromatography (HPLC)

1. Adjust the sample volume (preTAILS or TAILS samples) to 1 ml with SCX buffer A and adjust the pH to ≤2.7 by adding 50% H_3PO_4.
2. Centrifuge the sample for 7 min at 13,000 × *g* and room temperature and inject supernatant.
3. For fractionation use a linear gradient elution (0 min 0% SCX buffer B; 60 min 0% SCX buffer B; 65 min 5% SCX buffer B; 100 min 35% SCX buffer B; 110 min 100% SCX buffer B) at a flow rate of 0.2 ml/min.
4. Monitor eluate absorbance at 214 and 280 nm.
5. Collect ~27 peptide fractions with a volume of 0.5 ml along the gradient.
6. Pool fractions containing peptides according to their absorbance intensities (high absorbance intensity: pool only two fractions; low absorbance intensity: pool up to 4–5 fractions) to ~8 final samples and desalt (*see* below) for LC-MS/MS analysis.

3.5 Desalting, Pooling, and Cleanup of Fractionated Peptides

1. Reduce volumes of peptide fractions under vacuum to ≤100 μl.
2. (Optional) If samples are dried or their volumes are too low (≪100 μl), resuspend samples and adjust volumes to 100 μl by adding equilibration solution.
3. Wet the C18 OMIX tip by aspirating twice 100 μl 50% acetonitrile.

4. Equilibrate the C18 OMIX tip twice by aspirating twice 100 μl equilibration solution.
5. Bind peptides from multiple fractions to the same C18 OMIX tip by pipetting each fraction solution ten times up and down.
6. Wash the C18 OMIX tip seven times with 100 μl equilibration solution.
7. Elute peptides by pipetting ten times up and down in 100 μl elution solution.
8. Dry pooled peptides under vacuum in a SpeedVac.
9. For LC-MS/MS analysis, resuspend peptides in 20 μl 3% acetonitrile containing 0.1% FA.

3.6 LC-MS/MS Analysis

Analyze pooled peptide fractions for the preTAILS and TAILS samples by tandem mass spectrometry on a hybrid mass spectrometer coupled to a C18 in-line chromatography system. Settings will depend on the mass spectrometer used, but in general we apply the same parameters as for standard iTRAQ-based analyses. For example, results from our test dataset (*see* Subheading 3.7) were obtained on a Thermo Scientific Orbitrap Velos instrument coupled to an Eksigent-Nano-HPLC system utilizing a self-made tip column (75 μm × 150 mm; C18 material: AQ, 3 μm 200 Å) and the following settings: 1 μg peptide loaded; flowrate of 250 nl/min, gradient 0 to 35% acetonitrile in 62 min; full scan: 300–1700 *m/z* with resolution of 30,000 at 400 *m/z*, target value 1E6; collision-induced dissociation (CID) scan: eight most intense precursors, normalized collision energy of 35%, activation time of 10 ms; higher energy collision dissociation (HCD) of same precursors for iTRAQ: normalized collision energy of 42%, resolution of 7500 at 400 *m/z*, exclusion time of 45 s, exclusion window of 20 ppm (*see* **Note 8**). The detailed settings for a Thermo Scientific Q Exactive instrument have been published previously [12].

3.7 Data Analysis

In general, time-resolved TAILS datasets are analyzed using the CLIPPER analysis pipeline [9] according to a previously described procedure [12]. Here, we will describe modifications needed for time-resolved studies and focus the detailed discussion on steps required to exploit time-resolved information for effective discrimination of cleavage from non-cleavage events and the classification of cleavages by efficiencies. Central to our analysis strategy is the use of soft clustering to stringently filter out true cleavage events from true non-cleavage events and non-interpretable peptide abundance patterns. In contrast to hard clustering (e.g., k-means) soft clustering algorithms (e.g., fuzzy c-means in the Mfuzz R package [11]) do not definitively assign each object to a single cluster, but calculate a membership value alpha (0 < alpha < 1) that defines to what extent an object belongs to a cluster.

Consequently, the number of clusters determines the minimum alpha value that is indicative for assignment to a specific cluster, i.e., with two clusters an alpha of 0.5 indicates equal probability for both clusters and thus no discriminative pattern. In our analysis, we exploit this value to discriminate between peptides whose abundance increases over time and thus are derived from cleavage events of the test protease from those that remain constant in abundance and thus represent non-cleavage events or follow a pattern not suitable for meaningful interpretation. Cleavage events that are specific to the test protease are then further selected by applying abundance cutoffs to control conditions. Finally, in a second round of fuzzy clustering filtered cleavage events are extracted and assigned to subclusters comprising cleavages of different efficiencies. Several critical parameters that can be adjusted to optimize filtering and subclassification of true cleavages will be discussed in more detail below.

Fuzzifier value *m*: This value defines the fuzziness of the clusters and influences the degree to which an element can belong to multiple clusters. We have found that the function mestimate() that was introduced by Schwaemmle and Jensen [13] and integrated into the Mfuzz package gives the best results both for filtering and for subclustering of cleavage events.

Number of clusters: Determining the optimal number of clusters can be very difficult. However, in our experience restricting clusters to only two allows for the most effective filtering of cleavages related to the test protease from time-resolved TAILS datasets. Applying abundance cutoffs to control conditions to filter for high-confidence cleavages of the test protease appeared to be more efficient than generating more clusters at this stage of the analysis. The number of subclusters highly depends on the underlying abundance profiles of cleavage events. In general, these vary between two and three [5, 8], but it is advisable to test higher numbers to determine which gives the best results.

Membership value alpha: Varying the threshold cutoff for the alpha membership value allows adjusting the stringency of event classification. A low alpha value indicates strong deviation of the abundance profile of a specific event from a cluster core. Thus, using a high alpha value as a cutoff for filtering cleavage events from non-cleavage events and noise ensures high stringency. We recommend values of 0.8–0.9 in the first round of fuzzy clustering with only two clusters (cleavages and non-cleavages). The cutoff for subclusters of cleavage events will depend on the number of clusters, but as a rule of thumb, we obtained good results with an alpha value cutoff of 0.8 for two [5] and 0.7 for three [8] subclusters.

Below, we give an example of the procedure using a dataset that was extracted from [5] and that is available at http://www.mhs.biol.ethz.ch/research/werner/group-members-werner/

aufdemkeller.html. To aid in performing this part of the analysis we have provided the CLIPPER output file "MMP10_dataset_merge.annotate.clip.csv" and the underlying R script ("TAILS_time_analyzer.R") which can be adapted to your data. In addition, you will need to install the Trans-Proteomic Pipeline (TPP v7.1 POLAR VORTEX) [10] and an updated version of CLIPPER (for TPP_v_4_7_1).

1. Use the TPP for secondary validation of search results, combining SCX fractions and quantification of iTRAQ reporter signals as detailed in [9] and in the CLIPPER Manual. Since we include internal tryptic peptides and N termini in our analysis, pep.xml files from both preTAILS and TAILS LC-MS/MS runs have to be combined at the "xinteract" step in a single file.
2. Open "interact.pep.xml" output in PepXML Viewer, apply probability cutoffs, include all columns and export as a .xls tab-delimited text file.
3. Process the PepXML Viewer .xls output with CLIPPER and check the following options: "Leave unlabeled N-termini" (important to include tryptic peptides from preTAILS dataset), "Merge multiple spectra for same peptide," "Annotate peptides." This will generate a file named "_merge.annotate.clip.csv" as input for the "TAILS_time_analyzer.R" script.
4. Open "TAILS_time_analyzer.R" in a text editor or in an integrated R environment (*see* **Note 9**) and adjust the following user-defined variables:

   working_directory: full path to the directory containing the CLIPPER output file.

   data_file: file name for the CLIPPER output file (extension must be ".csv").

   max_expression_start: relative abundance cutoff for zero time point control (default: 0.25).

   max_expression_control: relative abundance cutoff for end time point control (default:0.25).

   alpha_cutoff_main: alpha membership cutoff for cleavage event filtering (default: 0.9).

   alpha_cutoff_sub: alpha membership cutoff for cleavage event subclustering.

   clipper_unique_id: column used as unique identifiers in CLIPPER output (default: "spectrum").

   time_point_controls: columns for zero and end time point controls (defaults: "libra1, libra8").

   time_point_labels: time point labels as vector.

   time_point_positions: time points in minutes as a numeric vector; the number for end time point control should be slightly higher than the maximal time point.

   num_subclusters: number of subclusters to be created from cleavages.

5. Run 'TAILS_time_analyzer.R' to generate the following output files for inspection:

   "_main_clusters_a[alpha_cutoff_main].pdf": figure of main clusters (*see* Fig. 4C in ([5])).

   "_cleavages_subclusters_a[alpha_cutoff_sub].pdf": figure of cleavage subclusters (*see* Fig. 5A in ([5])).

   "_clustering.csv": CLIPPER output file augmented with columns for alpha values of main and cleavage subclusters; filtering for "main_clusters_[1,2]" > [alpha_cutoff_main] will display peptides assigned to corresponding clusters in main clusters; filtering for "main_clusters_1" > [alpha_cutoff_main] and "sub_cluster_[0-9]" > [alpha_cutoff_sub] will display cleavage events assigned to corresponding subclusters.

4 Notes

1. Some MMPs can be auto-activated by incubation at 37 °C, but most need APMA for chemical induction of activation.
2. Cells that are deficient for the MMP (MMP KO) of interest produce a secretome that has never seen the test protease and thus is particularly suitable for substrate discovery.
3. Incubation times will highly depend on the ability of the cells to grow in serum-free conditions that are mandatory to avoid contamination with exogenous proteins and MMPs. Since serum-free conditions are often detrimental to cells and cause the release of intracellular proteins due to the death and lysis of cells, we recommend optimizing the incubation time by performing test experiments using assays that monitor cell death, e.g., lactate dehydrogenase (LDH) release or MTT ((3-(4,5-dimethylthiazol-2-yl)-2,5-diphenyl tetrazolium bromide) assay, to reduce cell death to a minimum.
4. To avoid APMA-induced activation of endogenous MMPs during test protease treatment, remove APMA after activation of the recombinant MMP by buffer exchange using spin columns.
5. Incubation times depend on the activity of the test protease. For MMPs, we have had good results for monitoring cleavages up to 16 h.
6. In our experience, tryptic digests are very efficient and after overnight incubation almost complete. However, monitoring digestion of proteins by SDS-PAGE should be performed for troubleshooting and especially, if the protocol is run for the first time.
7. The actual amount depends on the HPG-ALD concentration and binding capacity of the specific batch you purchased. Please check package insert for details.

8. In our experience, recording both CID and HCD spectra on a Thermo Scientific Orbitrap Velos mass spectrometer gave better results for identification and iTRAQ-based quantification than only HCD. In this case, spectra from both collision types (CID and HDC) have to be merged prior to data analysis. This can be achieved using Mascot Distiller (Matrix Science) or any other appropriate software package.
9. There are many ways to execute R scripts and it depends upon your particular R installation. For details, refer to the extensive documentation available at www.r-project.org. Open the directory that contains the file named "dataset_merge.annotate.clip.csv" and run the command "R < TAILS_time_analyzer.R --no-save." Alternatively, use an integrated R-environment such as RStudio (www.rstudio.com) [14] that will allow easy editing and execution.

Acknowledgments

A special thanks goes to S. Werner (ETH Zurich) for continuous support of our research. We want to thank Paolo Nanni, Tobias Kockmann, and the whole proteomics team of the Functional Genomics Center Zurich (FGCZ) for excellent support in mass spectrometry. This work was funded by grants from the Swiss National Science Foundation (31003A_140726 and 31003A_163216), the European Commission (Marie Curie International Reintegration Grant; FP7-PEOPLE- 2010-RG/SkiNterminomics) and the Novartis Foundation for Medical-Biological Research to U.a.d.K., and by funds from the ETH Zurich.

References

1. Schlage P, auf dem Keller U (2015) Proteomic approaches to uncover MMP function. Matrix Biol 44–46:232–238
2. Rogers LD, Overall CM (2013) Proteolytic post-translational modification of proteins: proteomic tools and methodology. Mol Cell Proteomics 12:3532–3542
3. Plasman K, Van Damme P, Gevaert K (2013) Contemporary positional proteomics strategies to study protein processing. Curr Opin Chem Biol 17:66–72
4. auf dem Keller U, Prudova A, Gioia M et al (2010) A statistics-based platform for quantitative N-terminome analysis and identification of protease cleavage products. Mol Cell Proteomics 9:912–927
5. Schlage P, Egli FE, Nanni P et al (2014) Time-resolved analysis of the matrix metalloprotease 10 substrate degradome. Mol Cell Proteomics 13:580–593
6. Plasman K, Van Damme P, Kaiserman D et al (2011) Probing the efficiency of proteolytic events by positional proteomics. Mol Cell Proteomics 10(M110):003301
7. Prudova A, auf dem Keller U, Butler GS et al (2010) Multiplex N-terminome analysis of MMP-2 and MMP-9 substrate degradomes by iTRAQ-TAILS quantitative proteomics. Mol Cell Proteomics 9:894–911
8. Schlage P, Kockmann T, Sabino F et al (2015) Matrix metalloproteinase 10 degradomics in keratinocytes and epidermal tissue identifies bioactive substrates with pleiotropic functions. Mol Cell Proteomics 14:3234–3246
9. auf dem Keller U, Overall CM (2012) CLIPPER-An add-on to the Trans-Proteomic

Pipeline for the automated analysis of TAILS N-terminomics data. Biol Chem 393: 1477–1483
10. Deutsch EW, Mendoza L, Shteynberg D et al (2010) A guided tour of the Trans-Proteomic Pipeline. Proteomics 10:1150–1159
11. Kumar L, Futschik ME (2007) Mfuzz: a software package for soft clustering of microarray data. Bioinformation 2:5–7
12. Kockmann T, Carte N, Melkko S et al (2015) Identification of protease substrates in complex proteomes by iTRAQ-TAILS on a Thermo Q Exactive instrument. In: Grant J, Li H (eds) Analysis of post-translational modifications and proteolysis in neuroscience, Neuromethods. Springer Protocols, New York
13. Schwammle V, Jensen ON (2010) A simple and fast method to determine the parameters for fuzzy c-means cluster analysis. Bioinformatics 26:2841–2848
14. RStudio Team (2015) RStudio: Integrated Development for R. RStudio, Inc., Boston, MA

Chapter 10

Identification of Protease Cleavage Sites by Charge-Based Enrichment of Protein N-Termini

Zon W. Lai and Oliver Schilling

Abstract

Differential proteolytic processing, for example by matrix metalloproteases (MMPs), has been recognized as an important hallmark in numerous pathological conditions. One crucial challenge in the present studies of proteases is system-wide identification of endogenous biological substrates. In this chapter, we highlight a robust method for the identification of bioactive substrates and their sites of MMP cleavage, as well as by other proteases and peptidases, in a system-wide manner. This approach enriches for putative protein N-termini by removal of internal peptides using a charge reversal strategy. In addition, this straightforward method can be used in combination with gel-based pre-separation of proteins to allow better estimation of the molecular weight of the identified cleavage product of a given bioactive substrate.

Key words Protease, Cleavage, Degradome, Substrate, Proteomics

1 Introduction

Dysregulated proteolysis is a hallmark of numerous pathological processes. However, very little is known regarding in vivo cleavage sites and degradation targets for most proteases (including MMPs) [1]. In recent years, investigations into elucidating endogenous proteases substrates have gained significant impetus. As such, a number of proteomic-based approaches have been developed to allow for system-wide substrate identification [2–6]. These approaches focus on the enrichment of peptides corresponding to protein N-termini, which includes stable cleavage fragments stemming from a substrate cleaved by a protease. Analysis of this N-terminome will unequivocally yield valuable information of protease cleavage sites and cleavage specificities, ultimately leading to identification of novel bioactive substrates. However, many of these strategies often involve laborious sample preparation followed by elaborate enrichment steps. To circumvent this, we have recently proposed a cost-effective, fast, and robust method to enrich for these N-terminal peptides via direct removal of internal

Charles A. Galea (ed.), *Matrix Metalloproteases: Methods and Protocols*, Methods in Molecular Biology, vol. 1579, DOI 10.1007/978-1-4939-6863-3_10,

peptides using a charge reversal strategy [7]. First, N-termini of proteins and polypeptides chains, including stable cleavage products stemming from bioactive substrates, are protected using chemical dimethylation. Next, these proteins and polypeptide chains are digested using trypsin, resulting in the generation of internal peptides with unmodified neo-N-termini. These unmodified neo-N-termini are readily susceptible to coupling (through Schiff–base reduction) to a commercially available, cost-effective compound containing two disulfonate groups. These modified internal peptides are subsequently removed using strong-cation exchange chromatography, while peptides reflecting native and protease-generated protein N-termini are retained for analysis using high resolution mass spectrometry to elucidate precise cleavage sites of endogenous bioactive substrates. A general outline of the workflow is illustrated in Fig. 1.

Proteome
e.g. cell lysate or tissue extract

↓

Block all primary amines by chemical dimethylation

↓

Trypsin digestion

↓

Modification of internal peptides using disulfonate groups

↓

Removal of modified internal peptides using strong cation-exchange chromatography

↓

Desalting using C18 solid phase exchange chromatography

↓

Mass spectrometry

↓

Data analysis

Fig. 1 Workflow for the enrichment of protein N-termini for protease cleavage site identification. Unmodified primary amines of protein N-termini and lysine side chains are first protected by chemical dimethylation followed by digestion with trypsin. Internal peptides bearing unmodified neo-N-termini are chemically modified using sodium 4-formylbenzene-1,3-disulfonate. Endogenously modified (e.g., acetylated Protein N-termini) and dimethylated peptides are subsequently enriched by strong cation exchange chromatography and identified by mass spectrometry analysis

2 Materials

2.1 Chemical Dimethylation of Protein N-Termini and Lysine Side Chains

1. Recommended starting material: 1.5–3 mg total protein from tissue/cell lysates or cell conditioned medium.
2. Refrigerated benchtop centrifuge with swing-bucket.
3. Water-bath sonicator.
4. 0.5 M iodoacetamide stock solution.
5. 1 M 1,4-dithiothreitol stock solution.
6. 0.1 M sodium hydroxide solution (keep cold on ice).
7. 6 M guanidine hydrochloride stock solution.
8. 2 M HEPES (4-(2-hydroxyethyl)piperazine-1-ethanesulfonic acid) stock solution.
9. 37% (w/v) formaldehyde stock solution (12.2 M).
10. 1 M sodium cyanoborohydride stock solution (make fresh).
11. Optional, for quantitation: 20% (w/v) d_2-^{13}C formaldehyde stock solution (6.6 M).
12. Acetone Precipitation Buffer: 90% (v/v) ice-cold acetone, 10% (v/v) ice-cold methanol. Keep cold on ice.

2.2 Digestion of Protein Mixtures by Trypsin

1. Refrigerated benchtop centrifuge with swing-bucket.
2. Water-bath sonicator.
3. 0.1 M sodium hydroxide solution (keep cold on ice).
4. 1 M Tris (Tris(hydroxymethyl)aminomethane) stock solution.
5. 2 M HEPES (4-(2-hydroxyethyl)piperazine-1-ethanesulfonic acid) stock solution.
6. 1 mg/mL sequencing-grade trypsin stock solution.
7. Bicinchoninic acid assay kit.
8. Acetone Precipitation Buffer: 90% (v/v) ice-cold acetone, 10% (v/v) ice-cold methanol (keep cold on ice).

2.3 N-Terminal Modification of Internal Peptides

1. SDS-PAGE apparatus.
2. 10% SDS-PAGE acrylamide gel.
3. Silver staining kit.
4. Bicinchoninic acid assay kit.
5. Vacuum concentrator (e.g.SpeedVac™).
6. 2 M HEPES (4-(2-hydroxyethyl)piperazine-1-ethanesulfonic acid), pH 7.0 stock solution.
7. 1 M 4-formylbenzene-1,3-disulfonic acid disodium salt hydrate stock solution (prepare fresh prior to use).
8. 1 M sodium cyanoborohydride stock solution (prepare fresh prior to use).

9. C18 solid phase extraction columns.
10. C18 Equilibration Buffer: 80% (v/v) acetonitrile, 0.1% (v/v) trifluoroacetic acid.
11. C18 Wash Buffer: 0.1% (v/v) trifluoroacetic acid.
12. C18 Elution Buffer: 80% (v/v) acetonitrile.

2.4 Removal of Internal Peptides Using Strong-Cation Exchange HPLC

1. High Pressure Liquid Chromatography (HPLC) device.
2. High resolution mass spectrometer coupled to a nano-HPLC (e.g., LC-MS/MS).
3. Software for analysis of LC-MS/MS data.
4. Vacuum evaporator (e.g.SpeedVac™).
5. 1 M Tris stock solution.
6. C18 solid phase extraction columns.
7. Strong Cation Exchange Chromatography Buffer A: 5 mM KH_2PO_4, pH 2.7, 25% (v/v) acetonitrile.
8. Strong Cation Exchange Chromatography Buffer B: 5 mM KH_2PO_4, 0.5 M KCl, pH 2.7, 25% (v/v) acetonitrile.
9. C18 Equilibration Buffer: 80% (v/v) acetonitrile, 0.1% (v/v) trifluoroacetic acid.
10. C18 Wash Buffer: 0.1% (v/v) trifluoroacetic acid.
11. C18 Elution Buffer: 80% (v/v) acetonitrile.

3 Methods

3.1 Chemical Dimethylation of Protein N-Termini

1. Extract proteins from cell/tissue lysate or cell conditioned media.
2. Reduce extracted proteins using 5 mM dithiothreitol for 1 h at 70 °C. Adjust pH to 7.0.
3. Alkylate proteins using 15 mM iodoacetamide. Incubate in the dark at room temp for 30 min.
4. Add another 5 mM dithiothreitol and incubate at room temperature for 30 min to quench excess iodoacetamide. Adjust to pH 7.0.
5. Precipitate proteins using prechilled Acetone Precipitation Buffer (10:1 (v/v) buffer to sample ratio). Store at −80 °C for at least 1.5 h.
6. Pellet precipitated proteins by centrifugation at 4,000 × *g*, 4 °C for 1 h.
7. Wash protein pellet using 1 mL ice-cold methanol followed by centrifugation at 4,000 × *g*, 4 °C for 15 min. Repeat three times.

8. Dissolve protein pellet in 0.5–1 mL prechilled 0.1 M NaOH/2 M guanidine hydrochloride. If needed, sonicate in a water-bath sonicator until proteins are completely dissolved (keep samples cold by adding ice into water-bath sonicator).
9. Adjust protein mixture to pH 7.5–8.0 by bringing to 0.2 M HEPES by using 2 M HEPES stock solution.
10. Determine protein concentration using bicinchoninic acid assay (adjust concentration to approximately 1 mg/mL).
11. Heat the samples at 70 °C for 10 min. Let it cool.
12. Add 40 mM sodium cyanoborohydride and 40 mM formaldehyde (*see* **Note 1** for experiments using multiplex labeling for relative quantitation). Mix and adjust to pH 7.0. Incubate at 37 °C for 16 h.

3.2 Digestion of Protein Mixtures by Trypsin

1. Quench excess formaldehyde by adding 50 mM Tris. Incubate at 37 °C for 1.5 h.
2. If different formaldehyde isotopes are used for comparative study, mix labeled samples in an equal ratio.
3. Precipitate proteins using ice-cold Acetone Precipitation Buffer (10:1 (v/v) buffer to sample ratio) for 1.5 h at −80 °C. Centrifuge precipitated proteins at 4,000 × *g*, 4 °C for 1 h.
4. Wash protein pellet using 1 mL ice-cold methanol followed by centrifugation at 4,000 × *g*, 4 °C for 15 min. Repeat three times, each time removing the fluid completely.
5. Dissolve protein pellet in 0.5–1 mL prechilled 0.1 M NaOH. If needed, sonicate in a water-bath sonicator until proteins are completely dissolved (keep samples cold by adding ice into water-bath sonicator).
6. Adjust protein mixture to pH 7.5–8.0 by bringing to 0.2 M HEPES using 2 M HEPES stock solution.
7. Determine protein concentration using bicinchoninic acid assay and save a small aliquot for SDS-PAGE (*see* Subheading 3.3, **step 1**).
8. Add 1:100 (w/w) ratio of sequencing-grade trypsin to protein and incubate at 37 °C for 16 h (*see* **Note 2**).

3.3 N-Terminal Modification of Internal Peptides (i.e., Trypsin Cleavage Sites)

1. Check tryptic digest by running 10% SDS-PAGE and follow silver staining procedure to compare samples from pre- and post-digest. No protein bands (except trypsin) should appear in the digested sample.
2. Add 20 mM (final concentration) sodium cyanoborohydride and 20 mM (final concentration) 4-formyl-1,3-benzenedisulfonic acid.

3. Adjust pH to 7.0 and incubate for 1 h at 37 °C. Repeat twice to achieve final concentration of 60 mM sodium cyanoborohydride and 60 mM 4-formyl-1,3-benzenedisulfonic acid. Incubate for 16 h at 37 °C.

3.4 Removal of Internal (Tryptic) Peptides Using Strong-Cation Exchange HPLC

1. Add 0.1 M Tris (final concentration) for 1.5 h at 37 °C.
2. Acidify digested peptides using 0.3% (v/v) trifluoroacetic acid to achieve pH of less than 3.0. If needed, hydrochloric acid can also be used.
3. Desalt digested peptides using C18 solid phase extraction column (55–105 μm particle size and 125 Å pore size, up to 3 mg binding capacity, e.g., Waters Sep-Pak C18 Cartridges):
   (a) Equilibrate C18 column using one volume of 80% (v/v) acetonitrile.
   (b) Wash C18 column using three volumes of 0.1% (v/v) trifluoroacetic acid.
   (c) Apply digested peptides into C18 column. Reapply the flow-through once to ensure all peptides are captured.
   (d) Wash C18 column using three volumes of 0.1% (v/v) trifluoroacetic acid.
   (e) Elute captured peptides using one volume of C18 elution Buffer.
4. Evaporate acetonitrile using vacuum evaporator.
5. Determine peptide concentration using bicinchoninic acid assay. Adjust peptide concentration to approximately 1 mg/mL using Strong Cation Exchange Chromatography Buffer A.
6. Resolve peptide mixture by strong cation exchange high performance liquid chromatography (HPLC). Recommended column specification: 5 μm particle size and 300 Å pore size, column length 100 mm, inner diameter 4.6 mm, e.g., PolyLC Polysulfoethyl A. Elute peptides using a linear gradient with increasing concentration of Strong Cation Exchange Chromatography Buffer B up to 45% (v/v) over a period of 90 min. (Due to a relatively large volume of samples, a larger injection loop (>2.0 mL) is recommended. *See* **Note 3**.)
7. Collect between 10–14 peptide fractions and evaporate acetonitrile using vacuum evaporator.
8. Desalt fractionated peptides using micro C18 solid phase extraction column or self-packed C18 STAGE-tips [8].
9. Analyze fractionated peptides using high resolution LC-MS/MS mass spectrometry analyzer (*see* **Note 4**).

3.5 LC-MS/MS Data Analysis

1. Convert raw LC-MS/MS data to a format compatible with the X! Tandem search engines. Vendor-specific converters as

described in detail at: http://tools.proteomecenter.org/wiki/index.php?title=Formats:mzXML.

2. Perform spectrum-to-sequence assignment searches using the following parameters: semi Arg-C for enzyme specificity with up to one missed cleavage site; static modification including carboxyamidomethylation of cysteine residues (+57.02 Da), N-terminal acetylation (+42.01 Da), and chemical dimethylation of primary amines at protein N-termini and lysine side chains (+28.03 Da) (*see* **Note 4**). If isotopic labeling with d_2-^{13}C formaldehyde was employed, dimethylation of primary amines at protein N-termini and lysine side chains (+34.06 Da) should be included. Exemplary mass tolerances for a Q-Exactive mass spectrometer are ±10 ppm for precursor ions and ±20 ppm for fragment ions.
3. Perform secondary validation of search results with the PeptideProphet [9] algorithm provided as part of the Trans Proteomic Pipeline (TPP) [10]. Convert search result files to pep.xml format (http://tools.proteomecenter.org/wiki/index.php?title=Formats:pepXML) and analyze peptides by PeptideProphet, allowing for an error (false discovery) rate of 0.05.
4. Optional step for quantitation: relative quantitation for each peptide can be analyzed using the XPRESS [11] algorithm in TPP, using the following settings: mass tolerance of 0.015 Da and mass difference between isotopes of N-terminal and lysine residues of 6.0131 Da.
5. For each identified peptide, derive nonprime-side sequences (approximately ten amino acids preceding identified peptides) and generate input files for specificity heatmaps and sequence logos by processing peptide lists with WebPICS [12].

4 Notes

1. For quantitative analyses of multiple proteomes, multiplex comparison using different isotopes of formaldehyde (e.g., d_0-^{12}C, d_2-^{12}C, or d_2-^{13}C) and sodium cyanoborodeuteride (for Schiff base reduction) may also be used for chemical dimethylation of protein N-termini and lysine side chains.
2. For gel-based pre-fractionation of proteins: Separate chemically dimethylated proteins using SDS-PAGE. Stain gels using bromophenol blue, excise specific band of interest, and proceed with in-gel tryptic digestion.
3. Modified internal peptides lose their ability to bind the strong cation exchange column at acidic pH due to the addition of two negatively charged disulfonate groups at peptide N-termini.

Hence, these internal peptides readily flow through the column and are removed, while peptides containing true protein N-termini bind to the strong cation-exchange resin and are retained on the column. If insufficient material is available for separation by HPLC, use micro-SCX tips with several elution conditions [13].

4. For mass spectrometry analysis, disulfonate-modified peptides are not detected unless MS acquisition is performed in the negative ion mode. To analyze these peptides, search parameters should include +247.94 Da for single and +495.89 Da for double disulfonation of N-terminal residues.

Acknowledgment

This study was funded by a Marie Curie Fellowship for Career Development (PIIF-GA-2012-329622 GlycoMarker to Z.W.L.), Deutsche Forschungsgemeinschaft (SCHI 871/2 and SCHI 871/5, SCHI 871/6, GR 1748/6, INST 39/900-1 and SFB850-Project B8 to O.S.), European Research Council (ERC-2011- StG 282111-ProteaSys to O.S.), and the Excellence Initiative of the German Federal and State Governments (EXC 294, BIOSS to O.S.).

References

1. Overall CM, Blobel CP (2007) In search of partners: linking extracellular proteases to substrates. Nat Rev Mol Cell Biol 8:245–257
2. Kleifeld O, Doucet A, Keller UAD, Prudova A, Schilling O, Kainthan RK, Starr AE, Foster LJ, Kizhakkedathu JN, Overall CM (2010) Isotopic labeling of terminal amines in complex samples identifies protein N-termini and protease cleavage products. Nat Biotechnol 28:281-U1441
3. Staes A, Van Damme P, Helsens K, Demol H, Vandekerckhove J, Gevaert K (2008) Improved recovery of proteome-informative, protein N-terminal peptides by combined fractional diagonal chromatography (COFRADIC). Proteomics 8:1362–1370
4. Shen PT, Hsu JL, Chen SH (2007) Dimethyl isotope-coded affinity selection for the analysis of free and blocked N-termini of proteins using LC-MS/MS. Anal Chem 79:9520–9530
5. Venne AS, Solari FA, Faden F, Paretti T, Dissmeyer N, Zahedi RP (2015) An improved workflow for quantitative N-terminal charge-based fractional diagonal chromatography (ChaFRADIC) to study proteolytic events in *Arabidopsis thaliana*. Proteomics 15:2458–2469
6. Mommen GPM, van de Waterbeemd B, Meiring HD, Kersten G, Heck AJR, de Jong APJM (2012) Unbiased selective isolation of protein N-terminal peptides from complex proteome samples using phospho tagging (PTAG) and TiO_2-based depletion. Mol Cell Proteomics 11:832–842
7. Lai ZW, Gomez-Auli A, Keller EJ, Mayer B, Biniossek ML, Schilling O (2015) Enrichment of protein N-termini by charge reversal of internal peptides. Proteomics 15:2470–2478
8. Rappsilber J, Ishihama Y, Mann M (2003) Stop and go extraction tips for matrix-assisted laser desorption/ionization, nanoelectrospray, and LC/MS sample pretreatment in proteomics. Anal Chem 75:663–670
9. Keller A, Nesvizhskii AI, Kolker E, Aebersold R (2002) Empirical statistical model to estimate the accuracy of peptide identifications made by MS/MS and database search. Anal Chem 74:5383–5392
10. Keller A, Eng J, Zhang N, Li XJ, Aebersold R (2005) A uniform proteomics MS/MS analysis

platform utilizing open XML file formats. Mol Syst Biol 1:2005.0017
11. Han DK, Eng J, Zhou HL, Aebersold R (2001) Quantitative profiling of differentiation-induced microsomal proteins using isotope-coded affinity tags and mass spectrometry. Nat Biotechnol 19:946–951
12. Schilling O, Keller UAD, Overall CM (2011) Factor Xa subsite mapping by proteome-derived peptide libraries improved using WebPICS, a resource for proteomic identification of cleavage sites. Biol Chem 392:1031–1037
13. Rappsilber J, Mann M, Ishihama Y (2007) Protocol for micro-purification, enrichment, pre-fractionation and storage of peptides for proteomics using StageTips. Nat Protoc 2:1896–1906

Chapter 11

Mapping the Substrate Recognition Landscapes of Metalloproteases Using Comprehensive Mutagenesis

Colin A. Kretz

Abstract

Protease specificity is controlled by exosites, which capture and orient the substrate, and the active site, which binds substrate residues near the P1–P1′ scissile bond and catalyzes peptide hydrolysis. Techniques used to identify critical contact points between a protease and its substrate can be time consuming and labor-intensive. Screening tools such as phage display have been revitalized with the emergence of high-throughput sequencing technology, and can be used to interrogate protease substrate specificity. This article will outline a method for creating and screening a comprehensive mutagenesis substrate phage display library. High-throughput sequencing of uncleaved phage at various reaction time points enables k_{cat}/K_M determination for every possible single amino acid substitution at each position of the substrate, providing unprecedented resolution for the interaction between a protease and its substrate.

Key words High-throughput sequencing, Phage display, Protease, Metalloprotease, Mutagenesis, Substrate

1 Introduction

Phage display is a powerful screening tool that has been successfully applied to antibody development [1], peptide engineering [2, 3], and protease biology [4, 5]. Traditional approaches to phage display utilize the replicative capacity of bacteriophage to iteratively reduce the complexity of a library to a subset of the best responders. Although phage display holds promise for countless applications, it has been limited by technical challenges (Table 1).

Despite these limitations, M13 filamentous substrate phage display has played an important role in understanding metalloprotease specificity, with efforts primarily focused on describing substrate residues surrounding the scissile bond. In this method, random peptides are displayed as fusions to minor coat protein PIII with an NH_2-terminal epitope tag. Phage-displaying peptides that are cleaved by the protease are separated from the epitope tag, allowing for purification and selective re-amplification of cleavable

Charles A. Galea (ed.), *Matrix Metalloproteases: Methods and Protocols*, Methods in Molecular Biology, vol. 1579,
DOI 10.1007/978-1-4939-6863-3_11,

Table 1
Traditional vs high-throughput phage display

Traditional phage display	High-throughput sequencing phage display
Multiple rounds of selection/amplification	Single round selection
Interference of displayed protein with infective domain of phage can bias clone distribution	No infection step required
Toxic protein expression can lead to depletion or loss of specific clones	Enrichment is calculated for each clone individually, correcting for initial abundance
Individual validation of selected clones	Replicates performed in parallel allow statistical interpretation of responders and non-responders
Sanger sequencing only practical for a limited number of individual clones (20–100)	High-throughput sequencing > 10^8 sequences

clones (Fig. 1a). Following the final panning reaction, individual phage clones are isolated and tested in functional assays to identify cleavable peptides, which are ultimately identified by Sanger sequencing (Fig. 1b). Through these and other studies, matrix metalloproteases have been found to have broad and overlapping substrate recognition preferences [4, 6–8]. Other metalloproteases in the human genome, such as the Adamalysins [9], have been less well characterized [5]. Through these studies, it is hypothesized that metalloproteases utilize exosites (substrate-binding sites distinct from the active site) to control substrate specificity.

Bacteriophage provides a direct physical link between a protein and the DNA encoding it. As a result, recent advances in high-throughput sequencing (HTS) technology [10] have enabled comprehensive and quantitative profiling of every clone in a phage display library following a single round of selection (Fig. 1c) [11–15]. This approach alleviates many shortcomings of the traditional approach to phage display, and greatly expands its potential uses to interrogate biological systems. For the first time, selection replicates performed in parallel can be combined for statistical interpretations of enrichment and depletion [16]. Substrate phage display affords a unique opportunity to leverage HTS technology because the stringency of the selection procedure can be controlled by varying incubation periods with the enzyme.

This article outlines a protocol for cloning, screening, and analyzing a mutagenesis substrate phage display library using HTS [15]. The size of the peptides can be engineered to include critical exosite-binding residues, allowing comprehensive profiling of residues beyond the scissile bond. Furthermore, carrying out the selection at various reaction time points allows for the quantification of the change in mutation burden as a function of time, and an estimation of the k_{cat}/K_M value for every single amino acid residue at each position of the peptide.

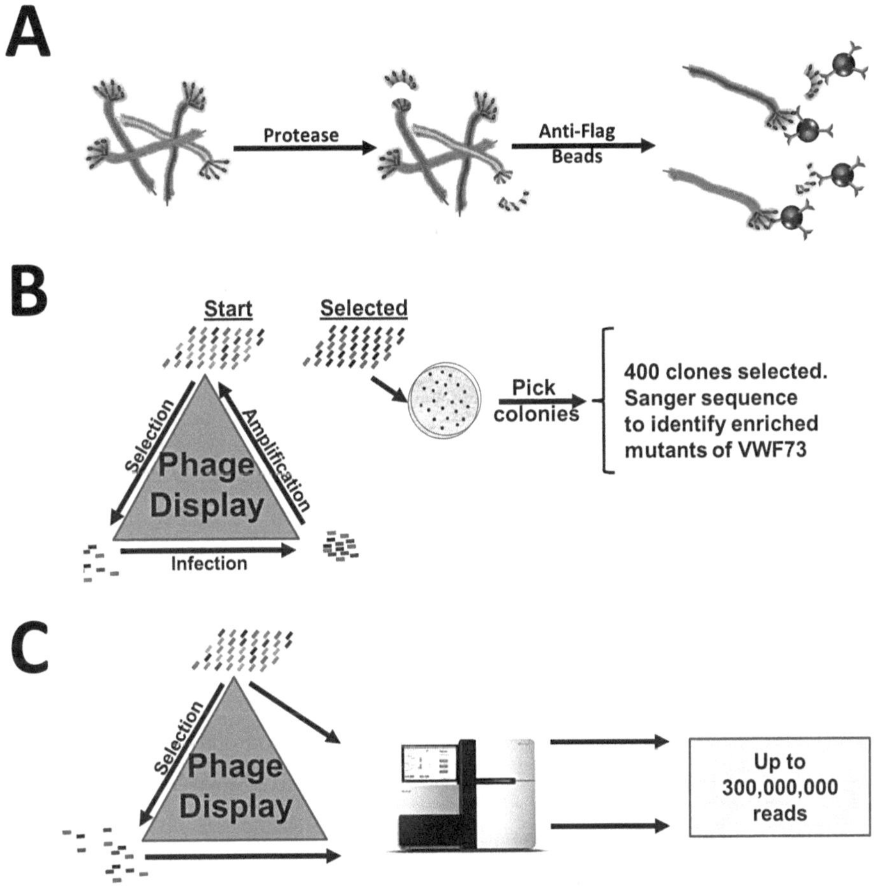

Fig. 1 High-throughput substrate phage display. (**a**) In substrate phage display, a random peptide library is displayed as an NH_2-terminal fusion to the PIII protein along with an epitope tag (such as FLAG). Phage displaying a peptide that is cleaved by the protease is separated from the epitope tag, allowing purification of remaining uncleaved phage particles. (**b**) This procedure is performed for multiple rounds of selection. The selected phage is re-amplified in bacteria to iteratively evolve the library. Following multiple rounds of selection and amplification, individual colonies are picked off agar plates, and a number of clones are identified by Sanger sequencing. (**c**) High-throughput sequencing mitigates the need for multiple rounds of selection and amplification. Millions of reads are obtained from the starting library and from phage following a single round of selection

2 Materials

2.1 Designing Mutagenesis Library

The design of a mutant substrate phage display library should account for HTS read length. The Illumina HiSeq platform currently delivers up to 300 million sequences of 50, 100, or 150 base pair lengths, whereas the MiSeq platform currently delivers 30 million reads up to 300 base pairs in length. Both formats include a paired-end read option, thus allowing for longer sequence designs. Because the quality of sequence data decreases toward the end of longer reads, the library should be designed such that paired-end reads overlap, to improve confidence in these sequence reads for downstream data analysis. As a result of these considerations, *the mutant oligonucleotide library should currently be restricted to less than 250 bases in length*, when using the HiSeq platform. Other HTS platforms suffer from high sequence error rates, lower throughput, or do not offer paired end read options, which make them less suitable for this application [17]. Future advances in this rapidly evolving technology will undoubtedly allow for longer mutant libraries. Consult with your HTS service provider before designing the library to better understand current capacities.

The library is constructed using mutations introduced during oligonucleotide synthesis, such that each nucleotide substitution is an independent event (Fig. 2). This is achieved by spiking in a desired amount of each alternative nucleotide into the individual nucleotide solutions prior to synthesis. For example, to achieve a 6% mutation frequency at each G in a given sequence, 2% of A, C, and T is added to the G solution. Alternative approaches, such as mutagenic PCR [18], can also be used but can result in unequal distribution of mutations and incomplete coverage of the entire mutation space.

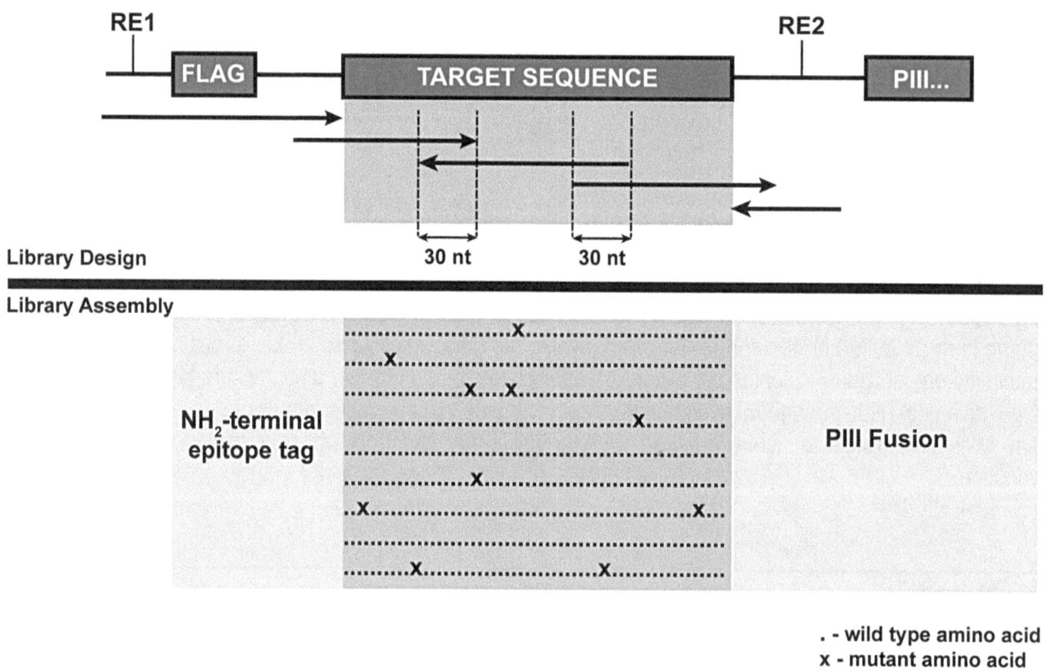

Fig. 2 Comprehensive mutagenesis library. To generate a comprehensive mutagenesis library, a series of overlapping oligonucleotides are designed spanning the target sequence. The oligonucleotides will contain both mutagenic sequences (*dark shaded*), and non-mutagenic sequences. Mutagenized oligonucleotides should contain not less than 30 nucleotides of overlap. Non-mutagenic oligonucleotides contain unique *BglI* restriction sites and an NH_2-terminal epitope tag for cloning into the FUSE55 phage display vector. Following assembly and cloning, each phage particle will contain up to five copies of a unique substrate variant displayed on its surface, with the DNA coding for that mutant peptide contained within the phage body

The required mutation frequency at each nucleotide position is calculated by the desired frequency of wild-type clones in the pool. For example, a 3% mutation frequency at each nucleotide position of a 210 base pair sequence results in 0.167% (0.97^{210}) of clones carrying no mutations. The resultant amino acid mutation frequency will vary between constructs because of codon usage, but a general guide is to design the library to contain 0.1–1% wild-type clones. Because nucleotide substitutions are independent events, the mutation frequency also dictates the fraction of clones that contain a single amino acid substitution or multiple amino acid substitutions. If multiple amino acid substitutions within a single clone are not desired, then mutation frequency should be decreased. Other options for creating mutagenized libraries with only a single amino acid substitution have recently been reported [12, 14], but will not be discussed here.

Once the library has been designed, contact your oligonucleotide synthesis provider for pricing and customization options. The wild-type sequence is typically indicated in upper case letters and sequences requiring the desired mutation frequency in lower case letters.

For example:

Primer 1: AAAAAAAAAAAAAAAAAAAAAAAAAAAAAAAAAAAAAAAAAA.

Mutant Oligo 1: AAAAAAAAAAAAAAAAAaaaaaaaaaaaaaaaaaaaaaaaaaaaaa.

Mutant Oligo 2: aaaaaaaaaaaaaaaaaaaaaaaaaaaaaaaaaaaaaaaaaaaaaaa.

Mutant Oligo 3: aaaaaaaaaaaaaaaaaaaaaaaaaaaaaaaaaAAAAAAAAAAAA.

Primer 2: AAAAAAAAAAAAAAAAAAAAAAAAAA.

2.2 Cloning Library

1. Gotaq green master mix (Promega).
2. Herculase II Fusion Enzyme kit (Agilent).
3. *BglI* (New England Biolabs).
4. T4 DNA ligase, HC (Promega).
5. FUSE55: M13 filamentous phage display vector from Smith lab [19] (University of Missouri) containing *BglI* cloning site 5′ to the gIII gene, and a single copy of each phage gene. As opposed to the phagemid system, FUSE55 does not require helper phage to supply the remaining phage proteins required for assembly and secretion. Importantly, all PIII proteins will display the recombinant protein fragment, allowing more precise control over the stoichiometry of substrate:enzyme ratios, and a better estimate of k_{cat}/K_M values for each mutant peptide.

6. Glycogen: UltraPure 20 μg/mL (Invitrogen). Working solution: 2 μg/mL in RNase-free water.
7. 3.5 M Sodium acetate, pH 3.
8. Isopropanol.
9. 70% Ethanol.
10. TE buffer: 10 mM Tris–HCl (pH 8.0), 1 mM EDTA.
11. LB-agar.
12. Tetracycline: Tetracycline-HCl is soluble in aqueous solutions whereas tetracycline is not. Prepare 30 mg/mL solution in water and 50% glycerol, stored at −20 °C and protected from light. Prolonged exposure to light has anecdotally been observed to limit growth or kill bacterial cultures.
13. MegaX DH10B electrocompetent *E. coli* (or other F− *E.coli*). These bacteria lack F-pili to prevent phage amplification bias. F+ bacterial strains are appropriate for helper-phage systems or for titering phage preps.
14. Cuvettes: 0.1 cm gap (BioRad).
15. Gene Pulser Xcell Electroporation system (BioRad).
16. SOC recovery media: 20 g/L Tryptone, 5 g/L yeast extract, 4.8 g/L $MgSO_4$, 3.603 g/L glucose, 0.5 g/L NaCl, 0.186 g/L KCl. Glucose, and $MgSO_4$ should be added after autoclaving. Final pH should be 6.8–7.0, and can be adjusted using sodium hydroxide. Alternatively, entire recipe can be assembled and sterile filtered through 0.22 μM filter.
17. 12 mL culture tubes.
18. Microtiter tubes.
19. Elutrap System (Whatman) which includes BT1 and BT2 membranes, electrophoresis chamber, and Elutrap device.
20. TBS-T: 20 mM Tris–HCl, pH 7.4, 150 mM NaCl, 0.05% Tw20.
21. PCR and Gel purification column Kits (QIAgen).
22. F55-S1 primer (5′–3′): CACCTCGAAAGCAAGCTGATAAACCG.
23. F55-AS1 primer (5′–3′): CGCCTGTAGCATTCCACAGACAGCCC.

2.3 Phage Preparation

1. 2800 mL Fernbach flask: aeration during bacterial growth improves phage yield as well as plasmid yields.
2. 2× YT media: 16 g/L Tryptone, 10 g/L Yeast extract, 5 g/L NaCl.
3. PEG/NaCl: 100 g polyethylene glycol-8000, 116.9 g NaCl, 475 mL water, total volume 600 mL (autoclave to sterilize and fully solubilize).

4. TBS: 20 mM Tris–HCl, 150 mM NaCl, pH 7.4.
5. K91 *E. coli*, or other F-pilus containing strain. Ensure that F-plasmid does not require tetracycline for maintenance.
6. LB agar plates containing 30 μg/mL tetracycline.

2.4 Panning

1. Protease.
2. 5× Reaction buffer: TBS, 5% bovine serum albumin, 25 mM $CaCl_2$, 50 μM $ZnCl_2$ (or other appropriate buffer for your protease).
3. Stop buffer: Reaction buffer + 50 mM EDTA (AEBSF can be used in place of EDTA for proteases other than metalloproteases).
4. M2 anti-FLAG agarose beads (Sigma).
5. 5 mg/mL 3× FLAG peptide (Sigma).

2.5 AlphaLISA Quantification of Phage Proteolysis

1. AlphaLISA buffer: 25 mM Hepes, 1 mg/mL dextran, 0.5% Triton-100, 0.5% bovine serum albumin, 0.5% Proclin-300.
2. 96-well half-well plate.
3. Anti-FLAG acceptor beads (Perkin Elmer).
4. Anti-Fd-tet biotinylated antibody (B2661-0.5ML, Sigma).
5. Streptavidin donor beads (Perkin Elmer).
6. EnSpire 2300 Multilabel Plate Reader (Perkin Elmer).

2.6 High-Throughout Sequencing

1. Proteinase K (10 mg/mL).
2. UltraPure Phenol/chloroform/isoamyl alcohol (25:24:1,v/v) (Invitrogen).
3. Herculase II Fusion Enzyme Kit (Agilent Technologies, prod uct #: 600677).
4. Agarose.
5. Illumina primers Table 2 (*see* **Note 1**).

3 Method

Mutagenized substrate phage display libraries have the potential to identify substrate residues that make important contacts with the protease. Individual substrate clones harboring amino acid substitutions that impair proteolysis will be consumed less efficiently in the reaction, and thus be disproportionately represented in the selected phage population. As a result, this method can not only identify substrate residues that interact with the protease active site, but also define residues that interact with the protease at sites distinct from the active site, called exosites. A well-constructed

Table 2
HTS library preparation primers

Primer	Sequence
NGS S1a	ACACTCTTTCCCTACACGACGCTCTTCCGATCTNNNNNNgaatc**TCAATGACAGGAGGACAACAAATG**
NGS S1b	ACACTCTTTCCCTACACGACGCTCTTCCGATCTNNNNNNtctga**TCAATGACAGGAGGACAACAAATG**
NGS S1c	ACACTCTTTCCCTACACGACGCTCTTCCGATCTNNNNNNcagtc**TCAATGACAGGAGGACAACAAATG**
NGS S1d	ACACTCTTTCCCTACACGACGCTCTTCCGATCTNNNNNNagtct**TCAATGACAGGAGGACAACAAATG**
NGS S1e	ACACTCTTTCCCTACACGACGCTCTTCCGATCTNNNNNNtgcaa**TCAATGACAGGAGGACAACAAATG**
NGS AS1a	CTCGGCATTCCTGCTGAACCGCTCTTCCGATCTNNNNNNgtcta**ACTTTCAACAGTTTCGGCCC**
NGS AS1b	CTCGGCATTCCTGCTGAACCGCTCTTCCGATCTNNNNNNcagag**ACTTTCAACAGTTTCGGCCC**
NGS AS1c	CTCGGCATTCCTGCTGAACCGCTCTTCCGATCTNNNNNNttcac**ACTTTCAACAGTTTCGGCCC**
NGS AS1d	CTCGGCATTCCTGCTGAACCGCTCTTCCGATCTNNNNNNagtct**ACTTTCAACAGTTTCGGCCC**
NGS AS1e	CTCGGCATTCCTGCTGAACCGCTCTTCCGATCTNNNNNNtcagg**ACTTTCAACAGTTTCGGCCC**
PESeq S1	CAAGCAGAAGACGGCATACGAGATCGGTCTCGGCATTCCTGCTGAACCGCTCTTCCGATCT
PESeq AS1	AATGATACGGCGACCACCGAGATCTACACTCTTTCCCTACACGACGCTCTTCCGATCT

mutagenized substrate phage display library can theoretically include every possible amino acid substitution at each position, offering a more comprehensive interrogation of protease/substrate interactions compared with other approaches.

3.1 Design, Synthesis, and Assembly of Mutagenized Substrate Phage Display Library and Wild-Type Substrate Displaying Phage

1. Design overlapping mutagenized oligonucleotides (contact supplier for details about oligonucleotide length limits) until entire length of desired target is covered (Fig. 2). Terminal 5′ and 3′ oligonucleotides should contain non-mutagenic adapter sequences for equal amplification following assembly as well as appropriate restriction sites for cloning into the vector (*see* **Note 2**). Internal oligonucleotides can be entirely mutagenized. Provide no less than 30 nucleotides of overlap between oligonucleotides to allow for mismatches during assembly.
2. In parallel, order non-mutageneic oligonucleotides for synthesis of a wild-type substrate phage clone.
3. Assemble an analytical PCR reaction: 1 nM of each internal oligonucleotide, 0.5 μM 5′ and 3′ terminal primers using Herculase II. Thermal profile: 95 °C (30 s), 55 → 66 °C gradient (30 s), 72 °C. Repeat for 15–20 cycles.
4. Identify lowest annealing temperature in Subheading 3.1, **step 3**. Repeat PCR at lowest optimal annealing temperature using Herculase II (lower annealing temperatures will reduce selection bias against nucleotide mismatches within the overlap region) (*see* **Note 3**).
5. Separate PCR product on 1.5% agarose gel, stain with ethidium bromide, and excise band.
6. Extract DNA from cut gel using EluTRAP system (*see* **Note 4**).
7. Precipitate DNA by adding 10 ng glycogen, 0.1:1 v:v sodium acetate, and 1:1 isopropanol. Incubate at −80 °C for 2 h. Centrifuge at 14,000 × *g* for 30 min at 4 °C. Wash pellet with 70% ethanol. Centrifuge at 14,000 × *g* for 20 min. Allow the pellet to airdry. Resuspend in 50 μL nuclease-free water.
8. Digest PCR product and vector with *Bgl*I, and purify (repeat Subheading 3.1, **steps 5–7**).
9. Assemble ligation reaction using DNA Ligation kit HC (Promega). Use a 3:1 molar ratio of insert to vector and 1 μg vector in a 50 μL ligation reaction (containing 5 μL 10× Ligase Buffer and 1 μL ligase) (*see* **Note 5**). Assemble the ligation in PCR tubes, and use a thermocycler to incubate the reaction at 16 °C overnight (*see* **Note 6**).
10. Heat-inactivate ligation reaction at 65 °C for 20 min and precipitate (Subheading 3.1, **step 7**). Resuspend ligation product in 10 μL nuclease-free water and use immediately.

11. Electroporate ligation product into DH10B *E.coli* using BioRad Gene Pulser. Add 1 μL ligation reaction to 30 μL cells in microcentrifuge tubes on ice. Add to prechilled 1 cm disposable cuvettes. Electroporation conditions: 2.0 kV, 200 Ω, and 25 μF (*see* **Note 7**). Recover each electroporation using 2 × 1 mL SOC in a 12 mL culture tube and place at 37 °C in shaker incubator for 1 h.
12. Pool cultures, remove 20 μL for titration, and add remaining volume to 1 L 2× YT in a Fernbach flask containing 30 μg/mL tetracycline. Place in 30 °C shaker incubator overnight. *Note total volume prior to addition to Fernbach flask.*
13. To assess library depth, perform a 1/10 (20 μL into 180 μL) serial dilution of culture into TBS-T. Plate 100 μL of each dilution onto LB Agar plates containing 30 μg/mL tetracycline. Incubate plates at 37 °C overnight.
14. Calculate total number of phage in recovered media: 0.01 × (CFU on plate) × (plate dilution factor) × (total electroporation volume in μL from Subheading 3.1, **step 12**).
15. Identify the number of phage clones with expected insert size. Pick 20–50 colonies into 50 μL autoclaved distilled deionized water to serve as a template in PCR reactions. Assemble a PCR reaction with 25 μL of 2× GoTaq. Green Master mix, 2 μL template solution, 0.2 μM F55-S1, 0.2 μM F55-AS1, nuclease-free water to 50 μL final volume. Thermal profile: (1) 95 °C (5 min), (2) 95 °C (30 s), (3) 60 °C (30 s), (4) 72 °C (30 s), (5) 72 °C (10 min), repeat 2–4 for 30 cycles.
16. Assess PCR reaction by running 10 μL on a 1–2% agarose gel. Count the frequency of PCR products with a band of the expected size for the insert and multiply this frequency by the phage depth estimate from Subheading 3.1, **step 14**. *This is the final library depth.*
17. Purify PCR products from reactions containing expected band size using QIAquick PCR purification. Submit for Sanger Sequencing using F55-S1 primer to validate mutation frequency.

3.2 Phage Preparation

1. From Subheading 3.1, **step 12**. Pellet bacteria using a two-step centrifugation. (**Step 1**) 10 min at 2400 × *g*, 4 °C. Transfer supernatant containing phage to fresh bottle. (**Step 2**) 10 min at 6200 × *g*, 4 °C.
2. Store aliquots of final library. Resuspend bacterial pellet in 50 × 1 mL aliquots of 2× YT with 20% glycerol, and store in −80 °C.
3. To the phage-containing culture supernatant, add 0.15× volume (150 mL for 1 L culture) of PEG/NaCl. Gently mix by inverting bottles several times (up to 100 times). Incubate at 4 °C overnight to precipitate.

4. Pellet phage by centrifugation at 6500 × *g* for 60 min at 4 °C.
5. Resuspend phage in 30 mL TBS. Incubate in an shaker incubator for 20 min to aid in solubilizing. Centrifuge at 12,000 × *g* for 10 min at 4 °C to pellet debris.
6. Repeat Subheading 3.2, **steps 3–4**. Resuspend phage in 10 mL TBS-T.
7. Titer phage. Perform a 1/10 serial dilution of phage solution (20 μL into 180 μL) into log growth phase (OD_{600nm} 0.3–0.5) K91 *E. coli* (or equivalent F-pilus expressing strain). Shake dilutions for 1 h at 37 °C. Plate 100 μL of each dilution onto LB Agar plates containing 30 μg/mL tetracycline. Incubate plates at 37 °C overnight.
8. Assess phage concentration:

   (a) 0.01 × (CFU on plate) × (plate dilution factor) = C_1 (phage/μL).

   (b) $(C_1 \times 10^6)/(6.022 \times 10^{23}) = C_f$ (M).

3.3 Initial Reaction Time Course with Phage Displaying Wild-Type Substrate and AlphaLISA

Experiments conducted using phage displaying the wild-type substrate are important for informing the appropriate conditions for the reaction between the protease and the mutant substrate phage display library.

1. Assemble 500 μL reaction containing 400 μL wild-type substrate displaying phage (from Subheading 3.2, **step 6**), 100 μL 5× reaction buffer. Remove 50 μL of solution into 50 μL stop buffer (0 min).
2. Initiate the reaction by adding remaining 450 μL of phage solution into 5 μL of 455 nM protease stock solution, achieving a final concentration of 5 nM protease (*see* **Note 8**).
3. Remove 50 μL reaction time points into 50 μL of 4 °C stop buffer. Repeat at each of the remaining reaction time points for a total of eight time points. Recommended initial time points (min): 0, 1, 5, 10, 50, 100, 500, overnight (*see* **Note 9**).
4. During reaction time course, perform a dilution series for a standard curve. Repeat Subheading 3.3, **step 1**, omitting protease. Perform a 1:2 dilution series, by adding 250 μL of phage to 250 μL 1× Reaction Buffer (1 mL 5× reaction buffer + 4 mL TBS). Add 250 μL Stop Buffer to each dilution (*see* **Note 10**).
5. Test the selected time course for rates of phage cleavage using AlphaLISA (*see* **Note 11**). Prepare a 1100 μL Master Mix solution of 20 μg/mL anti-FLAG coated Acceptor beads and 4 nM biotinylated anti-fdTET in AlphaLISA Buffer. Add 25 μL of Master Mix solution to 32 wells of a 96 half-well plate. In duplicate, add 5 μL of each standard curve dilution, or 5 μL of each reaction time point to each well. Incubate at room temperature for 1 h. Prepare a 700 μL solution of 40 μg/mL

streptavidin-coated donor beads in AlphaLISA Buffer, and add 20 μL to each well. Protect from light, and incubate at room temperature for 30 min (*see* **Note 12**). Read the plates on an EnSpire 2300 Multilabel Plate Reader at an excitation wavelength of 680 nm and emission wavelength of 615 nm.

6. Construct a standard curve of AlphaLISA signal as a function of the natural log of the molar phage concentration. Determine the concentration of uncleaved phage at each time point from the protease reacted phage, and calculate k_{cat}/K_M value.

3.4 Reaction of Phage Displaying Mutant Substrate Phage Display Library with Protease

1. Repeat Subheading 3.3, **steps 1–6** using mutant substrate phage display library.
2. Selection of reaction time points (t_0–t_5) for analysis by HTS should be based on time points exhibiting less than 75% substrate consumption (*see* **Note 13**). A single time point representing maximal library cleavage ($\Delta_{[substrate]} \rightarrow 0$) should also be taken ($t_{final}$).
3. Add 500 μL Stop Buffer and 50 μL anti-FLAG agarose beads to select mutant substrate phage library reaction time points. Incubate at 4 °C overnight with end-over-end mixing.
4. Centrifuge at 1,000 × *g* for 2 min to pellet beads. Wash beads 5× with Stop Buffer.
5. Elute phage from beads at each reaction time point by adding 500 μL of 0.15 mg/mL 3× FLAG peptide in Stop Buffer. Incubate 30 min at room temperature with end-over-end mixing.
6. Centrifuge at 1,000 × *g* for 2 min to pellet beads. Remove supernatant (containing selected phage) into fresh tube.
7. Add 5 μL of 10 mg/mL proteinase K. Incubate at 50 °C for 4 h (or overnight) to digest phage coat protein, exposing ssDNA.
8. Add 500 μL phenol/chloroform/isoamyl alcohol, and vortex. Centrifuge at 14,000 × *g* for 10 min at room temperature. Remove top aqueous layer into clean microcentrifuge tube. Add 200 μL chloroform to remove excess phenol, vortex, and then centrifuge for 10 min at 14,000 × *g*. Remove top aqueous layer into a fresh tube.
9. Precipitate and concentrate ssDNA as in Subheading 3.1, **step 7**.
10. Resuspend pelleted ssDNA in 20 μL TE Buffer.

3.5 Preparation of High-Throughput Sequencing Samples

1. Assemble a 50 μL PCR reaction for each ssDNA sample collected from reaction time points (Subheading 3.4, **step 10**) containing 5 μL template, 0.5 μM NGS-S1(a → e), 0.5 μM NGS-AS1(a → e), and Herculase reaction components according to manufacturer's instructions. The thermal profile is: (1) 98 °C (2 min), (2) 98 °C (20 s), (3) 60 °C (20 s), (4) 72 °C (30 s), (5) 72 °C (5 min), repeat 2–4 for 8–16 cycles (*see* **Note 14**).

2. Gel purify on a 2% agarose gel.
3. Assemble a 50 μL PCR reaction for each PCR product of reaction time points, containing 10 ng template (from Subheading 3.5, **step 2**), 0.5 μM PE1-seq, 0.5 μM PE2-seq, and Herculase reaction components according to manufacturer's instructions. The thermal profile is: (1) 98 °C (2 min), (2) 98 °C (20 s), (3) 60 °C (20 s), (4) 72 °C (30 s), (5) 72 °C (5 min), repeat 2–4 for 8–16 cycles (*see* **Note 15**).
4. Repeat Subheading 3.5, **step 2**.

HTS libraries can now be submitted for sequencing. To assess library quality (a) samples can be quantified using qPCR (Kapa BioSystems, product #: KK4835) to ensure proper Illumina adapter assembly, and (b) samples can be run on a BioAnalyzer to ensure proper fragment size. Paired-end sequencing is the most common format for these applications because it allows for coverage of longer sequences, and sequence redundancy in overlap regions can be used to correct for sequencing errors (*see* **Note 16**).

Each sample has a unique nucleotide sequence added to it (*see* primers for Subheading 3.5, **step 1**) that acts as a barcode for that sample. This allows many samples to be pooled and sequenced on the same HTS run, which saves cost and time. Sequencing data for each reaction time point is then separated bioinformatically based on the barcode sequence.

3.6 Data Analysis

Data analysis pipelines for high-throughput sequencing of phage display experiments are still largely dependent on in-house coding algorithms. Although some formal pipelines are available to calculate mutation frequency and assess enrichment [16, 20], they do not provide the flexibility for customized analysis. A general strategy is provided for analyzing the performance of mutant substrate sequences (Fig. 3), as quantified from high-throughput sequencing data.

3.6.1 Raw Data

1. Parse sequencing data into bins according to barcodes, represented by nucleotide reads 7–11. Each bin now contains the sequencing data collected from the uncleaved phage at each reaction time point. Check for comparable sequence read depth across reaction time points.
2. Trim sequences of the 6 N wobble, barcode, and primer, leaving only the mutant library.
3. FASTQ files from paired-end reads are aligned to wild-type sequence using Bowtie2 [21], yielding a paired SAM output file (http://bowtie-bio.sourceforge.net).
4. Assemble paired end reads into a single sequence. If paired reads overlap and alignments show discordant base calls, the base with the higher quality score is kept. If the discordant

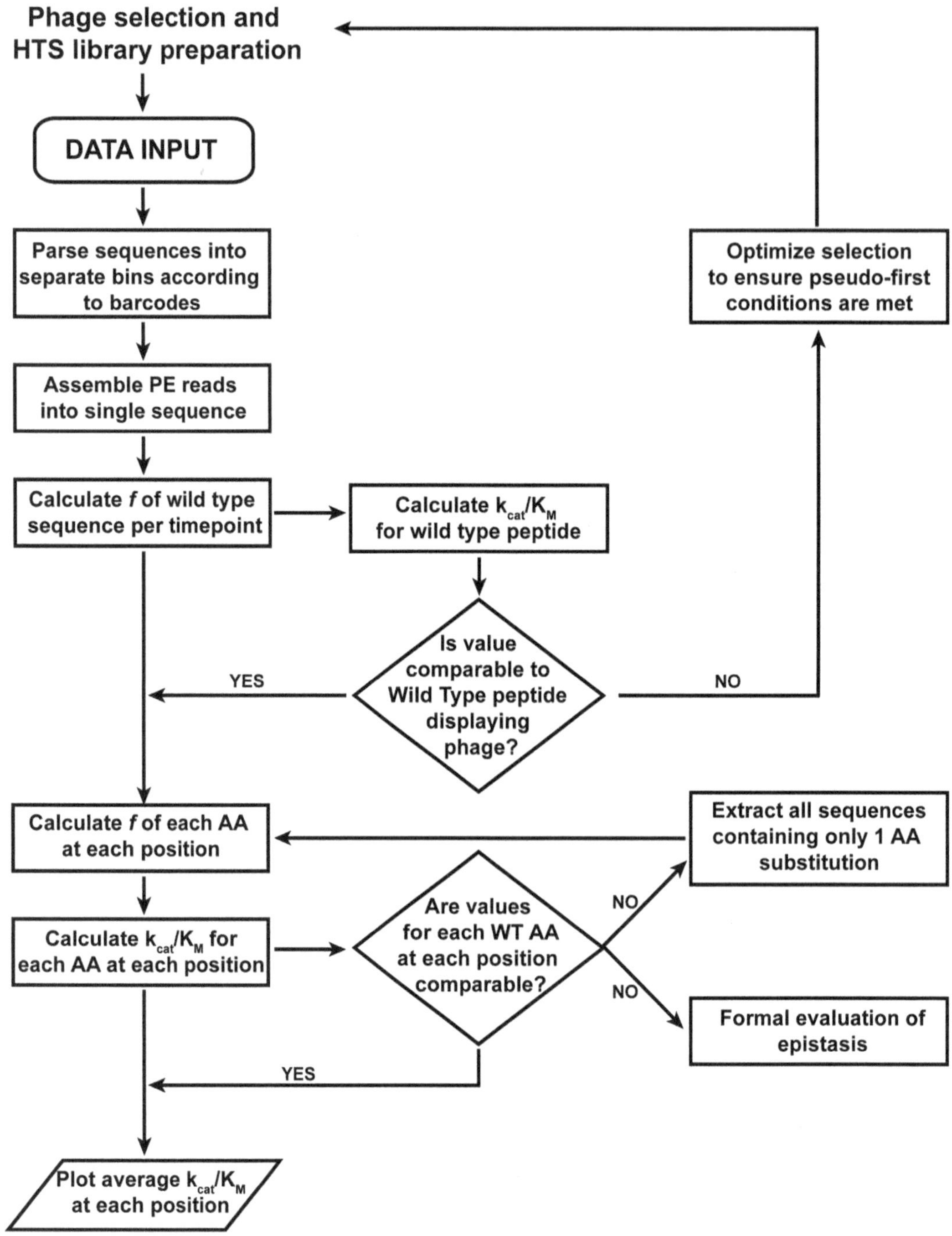

Fig. 3 High-throughput sequencing analysis scheme. An outline for the analysis of high-throughput sequencing data is shown. Following high-throughput sequencing of the uncleaved phage at various reaction time points, raw sequencing reads are separated into bins. Paired-end (PE) reads are assembled into single sequences, and the frequency (*f*) of wild-type sequences is initially analyzed. Next, the frequency of each amino acid (AA) at each position of the substrate is calculated by dividing the count by the total number of reads assigned to that bin. Alternatively, the frequency of each amino acid is only calculated from the subset of clones that only contain a single amino acid substitution. This can be used to mitigate the effects of multiple substitutions within a clone affecting the results. Ultimately, the frequency of each amino acid at each time point is multiplied by the concentration of uncleaved phage at that time point, before fitting the data to the first order rate law

base calls have identical quality scores, the forward read (read 1) should be preferred.

3.6.2 Analysis of Wild-Type Sequences Within the Library

1. Assess the frequency of wild-type substrate in each reaction time point. In the starting library (t_0), wild-type should be ~$(1 - \mu)n$, where μ is the mutation frequency at each nucleotide position and n is the number of nucleotides in the sequence (*see* **Note 17**).
2. Multiply the frequency of wild-type sequence, by the molar concentration of uncleaved phage at the corresponding reaction time point (from Subheading 3.4, **step 1**).
3. Fit the data to the first order rate law: $\ln([A]t) = -kt + \ln([A]_0)$. Perform a regression of the natural log of molar concentration of WT phage as a function of time. The slope, k, divided by the molar concentration of enzyme yields k_{cat}/K_M. Compare this value to previously determined values (from Subheading 3.3) to assess the reliability of the data (*see* **Note 18**).
4. Determine the effect of multiple substitutions within a clone on the rate of proteolysis measured for a given mutant at a position (i.e., epistasis), using the WT amino acid frequency at each position. These counts are derived from substrate clones containing multiple amino acid substitutions, but share the WT amino acid at this position in common. As in Subheading 3.6.2, **steps 2–3**, calculate the k_{cat}/K_M for each wild-type amino acid at each position. If the value for each WT amino acid at each position is comparable to each other, it suggests that the aggregate effect of multiple substitutions on a given peptide is mostly averaged across sequences sharing a given amino acid. Thus, strong epistasis is not common within the dataset. If needed, more formal and rigorous analysis of epistasis should be explored [15, 22].

3.6.3 Analysis of Mutant Sequences Within the Library

1. If Subheading 3.6.2, **step 4** indicates limited epistasis within the library, then the counts for each amino acid at each position of the substrate (Subheading 3.6.2, **step 3**, above) can be used to calculate the k_{cat}/K_M for each mutant sequence.
2. If Subheading 3.6.2, **step 4** suggests multiple interactions between mutations at certain positions within the library, then the mutation frequency calculated from Subheading 3.6.2, **step 3**, cannot reliably estimate the effect of a single amino acid substitution on the rate of cleavage in the library. Instead, the frequency of each amino acid substitution at each position of the peptide should be calculated from the subset of sequences that contain only a single amino acid substitution. This will only cover a fraction of the potential sequence space of the peptide. However, this subset of mutations will likely contain

the majority of amino acid changes requiring only a single nucleotide substitution within a codon, which accounts for the vast majority of point mutations that cause human disease (*see* Subheading 4).

3. Assemble a table for the counts of each 20 amino acid at each position of the substrate. Divide each count by the total number of sequence reads assigned to each time point. The wild-type amino acid should be the highest frequency amino acid at each position. Calculate the molar concentration of phage containing each amino acid by multiplying each frequency by the concentration of uncleaved phage at the corresponding reaction time point (from Subheading 3.4, **step 1**).
4. Exclude amino acid substitutions from the table that are covered by less than five sequencing reads in any time point.
5. If the slope of a regression line of concentration vs time is >0, then set k_{cat}/K_M value to 0. These values represent amino acid substitutions that prevent any detectable substrate cleavage in the conditions of the assay.
6. Calculate k_{cat}/K_M for each amino acid substitution at each position of the substrate to determine the effect of every possible amino acid substitution on proteolytic cleavage (as in Subheading 3.6.2, **step 3**).
7. Use the Benjamini-Hochberg [23] statistical model on reaction replicates to calculate p-values, adjusted for multiple observations, for every k_{cat}/K_M value (also called the False Discovery Rate, or FDR).

4 Conclusion

The capacity to investigate the entire sequence space of a protein can provide important insights into function. The method outlined in this report should serve as a useful guide for creating and screening mutagenized substrate phage display libraries. Coupling this approach with high-throughput sequencing allows the quantification of the impact of every possible amino acid substitution at each position of a substrate on the rate of its cleavage by the cognate protease.

In addition to defining important contact points between a protease and its substrate, this approach may be useful for interpreting the impact of genetic variants on protein function [24]. Whole genome and whole exome sequencing routinely identifies genetic mutations in both healthy individuals and patients. While frameshifts, stop-gain, and gene deletions can generally be expected to cause a loss of protein function, interpretation of the much larger set of mutations that result in an amino acid change is more difficult. These variants of uncertain significance are often assigned a functional score based on evolutionary sequence conservation using predictive computer algorithms. However, these computational tools exhibit high error rates [25], which limit their utility. High-throughput mutagenesis profiling methods, like the one described here, can generate empirical biochemical data for all possible amino acid substitutions of a given protein. Precise quantification of the impact of amino acid changes on protein function may better predict the link between genetic variation and disease. Ultimately, these techniques will facilitate the establishment of comprehensive missense mutation databases with accurate predictions of functional significance that can be used by investigators and clinicians to improve diagnosis and patient care.

5 Notes

1. These barcoding primers contain sequences that anneal up-stream or down-stream of the mutagenized library (bold), a barcode specific to each sample to allow for multiplexing many samples on a single sequencing lane (lowercase), and the first half of the Illumina adapter. Illumina sequencers use the first six nucleotide reads to establish optical coordinates for each cluster on the chip. As a result, it is often advisable to place a 6 N wobble upstream of the barcode to provide enough sequence diversity for each cluster to be distinguished from its neighbors.
2. In the case of FUSE55, these include unique *BglI* sites. In addition, the 5′ terminal primer should also add an epitope tag to the library, such as FLAG.
3. Be careful about the number of PCR cycles. Too many cycles can introduce biases in the library at an early stage and will not become evident until the first set of HTS data comes back. We aim for not more than 15–20 cycles.
4. Gel extraction techniques based on chaotropic salts to melt the gel and columns to capture DNA (such as QIAgen) can also be used, but can impair downstream steps. For example, ligation reactions are very sensitive to salt contamination.
5. To calculate the amount of insert to include in the ligation reaction, use the following equation: Insert Mass (ng) = Vector Mass × 3 × (Number of insert base pairs)/(Number of vector base pairs). It is generally recommended to perform multiple ligation reactions in parallel to optimize the ratio of insert:vector and the total volume of the ligation reaction.
6. Avoid freeze thaw cycles of the 10× ligation buffer, as this lowers the ATP yield in the buffer and negatively impacts ligation efficiency.

7. Much of the success of the library build comes from this step. Poor transformation efficiency can lead to poor library depth, and ultimately incomplete diversity. Typically, ten electroporations per library are sufficient to yield 1 × 10 [7] independent clones. However, more may be required. 30–40 electroporations can be needed to yield sufficient diversity. Keep an eye on the time constant, optimally at 4.2 ms, if drops below 4 ms may indicate salts in the ligation product that increase conductivity. Try additional 70% ethanol washes at room temperature (including a 20 min incubation) to extract excess salts from the precipitation reaction. Also, too much glycogen can impede transformation efficiency.
8. It is important to perform these reactions under pseudo-first order reaction conditions, where the concentration of substrate is $\ll K_M$. Leave lots of room here, in case mutations are discovered that enhance cleavage relative to WT. But an initial reaction setup to the conditions for a WT substrate will suffice for an initial analysis.
9. The reaction time course will need to be determined empirically if the interaction is not already well defined. Adjust protease concentration and time points accordingly.
10. Stop buffer contains EDTA, which we have found to interfere with the AlphaLISA signal. This must be accounted for in the standard curve.
11. AlphaLISA is a rapid alternative to standard ELISA protocols. However, if appropriate plate readers are not available, then standard ELISA can also be used. In this case, precoat plates with 10 μg/mL anti-fd-TET antibody (Sigma) and detection with 1/10,000 HRP-anti-FLAG antibody (Sigma). Detection is achieved using 1 Step Ultra TMB-ELISA substrate (Thermo Scientific #34028). Filamentous phage exhibits a high degree of nonspecific binding to plastic surfaces, which can be leveraged by capturing with an anti-phage antibody.
12. Donor beads are highly susceptible to photo-bleaching. Turn off overhead fluorescent lights when working with donor beads.
13. This may need to be determined empirically, as it depends on the K_M of the interaction. Dogmatically, initial rates of a reaction are determined with less than 10% substrate consumption. However, fitting the data to a first order rate law accounts for decreased rates of cleavage at later time points caused by substrate consumption.
14. The number of cycles should be determined empirically to avoid PCR recombination (product laddering) that can cause bias in library distribution unrelated to the protease selection. This phenomenon tends to occur when PCR reactions are

allowed to proceed beyond log-phase, and may be related to products themselves acting as primers. This is less of an issue with mutagenic libraries, and may be more relevant with whole genome sequencing libraries or other samples where products contain partially complementary overlapping sequences.

15. These are universal primers that can be used regardless of barcoding primer.
16. Illumina sequencing is currently limited by the size of DNA fragments that can reliably cluster on the chip surface. Typically, total fragment length of less than 700 bp is preferable. Check with sequencing services for more details. Discuss with sequencing service provider about your anticipated library diversity. Low mutational burden within the libraries can affect sequence quality scores. Increasing PhiX to 20% of total sample quantity is useful. Alternatively, multiplexing with RNASeq, ExomeSeq, or Whole genome sequencing can provide the necessary sample diversity for a successful sequencing reaction.
17. Some nucleotide substitutions will not alter the wild-type amino acid. This fact is not accounted for in this formula.
18. Because of the complexity of the reaction conditions, these values represent apparent rate constants.

Acknowledgments

I would like to thank David Ginsburg (University of Michigan) for critical review of the manuscript. Colin A. Kretz holds a McMaster University Department of Medicine Internal Career Award. This work was also supported by the Judith Graham Pool Fellowship from the National Hemophilia Foundation, awarded to Colin A. Kretz.

References

1. Chan CE, Lim AP, MacAry PA, Hanson BJ (2014) The role of phage display in therapeutic antibody discovery. Int Immunol 26:649–657
2. Chen S, Heinis C (2015) Phage selection of bicyclic peptides based on two disulfide bridges. Methods Mol Biol 1248:119–137
3. Whitney M, Crisp JL, Olson ES, Aguilera TA, Gross LA, Ellies LG, Tsien RY (2010) Parallel in vivo and in vitro selection using phage display identifies protease-dependent tumor-targeting peptides. J Biol Chem 285:22532–22541
4. Kridel SJ, Chen E, Kotra LP, Howard EW, Mobashery S, Smith JW (2001) Substrate hydrolysis by matrix metalloproteinase-9. J Biol Chem 276:20572–20578
5. Hills R, Mazzarella R, Fok K, Liu M, Nemirovskiy O, Leone J, Zack MD, Arner EC, Viswanathan M, Abujoub A, Muruganandam A, Sexton DJ, Bassill GJ, Sato AK, Malfait AM, Tortorella MD (2007) Identification of an ADAMTS-4 cleavage motif using phage display leads to the development of fluorogenic peptide substrates and reveals matrilin-3 as a novel substrate. J Biol Chem 282:11101–11109
6. Eckhard U, Huesgen PF, Schilling O, Bellac CL, Butler GS, Cox JH, Dufour A, Goebeler V, Kappelhoff R, Keller UA, Klein T, Lange PF, Marino G, Morrison CJ, Prudova A, Rodriguez D, Starr AE, Wang Y, Overall CM (2016)

Active site specificity of the matrix metalloproteinase family: proteomic identification of 4300 cleavage sites by nine MMPs explored with structural and synthetic peptide cleavage analyses. Matrix Biol 49:37–60
7. Chen EI, Kridel SJ, Howard EW, Li W, Godzik A, Smith JW (2002) A unique substrate recognition profile for matrix metalloproteinase-2. J Biol Chem 277:4485–4491
8. Ratnikov BI, Cieplak P, Gramatikoff K, Pierce J, Eroshkin A, Igarashi Y, Kazanov M, Sun Q, Godzik A, Osterman A, Stec B, Strongin A, Smith JW (2014) Basis for substrate recognition and distinction by matrix metalloproteinases. Proc Natl Acad Sci U S A 111(40):E4148–E4155
9. Huxley-Jones J, Clarke TK, Beck C, Toubaris G, Robertson DL, Boot-Handford RP (2007) The evolution of the vertebrate metzincins; insights from *Ciona intestinalis* and *Danio rerio*. BMC Evol Biol 7:63
10. Shendure J, Lieberman Aiden E (2012) The expanding scope of DNA sequencing. Nat Biotechnol 30:1084–1094
11. Tinberg CE, Khare SD, Dou J, Doyle L, Nelson JW, Schena A, Jankowski W, Kalodimos CG, Johnsson K, Stoddard BL, Baker D (2013) Computational design of ligand-binding proteins with high affinity and selectivity. Nature 501:212–216
12. Kitzman JO, Starita LM, Lo RS, Fields S, Shendure J (2015) Massively parallel single-amino-acid mutagenesis. Nat Methods 12:203–206
13. Fowler DM, Araya CL, Fleishman SJ, Kellogg EH, Stephany JJ, Baker D, Fields S (2010) High-resolution mapping of protein sequence-function relationships. Nat Methods 7:741–746
14. Starita LM, Young DL, Islam M, Kitzman JO, Gullingsrud J, Hause RJ, Fowler DM, Parvin JD, Shendure J, Fields S (2015) Massively parallel functional analysis of BRCA1 RING domain variants. Genetics 200:413–422
15. Kretz CA, Dai M, Soylemez O, Yee A, Desch KC, Siemieniak D, Tomberg K, Kondrashov FA, Meng F, Ginsburg D (2015) Massively parallel enzyme kinetics reveals the substrate recognition landscape of the metalloprotease ADAMTS13. Proc Natl Acad Sci U S A 112:9328–9333
16. Fowler DM, Araya CL, Gerard W, Fields S (2011) Enrich: software for analysis of protein function by enrichment and depletion of variants. Bioinformatics 27:3430–3431
17. Glenn TC (2011) Field guide to next-generation DNA sequencers. Mol Ecol Resour 11:759–769
18. Desch KC, Kretz C, Yee A, Gildersleeve R, Metzger K, Agrawal N, Cheng J, Ginsburg D (2015) Probing ADAMTS13 substrate specificity using phage display. PLoS One 10:e0122931
19. Scott JK, Smith GP (1990) Searching for peptide ligands with an epitope library. Science 249:386–390
20. Love MI, Huber W, and Anders S (2014). Moderated estimation of fold change and dispersion for RNA-seq data with DESeq2. Genome Biology 15:550
21. Langmead B, Salzberg SL (2012) Fast gapped-read alignment with Bowtie 2. Nat Methods 9:357–359
22. Araya CL, Fowler DM, Chen W, Muniez I, Kelly JW, Fields S (2012) A fundamental protein property, thermodynamic stability, revealed solely from large-scale measurements of protein function. Proc Natl Acad Sci U S A 109:16858–16863
23. Hochberg YBY (1995) Controlling the false discovery rate: a practical and powerful approach to multiple testing. J R Stat Soc Series B 57:289–300
24. Shendure J, Akey JM (2015) The origins, determinants, and consequences of human mutations. Science 349:1478–1483
25. Gnad F, Baucom A, Mukhyala K, Manning G, Zhang Z (2013) Assessment of computational methods for predicting the effects of missense mutations in human cancers. BMC Genomics 14(Suppl 3):S7

Part V

Detection of Matrix Metalloproteases

Chapter 12

Detection of Matrix Metalloproteinases by Zymography

Rajeev B. Tajhya, Rutvik S. Patel, and Christine Beeton

Abstract

Matrix metalloproteinases (MMPs) represent more than 20 zinc-containing endopeptidases that cleave internal peptide bonds, leading to protein degradation. They play a critical role in many physiological cell functions, including tissue remodeling, embryogenesis, and angiogenesis. They are also involved in the pathogenesis of a vast array of diseases, including but not limited to systemic inflammation, various cancers, and cardiovascular, neurological, and autoimmune diseases. Here, we describe gel zymography to detect MMPs in cell and tissue samples and in cell culture supernatants.

Key words Zymography, Zymogram, Protease, Detection, Semiquantitative

1 Introduction

Matrix metalloproteinases (MMPs) are zinc-containing endopeptidases responsible for extracellular matrix degradation via cleavage of internal peptide bonds of specific proteins. More than 20 MMPs have been identified. MMPs can be either membrane-bound or secreted; they can be segregated into six groups based on their substrate specificity and structure: collagenases, gelatinases, stromelysins, matrilysins, membrane type, and others.

MMPs are crucial to development and tissue repair and remodeling but are also important players in many disease processes. They have been suggested as either diagnosis tools or therapeutic targets in a broad range of conditions such as multiple sclerosis [1, 2], periodontal disease [3, 4], cancer [5], rheumatoid arthritis and osteoarthritis [6, 7], intestinal disease [8, 9], or cardiovascular diseases [10]. The accurate and reproducible detection of different MMPs is therefore critical for both basic and translational research.

A number of techniques have been developed to detect and identify MMPs in various samples. Enzyme linked immunosorbent assays (ELISA) and Western blots require the use of antibodies targeted to the MMP of interest. Here, we describe the procedure for gel zymography, which was first described in 1980 by Heussen

Charles A. Galea (ed.), *Matrix Metalloproteases: Methods and Protocols*, Methods in Molecular Biology, vol. 1579,
DOI 10.1007/978-1-4939-6863-3_12, © Springer Science+Business Media LLC 2017

and Dowdle [11], and continues to be extensively used to detect MMPs in many cell types and tissues and in most bodily fluids [2–4, 6].

2 Materials

Prepare all solutions using analytical grade reagents and ultrapure water. All solutions should be stored at room temperature in closed bottles unless otherwise specified and must be discarded after use following all waste disposal regulations in place.

2.1 Sample Preparation

1. Phosphate-buffered saline (PBS), pH 7.2: Add 800 mL water to a 1 L graduated cylinder. Weigh 8 g NaCl, 0.2 g KCl, 1.44 g Na_2HPO_4 (dibasic anhydrous) (*see* **Note 1**), and 0.24 g KH_2PO_4 (monobasic anhydrous) and add to the graduated cylinder. Mix and adjust pH with HCl. Add water to 1 L (*see* **Note 2**).
2. Protease inhibitor cocktail (100×): Add 16 mL water to a 20 mL graduated cylinder. Weigh 20 mg aprotinin, 4 mg leupeptin, 961 mg benzamidine and add to the graduated cylinder. Add water to 20 mL and mix. Aliquot into microcentrifuge tubes and store at −20 °C.
3. NP-40 lysis buffer, pH 7.5 (*see* **Note 3**): Add 80 mL water to a 100 mL graduated cylinder. Weigh 394 mg Tris–HCl and 584 mg NaCl, and add to the graduated cylinder. Add 1 mL Nonidet P-40 (NP-40) to the graduated cylinder and mix. Add water to 100 mL. Store at 4 °C. Immediately before use, add 10 μL of protease inhibitor cocktail (100×) (Subheading 2.1, **item 2**) to 1 mL of NP-40 lysis buffer.
4. Triton X-100 lysis buffer, pH 8.5 (*see* **Note 3**): Add 80 mL water to a 100 mL graduated cylinder. Weigh 315 mg Tris–HCl, 730 mg NaCl, 1 mL Triton X-100 (*see* **Note 4**) and add to the graduated cylinder. Add water to 100 mL and mix. Store at 4 °C. Immediately before use, add 10 μL of protease inhibitor cocktail (100×) to 1 mL of Triton X-100 lysis buffer.

2.2 Zymogram Gel

1. Separating gel buffer: Add 80 mL water to a 100 mL glass cylinder. Weigh 4.63 g ammediol•HCl and 0.02 g sodium azide and transfer to the cylinder. Mix and adjust pH to 8.96 with HCl. Add water to 100 mL. Store at 4 °C for up to a month.
2. Casein substrate solution (10×) (*see* **Note 5**): Add 100 mL 0.1 N NaOH to a glass beaker. Weigh 800 mg casein and transfer to the beaker. Heat the solution to 37 °C with occasional vortexing until full dissolution (*see* **Note 6**). Store at −20 °C for up to 6 months.

3. Gelatin substrate solution (10×) (*see* **Note 5**): Add 100 mL water to a glass beaker. Weigh 800 mg porcine skin gelatin. Heat in the microwave until the solution just boils (*see* **Note 7**). Swirl thoroughly to ensure homogeneous distribution (*see* **Note 8**). Store at 4 °C for 1 week or at −20 °C for up to 6 months.
4. Sucrose solution: Add 50 mL water to a 100 mL glass cylinder. Weigh 50 g sucrose and 0.02 g sodium azide and transfer to the cylinder. Add 30 μL toluene and make up to 100 mL with water. Store at 4 °C for up to a month.
5. Separating gel (*see* **Note 9**): Mix the components in the appropriate proportions (Table 1) for the choice of gel (*see* **Note 10**).
6. Stacking gel buffer: Add 80 mL water to a 100 mL glass cylinder. Weigh 3.51 g ammediol•HCl and 0.02 g sodium azide and transfer to the cylinder. Mix and adjust pH to 8.37 with HCl. Add water to 100 mL. Store at 4 °C for up to a month.
7. Stacking gel (*see* **Note 9**): Mix 142 μL acrylamide/bisacrylamide solution (37.5:1), 285 μL stacking buffer, 285 μL sucrose solution, 428 μL water, 14 μL 10% (weight/volume) ammonium persulfate, and 3 μL TEMED (*N,N,N′,N′*-tetramethylethylenediamine) (*see* **Note 11**).
8. 0.5 M Tris–HCl, pH 6.8: Add 400 mL water to a 1 L glass cylinder (*see* **Note 12**). Weigh 39.39 g Tris and transfer to the cylinder. Mix and adjust pH with HCl. Add water to 500 mL.
9. Sample buffer (2×): In a 10 mL glass cylinder, add 2.5 mL of 0.5 M Tris–HCl pH 6.8, 2.0 mL of glycerol, 10% (weight/

Table 1
Recipes for preparing zymogram separating gels with different percentages of acrylamide

	Final acrylamide concentration in gel			
Solution	**6%**	**7.5%**	**10%**	**12%**
Acrylamide/bisacrylamide solution 37.5:1	0.8 mL	1.0 mL	1.32 mL	1.61 mL
Separating gel buffer	1.0 mL	1.0 mL	1.0 mL	1.0 mL
10× casein or gelatin substrate solution	0.4 mL	0.4 mL	0.4 mL	0.4 mL
Sucrose solution	0.86 mL	0.86 mL	0.86 mL	0.86 mL
Water	0.94 mL	0.74 mL	0.42 mL	0.13 mL
10% (w/v) ammonium persulfate	14 μL	14 μL	14 μL	14 μL
TEMED[a]	1.5 μL	1.5 μL	1.5 μL	1.5 μL

The volumes are sufficient for a single mini-gel (9 cm × 6 cm × 0.75 cm)
[a]TEMED: *N,N,N′,N′*-tetramethylethylenediamine

volume) sodium dodecyl sulfate (SDS), and 0.1% (weight/volume) bromophenol blue. Add water to 10 mL.

10. Running buffer (10×): To prepare 1 L of running buffer, add 100 mL water to a 1 L graduated cylinder. Weigh 29 g of Tris Base, 144 g of glycine, and 10 g of SDS. Add all three to the graduated cylinder and add water to 1 L. Mix with a magnetic stir-bar until fully dissolved.

2.3 Protease Detection

1. Renaturing buffer (*see* **Note 13**): Add 50 mL of water to a 100 mL graduated cylinder. Add 2.5 mL of Triton X-100 to the cylinder (*see* **Note 4**). Add water to 100 mL and mix thoroughly. Do not store for more than a day.
2. Developing buffer (10×): To prepare 1 L of developing buffer, add 100 mL water to a 1 L graduated cylinder. Weigh 12.1 g Tris Base, 63 g Tris–HCl, 117 g NaCl, 7.4 g $CaCl_2$ and add all four to the cylinder. Add 2 mL Brij 35 to the cylinder (*see* **Note 14**). Add water to 1 L and mix thoroughly.
3. Coomassie Blue staining solution (*see* **Note 15**): Add 800 mL water to a graduated cylinder. Add 50 mL methanol and 100 mL acetic acid (*see* **Note 16**). Weigh 5 g Coomassie Blue R-250 and add to the graduated cylinder. Add water to 1 L.
4. Coomassie Blue destaining solution (*see* **Note 15**): Add 800 mL water to a 1 L beaker. Add 100 mL methanol and 50 mL acetic acid (*see* **Note 16**). Add water to 1 L.

3 Methods

Perform all procedures at room temperature unless otherwise specified and wear personal protective equipment.

3.1 Sample Preparation

3.1.1 Preparation of Cell Culture Supernatants

1. If the cells to be tested are adherent, plate in complete growth media and allow time to adhere (*see* **Note 17**).
2. Wash the cells three times with sterile PBS or serum-free culture media.
3. Incubate the cells at 37 °C with serum-free culture media (*see* **Note 18**) for the optimal duration dependent on the cells (*see* **Note 19**).
4. Collect the culture supernatants.
5. Proceed to measuring protein concentration (Subheading 3.2) or freeze at −80 °C until use (*see* **Note 20**).

3.1.2 Preparation of Cell Lysates

1. If the cells to be tested are adherent, plate in complete growth media and allow time to adhere (*see* **Note 21**).
2. Wash the cells three times with sterile PBS or culture media.

3. Incubate the cells at 37 °C with culture media for the optimal duration dependent on the cells (*see* **Note 19**).
4. Place a bottle of PBS and a bottle of NP-40 lysis buffer on ice for a minimum of 20 min to prepare ice-cold buffers.
5. Wash the cells twice with ice-cold PBS.
6. Add cold NP-40 lysis buffer at a volume of 2 mL per 150 mm dish.
7. If the cells to be tested are adherent, scrape them using a cell lifter.
8. Collect the cell lysate and incubate on ice for 15 min.
9. Centrifuge the cell lysate 16,000 × *g* for 20 min at 4 °C in a microcentrifuge.
10. Collect the supernatant.
11. Proceed to measuring protein concentration (Subheading 3.2) or freeze at −80 °C until use (*see* **Note 20**).

3.1.3 Preparation of Tissue Extracts—Technique 1: NP-40 Lysis Buffer

1. Place a bottle of NP-40 lysis buffer on ice for a minimum of 20 min to prepare ice-cold buffers.
2. Collect the tissue and process immediately.
3. Cut approximately 50 mg of tissue into small pieces and place in a 1.5 mL microcentrifuge tube.
4. Add 0.5 mL of cold NP-40 lysis buffer (*see* **Note 22**).
5. Homogenize the tissue with a pestle over ice.
6. Centrifuge the homogenate at 16,000 × *g* for 20 min at 4 °C in a microcentrifuge.
7. Collect the supernatant.
8. Proceed to measuring protein concentration (Subheading 3.2) or freeze at −80 °C until use (*see* **Note 20**).

3.1.4 Preparation of Tissue Extracts—Technique 2: Pulverization

1. Collect the tissue and immediately freeze with dry ice.
2. Pulverize the frozen tissue with a micro-dismembrator.
3. Weigh 50 mg of pulverized tissue and resuspend in 0.15 mL Triton X-100 lysis buffer under gentle rotation at 4 °C.
4. Centrifuge the homogenate at 16,000 × *g* for 20 min at 4 °C in a microcentrifuge.
5. Collect the supernatant.
6. Proceed to measuring protein concentration (Subheading 3.2) or freeze at −80 °C until use (*see* **Note 20**).

3.2 Measuring Protein Concentration

1. Remove the bovine serum albumin (BSA) standard reagent (*see* **Note 23**) from 4 °C storage and let it warm to room temperature (*see* **Note 24**). Invert a few times before use.

2. Prepare serial dilutions of 2 mg/mL BSA to create a calibration curve (linear) in the range of 200–1,000 μg/mL.
3. Pipet an aliquot of each standard or sample into separate clean cuvettes filled with 500 μL dH_2O and 500 μL Bradford reagent and vortex. Prepare a blank sample in a clean cuvette by mixing 500 μL dH_2O and 500 μL Bradford reagent.
4. Incubate at room temperature (*see* **Note 24**) for at least 5 min. Do not incubate longer than 1 h (*see* **Note 25**).
5. Warm up the spectrophotometer for 10 min. Zero the instrument with the blank sample. Measure the absorbance of the standards and the samples at 280 nm.
6. Create a standard curve by plotting the absorbance at 280 nm on the *y*-axis and concentration of the serial standard dilutions on the *x*-axis. Determine the slope of the standard curve.
7. Using the slope and absorbance at 280 nm calculate the concentration of each sample (*see* **Note 26**).

3.3 Gel Preparation

1. Assemble the electrophoresis plates (*see* **Note 27**).
2. Prepare the separating gel and immediately pour (*see* **Note 28**) between the electrophoresis plates up to 2 cm from the top.
3. Overlay with water-saturated *n*-butanol and allow the gel to polymerize for approximately 30 min at room temperature (*see* **Note 29**).
4. Prepare the stacking gel (*see* **Note 28**).
5. Decant the *n*-butanol from the gel and gently rinse with water (*see* **Note 30**).
6. Pour the stacking gel on the top of the separating gel.
7. Immediately insert the well comb (*see* **Note 31**).
8. Allow the stacking gel to polymerize for approximately 20 min at room temperature.

3.4 Loading and Electrophoresis

1. Remove the well comb from the gel and rinse gently with 1× running buffer.
2. Install the zymography gel in the electrophoresis setup.
3. Fill the inside chamber with 1× running buffer and the outside chamber with 500 mL 1× running buffer (*see* **Note 32**).
4. Place 10 μL of 2× sample buffer on a Parafilm membrane and add 10 μL of sample. Pipet a few times to mix and load the mixture into each well using gel loading pipet tips (*see* **Note 33**).
5. Load a protein molecular weight standard in at least one well per gel.
6. Load appropriate recombinant MMP or a control known to contain the MMP of interest to identify the sample MMPs (optional) (*see* **Note 34**).

7. Run the gel at 125 V for 90 min or until the loading dye reaches the bottom of the gel.

3.5 Gel Renaturation

1. Place 100 mL of 1× renaturing buffer in a container large enough to fit the gel.
2. Open the gel plates (or cassette in the case of a commercial precast gel).
3. Make a diagonal cut in a corner of the gel to mark its orientation.
4. Remove the gel from the plate or cassette (*see* **Note 35**).
5. Place the gel in the container containing the renaturing buffer.
6. Incubate for 30 min at room temperature with gentle agitation on a bench rocker.

3.6 Gel Development

1. Remove the renaturing buffer by gently pouring it into a waste container (*see* **Note 36**).
2. Add 100 mL of 1× developing buffer to the container with the gel.
3. Incubate for 30 min at room temperature with gentle agitation on a bench rocker.
4. Remove the developing buffer by gently pouring it into a waste container.
5. Add 100 mL of 1× developing buffer to the container with the gel. Close the lid tightly.
6. Incubate overnight (16–18 h) at 37 °C (*see* **Note 37**).

3.7 Gel Staining and Destaining

1. Remove the developing buffer by gently pouring it into a waste container.
2. Add 100 mL of water to the container containing the gel.
3. Incubate for 5 min at room temperature with gentle agitation on a bench rocker.
4. Remove the water by gently pouring it into a waste container.
5. Repeat **steps 2–4** twice.
6. Scan the gel to record the exact position of the protein standard bands as they will become difficult to see after gel staining and destaining.

3.7.1 Staining with Coomassie Blue R-250

1. Add 20 mL of Coomassie blue staining solution (*see* **Note 15**) to the container containing the gel.
2. Incubate for 1 h at room temperature with gentle agitation on a bench rocker (*see* **Note 38**).
3. Remove the staining buffer by gently pouring it into a waste container.

4. Add 100 mL of 1× destaining solution to the container containing the gel.
5. Incubate at room temperature with gentle agitation on a bench rocker until areas of proteolytic activity are clearly visible (*see* **Note 39**).

3.7.2 Staining with Colloidal Coomassie Blue (G-250)

1. Add 20 mL of ready-to-use commercial buffer with colloidal Coomassie Blue to the container with the gel.
2. Incubate for 1 h at room temperature with gentle agitation on a bench rocker (*see* **Note 38**).
3. Remove the staining buffer by gently pouring it into a waste container.
4. Add 100 mL of water to the container with the gel.
5. Incubate for 1 h at room temperature with gentle agitation on a bench rocker (*see* **Note 40**).

3.8 Data Acquisition and Analysis

1. Remove the gel from the water, place in a plastic sheet protector, and gently remove air bubbles.
2. Scan the gel with a digital scanner (*see* **Note 41**).
3. The integrative intensity of the MMP bands can be quantified with ImageJ (freely available from the National Institutes of Health) or similar software.

4 Notes

1. The 1.44 g of Na_2HPO_4 (dibasic anhydrous) can be replaced by 1.81 g $Na_2HPO_4{\bullet}2H_2O$ (dibasic dihydrate) or by 2.72 g $Na_2HPO_4{\bullet}7H_2O$ (dibasic septahydrate) in the preparation of PBS.
2. If PBS is to be kept more than a day, it should be sterilized to prevent bacterial growth by either autoclaving or sterile filtration (0.22 μm filter pores).
3. Choice of lysis buffer for protein extraction will depend on the sample and also on the target MMP. For example, MMP-23 can only be efficiently extracted using the nonionic detergent Triton X-100 [12].
4. Triton X-100 is very viscous. When pipetting Triton X-100, only place the very tip of the pipet into the liquid to avoid coating the outside of the pipet with the detergent. Apply suction very slowly to allow the liquid to reach the correct volume in the pipette. If Triton X-100 is visible on the outside of the pipet, remove it with a clean tissue. Empty the pipet very slowly into a receiving container containing water and wait until the pipet has completely emptied (can take a few minutes).

5. The choice of substrate in the gel will depend on the MMP of interest. For example, MMP-2 and MMP-9 can be detected using gelatin gels, as can MMP-1, MMP-8, and MMP-13 although gelatin is not their preferred substrate. MMP-1 and MMP-13 are best detected on collagen gels while casein is the preferred substrate for MMP-11 and also allows for the detection of MMP-1, MMP-3, MMP-7, MMP-12, and MMP-13 [13].
6. Full dissolution of casein is very important to ensure homogeneity of the separating gel during zymography. Lack of homogeneity will prevent data interpretation.
7. Do not allow the solution to boil over as that will change its concentration.
8. Full dissolution of the gelatin is very important to ensure homogeneity of the separating gel during zymography. Lack of homogeneity will prevent data interpretation.
9. Precast gels can be purchased from various vendors. These gels are provided in sealed pouches and have strict expiration dates; expired gels will not provide expected results as the substrate will have started degrading in a heterogeneous manner.
10. The correct acrylamide percentage to use will depend on the molecular weight of the MMP of interest. Proteins with high molecular masses resolve better on low percentage gels, whereas proteins with low molecular masses resolve better on high percentage gels. The most commonly used gels for zymography are 10% gels as they resolve proteins between 20 and 100 kDa.
11. The volumes given here are sufficient for a single mini-gel (9 cm × 6 cm × 0.75 cm).
12. Placing water at the bottom of the cylinder will help to rapidly dissolve the Tris. In addition, warming the water to 37 °C will hasten the dissolution of Tris. However, the solution should be brought to room temperature before adjusting the pH.
13. The renaturing buffer should be prepared fresh on the day it will be used.
14. Brij 35 is very viscous. When pipetting Brij 35, only place the very tip of the pipet into the liquid to avoid coating the outside of the pipet with the detergent. Apply suction very slowly to allow the liquid to reach the correct volume in the pipette. If Brij 35 is visible on the outside of the pipet, remove it with a clean tissue. Empty the pipet very slowly into the receiving container containing water and wait until the pipet has completely emptied (can take a few minutes).
15. Coomassie Blue R-250 can be replaced by a ready-to-use commercial solution containing colloidal Coomassie blue G-250.

An advantage of the colloidal Coomassie blue is that it only requires water (no methanol or acetic acid) for destaining.

16. Always add concentrated acid to a large volume of water; never add water to concentrated acid as this would result in a violent exothermic reaction.
17. The number of cells to use for preparing cell culture supernatants for zymography will vary depending on the type of cells and must be tested prior to assaying a new cell system. For example, MMP-2 was detectable in the culture supernatants of human fibroblast-like synoviocytes isolated from patients with rheumatoid arthritis when cells were plated at a density of 50,000 cells per well with 0.3 mL culture medium per well in a 24-well plate [14].
18. Serum contains MMPs (Fig. 1); it is therefore crucial to remove all serum for the cell culture and then maintain the cells in serum-free media during MMP production. Indeed, any serum present in the culture may obscure effects due to differential treatment of the cells (Fig. 2) [3, 6, 8, 14–16].
19. The duration of incubation will differ depending on the cell type and must be tested prior to assaying a new cell system. For example, when assessing MMP-2 levels in the culture supernatants of human fibroblast-like synoviocytes isolated from patients with rheumatoid arthritis or from rats in an animal model of rheumatoid arthritis, a 24 h incubation gave optimal results [2, 3, 14, 16]. In contrast, rabbit corneal fibroblasts were cultured for 72 h before collecting the supernatant [6, 17] whereas a 1 h incubation was sufficient to detect pro-MMP-9 in the culture supernatants of human neutrophils [2, 18].
20. It is crucial that MMPs are not denatured to remain functional. If the samples are to be frozen, they should be frozen immediately upon collection. In addition, every effort should be made to avoid repeated cycles of freezing and thawing of the samples.

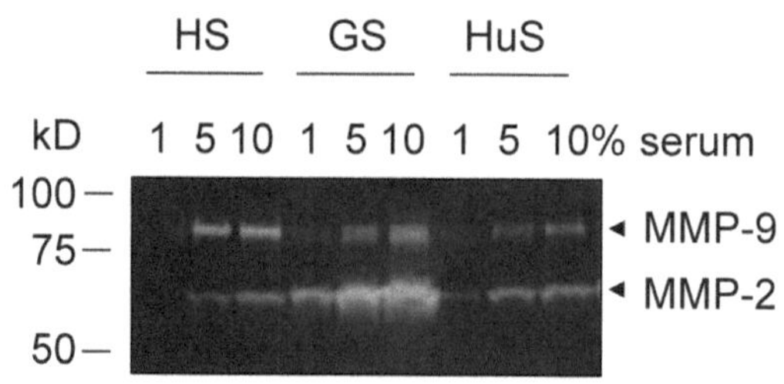

Fig. 1 Gelatin zymography gel loaded with different dilutions of horse serum (HS), goat serum (GS), and human serum (HuS) (1%, 5%, 10%). Stronger bands of MMP-2 (~65 kDa) and MMP-9 (~85 kDa) are visible with increasing concentrations of the different sera

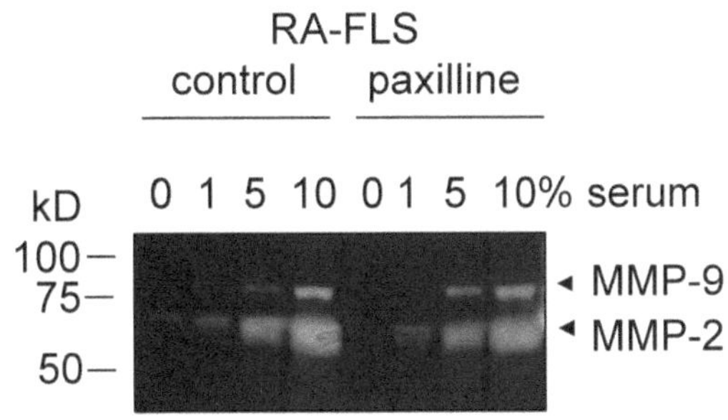

Fig. 2 Gelatin zymography gel comparing MMP-2 and MMP-9 production in the supernatant of untreated fibroblast-like synoviocytes isolated from a patient with rheumatoid arthritis (RA-FLS; control) and paxilline-treated RA-FLS incubated in the presence of 0%, 1%, 5%, or 10% of fetal bovine serum. Paxilline-treated RA-FLS exhibit a decrease in MMP-2 production when incubated without serum. This effect is masked by incubation with 1%, 5%, 10% serum

21. The number of cells to use for preparing cell culture supernatants for zymography will vary depending on the type of cells and must be tested prior to assaying a new cell system. For example, MMP-2 was detectable in the culture supernatants of human fibroblast-like synoviocytes isolated from patients with rheumatoid arthritis when cells were plated at a density of 50,000 cells per well with 0.3 mL culture medium per well in a 24-well plate [14].
22. The volume of lysis buffer to add will depend on the type of tissue to be homogenized as protein extractability varies with specimens.
23. While BSA is commonly used, any known protein standard can be used as a control.
24. Cold solutions can condense atmospheric moisture on the outside of the cuvette and scatter light giving erroneous readings.
25. Absorbance of protein-dye conjugate increases over time and affects the accuracy of protein concentration measurements.
26. If cell supernatants or tissue extracts contain low levels of gelatinases, a phase extraction with Triton X-114 is recommended [19].
27. The electrophoresis plates must be completely clean and dry at this time or the gel could incompletely polymerize or polymerize in a heterogeneous manner.
28. Once the polymerization agent (TEMED) is added to the gel, immediately mix and pour as the gel will start polymerizing rapidly. If the gel polymerizes too fast, reduce the ambient temperature as this will delay polymerization.
29. Ensure that the electrophoresis system is perfectly horizontal during gel pouring and polymerization.

30. It is crucial to remove all *n*-butanol to ensure full contact between the separating and stacking gels.
31. Place the comb carefully so as to not trap air bubbles.
32. Ensure that there is no leaking of the inside chamber buffer. If the inside chamber buffer leaks before sample loading, empty the gel box and unlock the gel. Wet the chamber gasket with running buffer and install the gel.
33. Volumes given here are for 12-well gels. These volumes can and should be increased when loading a gel with fewer wells.
34. Using recombinant MMP as a control can be very useful for semiquantification of MMPs in the samples. In addition, although MMPs are highly conserved between species in terms of molecular weight and protease activity, a positive control containing the MMP of interest in the species of choice would be beneficial.
35. Removing the gel from the plate or cassette can be difficult as the gel is very fragile and easy to damage. It is safer to slide the gel off one side of the cassette rather than lifting it.
36. When removing the buffer after incubating the gel, take precautions not to damage the gel. Wearing gloves, use two fingers to gently hold the gel to the bottom of the container while pouring the liquid out from one of the corners of the container.
37. We recommend an overnight incubation in developing buffer. If the bands are too strong the developing time can be reduced to 4 h. In contrast, if the bands are very faint, gels can be incubated in the developing buffer for up to 48 h.
38. The gel should be stained in Coomassie Blue staining solution for 1 h or until the gel is uniformly dark blue.
39. If no zones of digestion are evident, use a positive control (e.g., recombinant MMP-2 or MMP-9 for gelatin gels and MMP-1 for casein gels). If these controls are visible, the test samples may be devoid of proteases specific to the chosen substrate or the MMP concentration may be lower than expected, in which case increase the incubation time in developing buffer or concentrate the test samples. If the positive controls are not visible, check that (1) the samples were not boiled prior to gel loading as this denatures the proteins, (2) the sample loading buffer does not contain anything that inhibits the proteases (e.g., EDTA), and (3) the renaturing buffer contains sufficient amounts of Triton X-100.
40. Destaining the gel stained with a ready-to-use commercial solution containing colloidal Coomassie Blue for an hour may not be optimal in all situations. Further destaining can be achieved with a second incubation of 1 h in fresh water. The

gel can be scanned multiple times during the destaining process to catch the best contrast in the bands.

41. If the background contains blotches or streaks, this could indicate an uneven distribution of the substrate protein in the gel. Ensure that the gelatin of casein solution used to prepare the separating gel contains no undissolved materials.

Acknowledgment

This work was supported in part by funding from Baylor College of Medicine and National Institutes of Health grants AR059838 and NS073712 to C.B.

References

1. Baranger K, Rivera S, Liechti FD, Grandgirard D, Bigas J, Seco J, Tarrago T, Leib SL, Khrestchatisky M (2014) Endogenous and synthetic MMP inhibitors in CNS physiopathology. Prog Brain Res 214:313–351
2. Fainardi E, Castellazzi M, Bellini T, Manfrinato MC, Baldi E, Casetta I, Paolino E, Granieri E, Dallocchio F (2006) Cerebrospinal fluid and serum levels and intrathecal production of active matrix metalloproteinase-9 (MMP-9) as markers of disease activity in patients with multiple sclerosis. Mult Scler 12:294–301
3. Baeza M, Garrido M, Hernández-Ríos P, Dezerega A, García-Sesnich J, Strauss F, Aitken JP, Lesaffre E, Vanbelle S, Gamonal J, Brignardello-Petersen R, Tervahartiala T, Sorsa T, Hernández M (2015) Diagnostic accuracy for apical and chronic periodontitis biomarkers in gingival crevicular fluid: an exploratory study. J Clin Periodontol 43:34–45
4. Thomadaki K, Bosch J, Oppenheim F, Helmerhorst E (2013) The diagnostic potential of salivary protease activities in periodontal health and disease. Oral Dis 19:781–788
5. Cathcart J, Pulkoski-Gross A, Cao J (2015) Targeting matrix metalloproteinases in cancer: bringing new life to old ideas. Genes Dis 2:26–34
6. Cuéllar VG, Cuéllar JM, Kirsch T, Strauss EJ (2015) Correlation of synovial fluid biomarkers with cartilage pathology and associated outcomes in knee arthroscopy. Arthroscopy 32:475–485
7. Murphy G, Nagase H (2008) Reappraising metalloproteinases in rheumatoid arthritis and osteoarthritis: destruction or repair? Nat Clin Pract Rheumatol 4:128–135
8. Chang J, Wehner S, Schäfer N, Sioutis M, Bortscher S, Hirner A, Kalff JC, Bauer AJ, Overhaus M (2012) Iatrogenic extracellular matrix disruption as a local trigger for postoperative ileus. J Surg Res 178:632–639
9. Medina C, Radomski MW (2006) Role of matrix metalloproteinases in intestinal inflammation. J Pharmacol Exp Ther 318:933–938
10. Hopps E, Caimi G (2015) Matrix metalloproteases as a pharmacological target in cardiovascular diseases. Eur Rev Med Pharmacol Sci 19:2583–2589
11. Heussen C, Dowdle EB (1980) Electrophoretic analysis of plasminogen activators in polyacrylamide gels containing sodium dodecyl sulfate and copolymerized substrates. Anal Biochem 102:196–202
12. Pei D, Kang T, Qi H (2000) Cysteine array matrix metalloproteinase (CA-MMP)/MMP-23 is a type II transmembrane matrix metalloproteinase regulated by a single cleavage for both secretion and activation. J Biol Chem 275(43):33988–33997
13. Snoek-van Beurden PA, Von den Hoff JW (2005) Zymographic techniques for the analysis of matrix metalloproteinases and their inhibitors. Biotechniques 37:73–83
14. Hu X, Laragione T, Sun L, Koshy S, Jones KR, Ismailov II, Yotnda P, Horrigan FT, Gulko PS, Beeton C (2012) KCa1.1 potassium channels regulate key pro-inflammatory and invasive properties of fibroblast-like synoviocytes in rheumatoid arthritis. J Biol Chem 287: 4014–4022
15. Hu X, Beeton C (2010) Detection of functional matrix metalloproteinases by zymography. J Vis Exp 45:pii 2445
16. Tanner MR, Hu X, Huq R, Tajhya RB, Sun L, Khan FS, Laragione T, Horrigan FT, Gulko

PS, Beeton C (2015) KCa1.1 inhibition attenuates fibroblast-like synoviocyte invasiveness and ameliorates rat models of rheumatoid arthritis. Arthritis Rheumatol 67:96–106

17. Pan H, Chen J, Xu J, Chen M, Ma R (2009) Antifibrotic effect by activation of peroxisome proliferator-activated receptor-gamma in corneal fibroblasts. Mol Vis 15:2279–2286
18. Puente J, Jaque M, Carrasco C, Cruz C, Valenzuela M, Wolf M, Mosnaim A (2008) Triptan drugs, natural killer cell cytotoxicity, and neutrophils pro-matrix metalloproteinase-9 secretion. Headache 48:1482–1489
19. Toth M, Sohail A, Fridman R (2012) Assessment of gelatinases (MMP-2 and MMP-9) by gelatin zymography. Methods Mol Biol 878:121–135

Chapter 13

Imaging Matrix Metalloproteases in Spontaneous Colon Tumors: Validation by Correlation with Histopathology

Harvey Hensley, Harry S. Cooper, Wen-Chi L. Chang, and Margie L. Clapper

Abstract

The use of fluorescent probes in conjunction with white-light colonoscopy is a promising strategy for improving the detection of precancerous colorectal lesions, in particular flat (sessile) lesions that do not protrude into the lumen of the colon. We describe a method for determining the sensitivity and specificity of an enzymatically activated near-infrared probe (MMPSense680) for the detection of colon lesions in a mouse model ($APC^{+/Min\text{-}FCCC}$) of spontaneous colorectal cancer. Fluorescence intensity correlates directly with the activity of matrix metalloproteinases (MMPs). Overexpression of MMPs is an early event in the development of colorectal lesions. Although the probe employed serves as a reporter of the activity of MMPs, our method can be applied to any fluorescent probe that targets an early molecular event in the development of colorectal tumors.

Key words Colorectal cancer, Mice, Min mice, Imaging, Fluorescence imaging, Matrix metalloproteinases, Early cancer detection, Colonoscopy, Activated imaging probes, Near-infrared imaging

1 Introduction

Colonoscopy, the standard procedure for the detection of lesions in the colon, employs broad-spectrum visible light ("white light") for illumination, with lesions identified based on morphological features. Although effective, it is estimated that this method fails to detect up to 20% of lesions [1, 2]. Flat, sessile lesions are particularly difficult to detect because they do not protrude from the colon wall. Efforts are underway to develop optical methods for detecting these lesions more reliably, including the use of fluorescent probes [3]. Fluorescent probes produce optical contrast that reveals regions of biological activity that are not distinguishable from adjacent normal tissue on gross examination. Furthermore, fluorescence imaging can be employed simultaneously with white light imaging, using a suitably modified endoscope. In a preclinical setting and following intravenous injection, the distribution of

Charles A. Galea (ed.), *Matrix Metalloproteases: Methods and Protocols*, Methods in Molecular Biology, vol. 1579, DOI 10.1007/978-1-4939-6863-3_13, © Springer Science+Business Media LLC 2017

fluorescent probes in experimental animals can be determined by in vivo epifluorescence, fluorescence endoscopy, fluorescence molecular tomography, or ex vivo examination at necropsy [4].

A unique animal model ($Apc^{+/Min\text{-}FCCC}$ mouse) has been established by this group that spontaneously develops multiple colorectal adenomas (3.7 ± 0.3, mean ± SEM) [5]. Due to the small size of the lesions and their location deep within the abdomen, their detection is not feasible using the external, whole body imaging protocols that are available currently. While fluorescence endoscopy is possible, only the distal portion of the colon can be interrogated using currently available Hopkins telescopes. In addition, the presence of a large lesion in the colon often impedes full insertion of the endoscope, preventing the detection of more caudal lesions. In contrast, a number of instruments are commercially available to interrogate the entire colon and acquire accurate ex vivo measurements. Such experiments permit detailed, high resolution imaging and provide valuable proof-of-principle information. Crucial to the evaluation of any fluorescent imaging probe is validation of the resulting data. This process entails correlating regions of probe retention, as measured by fluorescence detection, with histopathologically confirmed biological features.

Numerous fluorescent probe-based strategies have been proposed and investigated to improve the early detection of colon lesions [6]. Among these, matrix metalloproteinases (MMPs), calcium-dependent zinc-containing endopeptidases, represent attractive targets for fluorescence imaging. MMPs play multiple important roles in the transformation of the normal human colorectal mucosa to a neoplastic state, in both the initial progression of adenomatous polyps to invasive colorectal cancers, and their eventual transition to metastatic disease [7, 8]. Of particular interest, overexpression of MMP-7 is found in 85% of colorectal adenocarcinomas [9], and occurs as a result of mutation of the adenomatous polyposis coli (APC) gene; the "gatekeeper" of sporadic colorectal carcinogenesis [10]. Among the probes available for detecting MMPs is MMPSense680 (PerkinElmer, Hopkinton, MA), a bioactivatable probe that consists of a number of fluorophores attached in close proximity to an MMP-cleavable backbone. In the uncleaved state, the fluorophores quench each other, rendering the probe optically silent. In the presence of specific MMPs, the backbone is cleaved releasing the fluorophores.

Rigorous validation of results obtained using probes can be challenging. This is accomplished by: (1) careful identification of regions of fluorescence where the intensity is above that of the background mucosa; (2) correlating areas positive for fluorescent signal with the presence of gross lesions identified at necropsy; and most importantly, (3) histopathological confirmation of areas that are both neoplastic and positive for fluorescent signal. Pathological analyses are performed on the entire colon, following tissue fixation, processing, embedding, and sectioning, while maintaining the location of each tissue sample as it relates to the colon structure.

2 Materials

2.1 Animal Treatment, Imaging, and Tissue Processing

1. Apc+/Min-FCCC mice, or other animal model that develops colorectal tumors.
2. Fluorescence imaging system (IVIS Spectrum or a similar system with an appropriate illumination system, filters, and sensitive, thermo-electrically cooled CCD camera).
3. Image processing system (computer and software—Matlab with image processing toolbox).
4. Plastic sheet with printed ruler (made by copying a plastic ruler onto transparency sheets using a standard photo-copy machine).
5. MMP probe (MMP680) (PerkinELmer, Hopkinton, MA).
6. PBS: Isotonic phosphate buffered saline.
7. Insulin syringe with 30-gauge needle (Kendall Monoject 8881600700) (Tyco Healthcare Group).
8. 10% normal buffered formalin.
9. Standard supplies, reagents, and equipment for paraffin embedding and tissue sectioning.
10. Hematoxylin and eosin stain.
11. Dissecting tools (forceps, Metzenbaum scissors, Mayo scissors, scalpels).
12. 70% Ethanol in a spray bottle.
13. Histowrap.
14. 50 ml conical centrifuge tubes or other container for fixing individual colons.
15. Tissue cassettes.
16. Microscope slides.

2.2 Immunohistochemical Validation of MMP 7 and MMP 9

1. Rat anti-human MMP-7 (clone 338; Vanderbilt University).
2. Goat anti-rat IgG (Vector Laboratories).
3. Rabbit anti-MMP-9 (catalog no. 38898; Abcam).
4. Goat anti-rabbit IgG (EnVision Kit; Dako).
5. Cell conditioning 1 (CC1) (Ventana Medical Systems).
6. Ventana XT automated immunostainer (Ventana Medical Systems).

2.3 Probes

Although this review focuses on our work using MMPsense680 as probe, the procedures described can be used to study any fluorescent probe in the mouse colon, or any similar organ in small animals.

We have imaged a variety of probes with fluorophores in the red or near-infrared spectrum including MMPSense680, ProSense680, ProSense750, IntegriSense680, and IntegriSense750 (PerkinElmer, Hopkinton, MA), as well as PSVUE794 (Molecular Targeted Technologies, Malvern, PA) using the following methodology.

2.4 IVIS Imaging System

Images are captured on an IVIS Spectrum (PerkinElmer). This system consists of a highly sensitive, cooled (−80 °C) CCD camera, and a versatile filter set for imaging fluorophores. The system also contains a set of LEDs for white light illumination, permitting the acquisition of both fluorescent and conventional (black and white) images, and displays the fluorescence intensity as an overlay on the white light image. Exposure time, binning, fstop, light illumination level, imaging field of view, focal point, and excitation and emission filters can be adjusted to optimize image acquisition.

2.5 Image Processing

IVIS images are processed with routines written in the Matlab (Mathworks) programming environment with the Image Processing Toolkit. A large number of free and commercial image processing software programs are available. However, customized routines are necessary to implement a convenient method for precisely pairing each fluorescent image with a corresponding paraffin embedded section.

3 Methods

3.1 Injection of MMP Probe

To maximize the use of the probe and ensure the ethical treatment of animals, endoscopy, MRI, CT, or other noninvasive imaging methods should be employed prior to injection to confirm the tumor-bearing status of the experimental animals. If at least one colon lesion has been identified, inject the mouse intravenously (via either the tail vein or retro-orbital sinus) with an appropriate amount of the fluorescent imaging agent (~2 nM for MMPSense680) in 100–150 μl of isotonic PBS (*see* **Note 1**). The inoculation time in vivo will differ depending upon the pharmacokinetic properties (plasma half-life) of the probe to be imaged. Time course analyses revealed 66 h post-injection as the optimal time to image MMPSense680 ex vivo in C57BL/6J $Apc^{+/Min\text{-}FCCC}$ mice. At the predetermined time point, euthanize the mice according to the American Veterinary Medical Association Guidelines for Euthanasia (slow inhalation of CO_2, followed by confirmation of death).

3.2 Preparation of the Mouse Colon for Ex Vivo Imaging

1. Immediately following euthanasia, place the mouse in a supine position and dampen its abdominal fur with 70% ethanol. Using steel dissecting scissors, cut a longitudinal slit through the ventral skin starting near the anus, moving through the pelvis, and finishing at the sternum to expose the colon in its entirety. Care should be taken not to cut through the colon, any internal organs, or major blood vessels.
2. Remove the small intestine (anterior to the colon) from the abdominal cavity to reveal the colon as a long tubular structure, approximately 3 mm in diameter, that extends from the anus to the mouse's left kidney, and then runs transverse across the body to connect to the cecum, a large, sac-like structure.
3. Insert Metzenbaum scissors (large), closed, into the pelvis area, and open scissors to crack through the pelvic bone. Remove the Metzenbaum scissors.
4. Gently lift the colon with forceps, and with the smaller Mayo scissors, dissect it free by cutting any attached mesenteries or other connective tissues that are securing it to other portions of the colon or other organs within the peritoneum.
5. Make a further cut between the proximal end of the colon and the cecum, completely separating the two structures.
6. With the Mayo scissors, cut the distal colon as close as possible to the anus to separate the distal section of the colon from the body cavity. This will permit the separation of the distal end of the colon from the base of the tail and surrounding skin.
7. Once the colon has been completely cut free, remove it from the abdominal cavity, place the entire colon in a weigh boat containing 1× PBS, and rinse. The colon must be handled with forceps at all times after its removal from the abdominal cavity.
8. Place the colon on a paper towel and make a longitudinal incision along its entire length with fine dissecting (Mayo) scissors.
9. Return the colon to the weigh boat and vigorously wash it in 1× PBS to remove all fecal matter.
10. Lay the colon flat (luminal side up) on a preprinted plastic ruler made by copying a plastic ruler onto transparency sheets using a standard copy machine. Place the distal end of the colon to the left or right of the origin of the ruler, with the gradations and numbers visible (*see* **Note 2**). Write the mouse ID on the plastic sheet with a black permanent marker (*see* **Note 3**), at a peripheral location that will not mask the image.

3.3 Ex Vivo Imaging

1. Place the transparent plastic sheet containing the excised colon in the imaging system on a black plastic mat (*see* **Note 4**). Image settings for our work [4] employing the IVIS Spectrum

system were: excitation wavelength = 640 nm, detection wavelength = 700 nm, exposure time = 5 s, F-stop = 2, binning = 4 ("small" setting), field of view = 13.6 cm (setting "C"). (*see* **Note 5**). The "small" binning section corresponds to a binning factor of 4 and producs images with 230,400 pixels (480 × 480), making each pixel (28 μm)^2. For higher resolution images , binning factors of 1 can be used to successfully create images with pixels of (7 μm^2), with appropriate increases in image acquisition time applied. Adjust the camera binning setting for the white light photographic image (acquired immediately prior to the acquisition of the fluorescence image in the IVIS Spectrum) to be the same as that for the fluorescence image (*see* **Note 6**).

2. Place the colon, while attached to the transparent plastic sheet, in a 50 ml conical centrifuge tube with 10% normal buffered formalin for fixation at 4 °C overnight.

3.4 Tissue Processing and Scoring of Images

3.4.1 Preparation of the Colons for Embedding and Documentation of the Position of Each Tissue Segment

1. Following fixation overnight on the transparent plastic sheet, process the tissue for paraffin embedding under a fume hood.
2. Using the white-light image as a guide, cut the entire colon at 2 mm increments (*see* **Note 7**).
3. Assign each piece a consecutive identifier and place it in a unique tissue cassette after wrapping it in histowrap dampened with formalin.
4. Record the precise location of each excised piece relative to the ruler markings by drawing a boxed area on a hard copy of the fluorescence image/white light image overlay (Fig. 1) that corresponds to the position of each excised piece of colon on the ruler.
5. Embed all tissues in paraffin, cut 5 μm sections and stain them with hematoxylin and eosin for histopathological review.

3.4.2 Identification of Tissue Sections with Fluorescence Image Features and Image Display

1. Using the locations marked on the hard copy printout as a guide (see **step 4** of Subheading 3.4.1), identify the position of each 2 mm segment of the colon on both white light and fluorescence/white light image overlays as displayed on a computer screen.
2. Mark a rectangular region of interest on the images as displayed on the computer with an image processing software program.
3. Since up to 60 separate regions of interest (ROIs) must be placed on the images to cover the entire colon, we have found it convenient to write routines to help do this using the MATLAB Imaging Processing Toolbox (MathWorks, Inc., Natick, MA). 30–60 ROIs (2 mm in width; the number depending on the length of the particular colon) are placed on

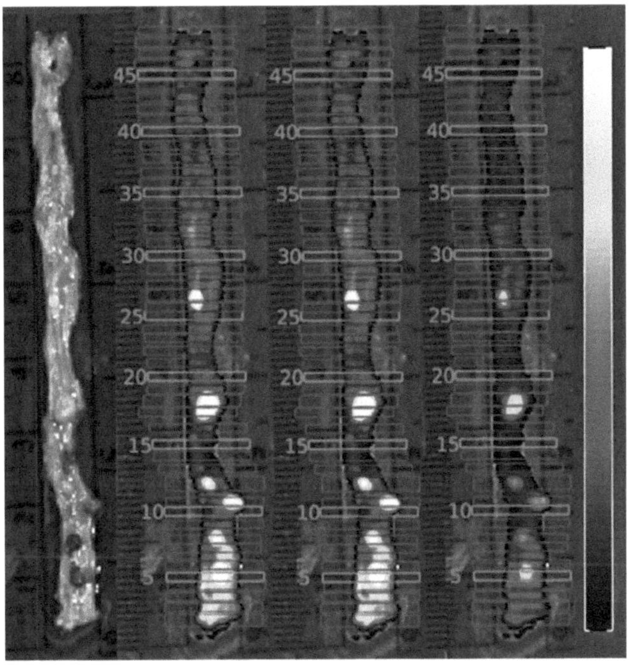

Fig. 1 Images of MMPSense680 fluorescence within the mouse colon, coded according to intensity with a "hot" colormap (far right). Each image represents the same colon, with the fluorescence intensity scaled to display a different color range. In the fluorescence overlay image on the left, a distinct region of strong signal intensity can be discerned (*section 47*), relative to local background, while regions of the distal colon exhibit more intense signal (*sections 1–8*), but display no discernable contrast. After adjusting the contrast (see the fluorescence overlay image at the far right), a discrete lesion becomes apparent in *section 5*. The grayscale white-light image, acquired immediately preceding the acquisition of the fluorescence images, is shown on the far left. Note the presence of other lesions (*sections 11*, *13*, *18*, and *26–27*) visible on all fluorescence overlay images

the image (Fig. 1). The ROI positions are then adjusted individually according to the precise position of each segment on the ruler, which were recorded when the colons were cut into segments for embedding into paraffin.

4. Display the fluorescence images as overlays on the white light images, with the white light intensities set to a grayscale colortable, while the fluorescence image is set to a different color lookup table. We employed a "hot colortable" for fluorescence intensities (Fig. 1), with the most intense signals corresponding to a display color of white. The least intense is set to a minimum indicative of the signal background level in the colon and presented as dark red, while the intermediate values display as a range of shades of red to yellow (Fig. 1). Images were acquired with a 16 bit CCD camera, producing a total of 65,536 intensity levels. In order to take advantage of this large

dynamic range, and given the large range of fluorescence intensities in the colon, display the fluorescence intensities in at least two different color mappings (Fig. 1), so that both low and high intensity regions can be visualized. This method is preferable to the alternative of a nonlinear (for example logarithmic) mapping of fluorescence intensity to color (*see* **Note 8**).

5. Estimate lesion size by measuring the number of pixels that span the lesion on either the white light or fluorescence image, and converting the region to millimeters using the image acquisition field of view. Both images are useful for determining the lesion edges; however, they tend to be more distinct in the fluorescence images.

3.4.3 Use of a Training Set

Use of a subset of the data as a training set is essential to: (1) develop experience identifying lesions; (2) establish rules and guidelines for scoring images; and (3) identify features on fluorescence images that correspond to histologically confirmed adenomas.

1. Set aside 10 of the annotated fluorescence/white light images to be used as a training set.
2. Identify regions of high fluorescence intensity relative to the local background, and compare these regions with results from H&E stained sections as reviewed and annotated by a trained pathologist (Fig. 2).
3. Conversely, determine which features on the fluorescence images correspond to the presence of a pathologically confirmed lesion.
4. Identify which areas have pathologically confirmed lesions and determine whether or not there is an identifiable fluorescence signal on the image in that location.

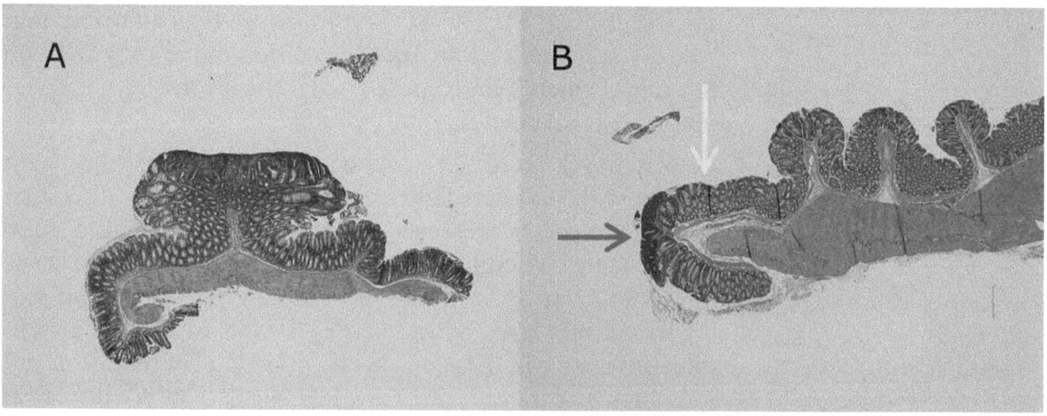

Fig. 2 Micrographs of colorectal lesions in an Apc$^{+/Min-FCCC}$ mouse. (**a**) 40× view of a peduncular adenoma that extends above the surface of the normal mucosa compared with a nonpolypoid lesion (**b**). (**b**) 40× view of a sessile adenoma (*blue arrow*) with a height less than twice that of the adjacent normal mucosa (*yellow arrow*)

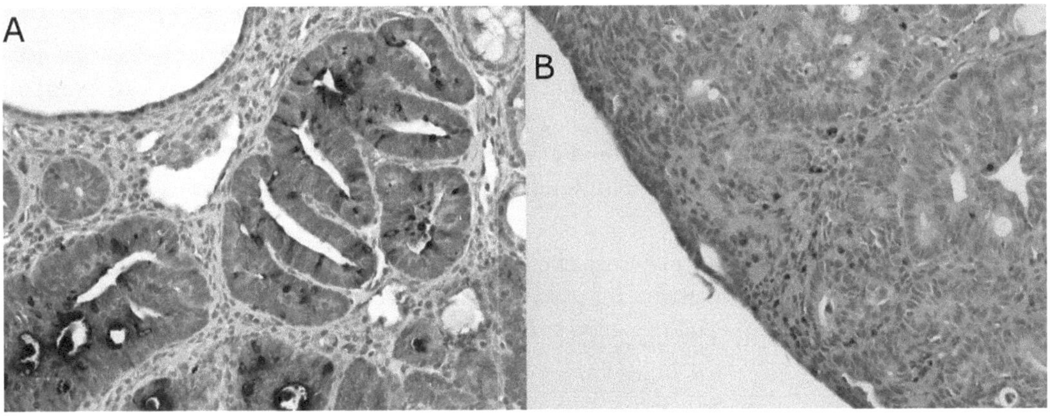

Fig. 3 Immunohistochemical staining of MMP-7 and -9 in colorectal lesions of Apc+/Min-FCCC mice. (**a**) 400× view of an adenoma stained with anti-MMP-7 antibodies. MMP-7 is strongly and diffusely expressed in the adenomatous epithelium. (**b**) 400× view of an adenoma stained with anti-MMP-9 antibodies. MMP-9 is strongly expressed in the polymorphonuclear leukocytes (PMNLs) in the stroma/lamina propria of the adenoma but not in the epithelial cells

5. After training is complete, at least two independent observers can conduct an analysis of an independent set of fluorescence images in a blinded manner to identify regions most likely containing atypical pathology or early lesions.

3.5 MMP-7 and MMP-9 Immuno-histochemistry

1. Employ standard immunohistochemical methods, as per the manufacturer's instructions, to determine the expression of MMP-7 and MMP-9 in paraffin embedded sections (Fig. 3).
2. Use rat anti-human MMP-7 antibody clone 338, at a dilution of 1:400 with sections incubated for 1 h at 37 °C, followed by goat anti-rat secondary antibody as the secondary antibody.
3. For MMP-9 staining, use rabbit anti MMP-9 at a dilution of 1:200, with sections incubated for 1 h at room temperature. A goat anti-rabbit secondary antibody is used with this primary MMP-9 antibody. For best results, sections should be stained using an automated immunostainer, such as the Ventana XT.

4 Notes

1. The presence of fluorescent substances in the most commonly used animal chows can lead to false-positive areas of strong fluorescence. The fecal matter of the mice will generally be fluorescent if all of these contaminants are not fully removed. The use of a non-fluorescent chow (Teklad 2019s, Envigo, East Millstone, NJ) for 1 week prior to imaging can be advantageous.

2. When laying the colon flat on the plastic ruler, be sure to keep track of the proximal and distal ends. This can be done by noting the distinct three-dimensional architecture (V-shaped) of the proximal colonic mucosa, and placing the colon so it is always oriented in the same direction. Always lay the colon with the luminal surface facing up, with no folds or twists.
3. Use a plain black colored marker when recording the animal number on the plastic sheets, as the dyes in some colored markers may have some fluorescent signal.
4. The black plastic mat is usually supplied by the manufacturer of the imaging system.
5. The focal distance on the IVIS Spectrum is set to a default of 1.5 cm for in vivo imaging. When making high spatial resolution images with a low f-number setting in excised organs, this setting must be decreased to approximately 1.5 mm and adjusted carefully.
6. Imaging times (from dissection to image acquisition) should not exceed 5 min to prevent dehydration.
7. The ruler markings photocopied onto the transparent sheet can be used as a guide for cutting the colon.
8. To increase the dynamic range of the images, an alternative to displaying multiple images with different contrast settings would be to employ a nonlinear mapping (e.g., logarithmic) of intensity to color. In practice, we find it preferable to examine multiple representations of the same data with a linear mapping of intensity to color.

Acknowledgments

This work was supported by grants CA-006927 and CA-124693 from the National Cancer Institute, by Fox Chase Cancer Center Keystone Initiative in Personalized Risk and Prevention, and by an appropriation from the Commonwealth of Pennsylvania. The authors have no conflicts of interest or financial disclosures to report.

References

1. Cornett D, Barancin C, Roeder B et al (2008) Findings on optical colonoscopy after positive CT colonography exam. Am J Gastroenterol 103:2068–2074
2. Heresbach D, Barrioz T, Lapalus MG et al (2008) Miss rate for colorectal neoplastic polyps: a prospective multicenter study of back-to-back video colonoscopies. Endoscopy 40:284–290
3. Burggraaf J, Kamerling IM, Gordon PB et al (2015) Detection of colorectal polyps in humans using an intravenously administered fluorescent peptide targeted against c-Met. Nat Med 21:955–961
4. Clapper ML, Hensley HH, Chang WC et al (2011) Detection of colorectal adenomas using a bioactivatable probe specific for matrix metalloproteinase activity. Neoplasia 13:685–691

5. Cooper HS, Chang WC, Coudry R et al (2005) Generation of a unique strain of multiple intestinal neoplasia (Apc(+/Min-FCCC)) mice with significantly increased numbers of colorectal adenomas. Mol Carcinog 44: 31–41
6. Weissleder R, Ntziachristos V (2003) Shedding light onto live molecular targets. Nat Med 9:123–128
7. Wagenaar-Miller RA, Gorden L, Matrisian LM (2004) Matrix metalloproteinases in colorectal cancer: is it worth talking about? Cancer Metastasis Rev 23:119–135
8. Zucker S, Vacirca J (2004) Role of matrix metalloproteinases (MMPs) in colorectal cancer. Cancer Metastasis Rev 23:101–117
9. Newell KJ, Witty JP, Rodgers WH et al (1994) Expression and localization of matrix-degrading metalloproteinases during colorectal tumorigenesis. Mol Carcinog 10:199–206
10. Kinzler KW, Vogelstein B (1996) Lessons from hereditary colorectal cancer. Cell 87:159–170

Part VI

Matrix Metalloprotease Inhibitors

Chapter 14

Virtual High-Throughput Screening for Matrix Metalloproteinase Inhibitors

Jun Yong Choi and Rita Fuerst

Abstract

Structure-based virtual screening (SBVS) is a common method for the fast identification of hit structures at the beginning of a medicinal chemistry program in drug discovery. The SBVS, described in this manuscript, is focused on finding small molecule hits that can be further utilized as a starting point for the development of inhibitors of matrix metalloproteinase 13 (MMP-13) via structure-based molecular design. We intended to identify a set of structurally diverse hits, which occupy all subsites (S1′–S3′, S2, and S3) centering the zinc containing binding site of MMP-13, by the virtual screening of a chemical library comprising more than ten million commercially available compounds. In total, 23 compounds were found as potential MMP-13 inhibitors using Glide docking followed by the analysis of the structural interaction fingerprints (SIFt) of the docked structures.

Key words Matrix metalloproteinase, Structure-based virtual screening, Docking, Structural interaction fingerprints, Zn-chelating inhibitor

1 Introduction

The identification of a lead structure usually initiates a medicinal chemistry program in drug discovery. Traditionally, high-throughput screenings (HTS) of large compound libraries are used to deliver lead compounds for a certain protein target. The resultant small molecule leads need to be optimized through iterative analog synthesis efforts. The high costs and low hit rates associated with HTS screening campaigns as well as a steadily increasing number of new drug targets accelerated the development of cheaper and faster computational alternatives starting in the early 1990s [1, 2]. Nowadays, virtual screening (VS) methods are broadly used in early-stage drug discovery for hit identification by analyzing chemical databases.

There are two different approaches, ligand- and receptor-based VS, used to prioritize compounds for synthesis and evaluation. The ligand-based approach tries to find molecules with similar physical

Charles A. Galea (ed.), *Matrix Metalloproteases: Methods and Protocols*, Methods in Molecular Biology, vol. 1579,
DOI 10.1007/978-1-4939-6863-3_14, © Springer Science+Business Media LLC 2017

and chemical properties to already known protein ligands. On the other hand, the receptor-based approach, also known as structure-based VS (SBVS), is not biased by known protein interaction partners. The screening starts with the 3-D structure of a target protein and a 3-D database of ligands. After virtual filtering of the compound library, the predictive binding mode of the lead structures is found by docking and the hits are scored by evaluating their binding affinities to the protein target [3].

Within this manuscript we describe a SBVS aimed at finding new lead compounds capable of inhibiting the zinc-dependent matrix metalloproteinase 13 (MMP-13). MMP-13 is an extremely promising drug target for the treatment of osteoarthritis (OA). This protease is mainly responsible for the cleavage of collagen type II, which leads to the destruction of articular cartilage, one of the main symptoms of OA [4]. MMP-13 is also involved in cancer progression. The proteolytic degradation of the extracellular environment in melanoma invasion and metastasis is dependent on the stromal expression of MMP-13 [5]. In breast cancer, metastasis to the bone is a very common mechanism for secondary tumor growth. Breast cancer cells manipulate signaling pathways in osteoblasts, leading to increased MMP-13 release, which promotes metastatic invasion to bone tissue [6].

1.1 Matrix Metalloproteinases as Drug Targets

MMP activity was described for the first time in 1962 as an enzymatic player in the metamorphosis process of tadpoles. Since then, 24 structurally and functionally related MMPs have been found in mammals. MMPs are characterized by a Zn^{2+}-cation in their enzyme active site, which is coordinated by three histidines followed by a conserved methionine residue. Under healthy conditions, MMPs are very important regulators of cellular activities and physiological processes including reproduction, tissue remodeling, embryogenesis, and angiogenesis via the degradation of extracellular matrix (ECM) components. MMPs can also influence the immune system by directly cleaving signaling molecules including the tumor necrosis factor (TNF) and other cytokines. MMP activity is very low under healthy conditions but can be detected during repair or remodeling processes including angiogenesis, bone development, or wound healing, as well as in inflamed tissue [7].

Originally, MMPs attracted the attention of numerous research groups due to their involvement in tumor angiogenesis and metastasis. First-generation MMP inhibitors, developed in the early 1990s, were generally not selective for any particular MMP. Drug candidates resultant from these early programs were designed to mimic peptide sequences of collagen. These peptidomimetics also consisted of a functional group capable of chelating the active site zinc within the MMP active site in a competitive manner. Zinc chelating functionalities include the carboxylate-, carbamate-, thiol-, and hydroxamate group as well as phosphoric acid derivatives. The broad-spectrum MMP inhibitors and hydroxamic acid containing peptidomimetics, Batimastat (**1**) and Marimastat (**2**), were the first MMP inhibitors that entered clinical testing for cancer therapy. Clinical trials were suspended in phase III due to musculoskeletal toxicities resultant from nonselective binding of the inhibitors within the MMP family [8].

Broad-spectrum MMP inhibition is common among MMP inhibitors due to the high structural homology among the active sites between MMP isoforms (Fig. 1). To circumvent this difficulty, recent research has switched to target alternative, less conserved allosteric sites with non-zinc chelating small molecules. The reports of co-crystal structures of MMPs with bound ligands could demonstrate that the significant structural differences within this highly conserved protein family lie in the conformation of surface loops (S1–S3 and S1′–S3′ subsites), which surround the catalytically active zinc ion [9]. Additionally, the development of isoform selective MMP inhibitors was aided by the discovery of the $S_1'^*$ specificity loop, which is a hydrophobic pocket located adjacent to the S1′ subsite that shows the highest sequence variability within the MMP family. In the case of MMP-13, the S1′/S1′* subsite forms a very deep hydrophobic pocket that is enclosed at one end by a lysine residue [10, 11].

1.2 MMP-13 in Drug Discovery

During the past 15 years, several highly potent and selective non-zinc chelating inhibitors targeting the S1′/S1′* subsite of MMP-13 were developed and several have entered clinical trials for the treatment of osteoarthritis. In the early 2000s, the first highly selective non zinc chelating MMP-13 inhibitors, **3** and **4** (Fig. 2), were identified as part of a HTS campaign followed by an extensive structure activity relationship (SAR) study [12]. X-ray co-crystal structures revealed that the dynamic movement of the S1′ pocket

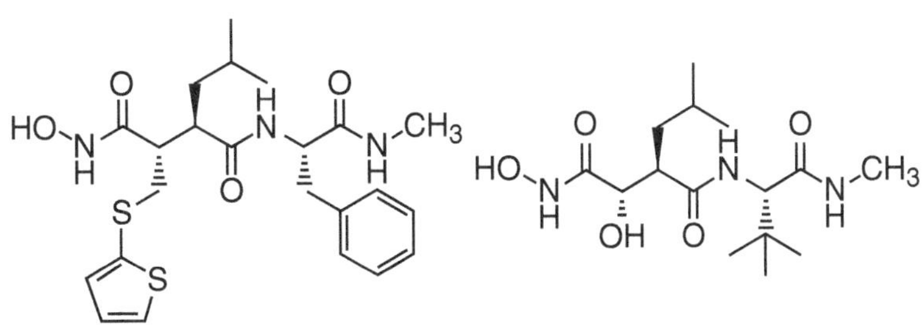

1, IC_{50}
MMP-1 = 3 nM
MMP-2 = 4 nM
MMP-3 = 20 nM
MMP-7 = 6 nM
MMP-9 = 4 nM

2, IC_{50}
MMP-1 = 5 nM
MMP-2 = 6 nM
MMP-3 = 230 nM
MMP-7 = 16 nM
MMP-9 = 3 nM

Fig. 1 Broad-spectrum MMP inhibitors

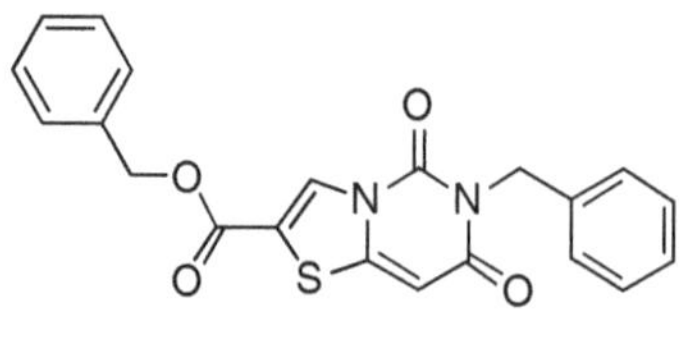

3, IC_{50}
MMP-13 = 30 nM
MMP-1, -2, -3, -7, -9, -12, -14
>100,000 nM

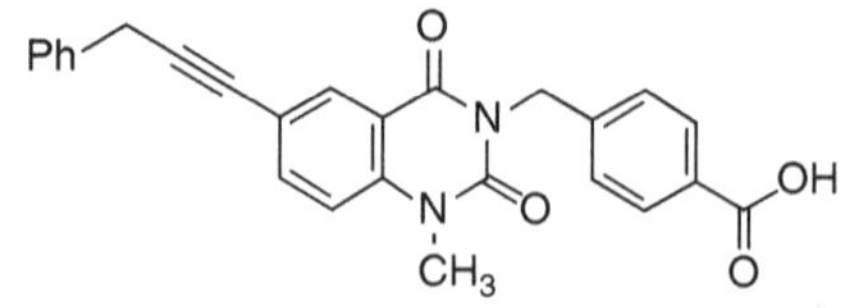

4, IC_{50}
MMP-13 = 0.67 nM
MMP-1, -2, -3, -7, -9, -12, -14
>30,000 nM

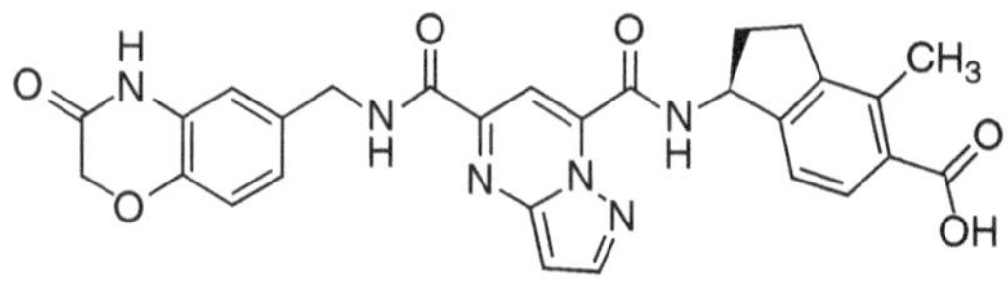

5, IC_{50}
MMP-13 = 0.03 nM
MMP-1, -2, -3, -7, -8, -9, -12, -14 > 20,000 nM

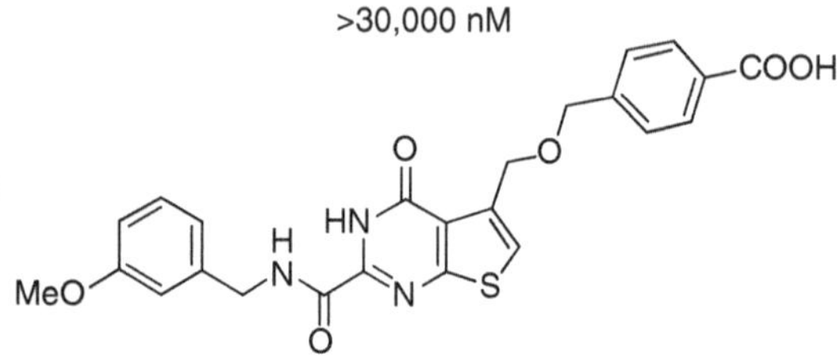

6, IC_{50}
MMP-13 = 3.9 pm
MMP-8 = 720 nM
MMP-10 = 160 nM
MMP-1, -2, -3, -7, -9, -14 > 4000 nM

Fig. 2 Non-zinc chelating MMP-13 inhibitors

seems essential for enzymatic turnover and the two compounds block enzymatic activity simply by inducing rigidity in the S1′ subsite and its adjacent S1′* specificity loop. The highly potent inhibitors **3** and **4** entered preclinical trials but had to be discontinued unexpectedly because both compounds induced renal toxicity in cynomolgus monkeys [13, 14].

In 2011, Baragi *et al.* from Alantos Pharmaceuticals reported the discovery and evaluation of compound **5** for intra-articular (IA) treatment of OA. In a manner analogous to compounds **3** and **4**, this compound binds deep within the S1′ pocket and additionally interacts with the S1′* specificity loop. An impressive 20,000-fold selectivity for **5** was observed after assaying the activity of the compound against a broad panel of MMPs (Fig. 2). Compound **5** also exhibited a remarkable half-life in rat knee joints after IA injection. The concentration of **5** in cartilage was >2 μM after 8 weeks following a single treatment [15].

Kori *et al.* from Takeda published the activity of compound **6**, which binds in the same binding mode as the compounds described before (**3**–**5**) and exhibits excellent potency and selectivity for MMP-13 relative to other MMPs. The oral bioavailability of compound **6** in different species (rats, guinea pigs, rabbits, beagle dogs, and cynomolgus monkeys) could be significantly improved by using the monosodium salt of the carboxylic acid functionality of this compound. Furthermore, the authors demonstrated that oral administration of the monosodium salt induced a significant reduction of a cartilage marker (C-telopeptide of type II collagen

(CTX-II)) of OA, after inducing OA symptoms by injecting monoiodoacetic acid (MIA) into the rat knee joint [16].

Despite these extremely promising preclinical results, no further improvements of compounds **5** and **6** or clinical trials using these compounds have been published in the literature and no MMP-13 inhibitor has achieved FDA approval thus far. The community continues to search for a disease-modifying anti-OA agent. However, the promising anti-cancer activities of MMP-13 inhibitors merit further investigation.

In the following section, a SBVS for MMP-13 is described. The SBVS is focused on the identification of small molecule hits that can be further utilized as starting points for the development of MMP inhibitors via structure-based molecular design. Therefore, we sought to identify structurally diverse hits, which occupy all subsites (S1′–S3′, S2, and S3) centering on the zinc containing binding site of MMP-13. We attempted to search for Zn-chelating structures within this SBVS, which could be additionally subjected to the structure-based design of selective MMP-inhibitors targeting different MMP isoforms. Although compounds interacting with the catalytically active Zn-ion lacked selectivity, the identified hits may be a useful resource for the structure-based design of non-Zn-chelating inhibitors targeting individual MMP isozymes.

2 Materials

2.1 Software

Two different programs were used in the SBVS focused on finding new inhibitors for MMP-13. The program Glide from the Schrödinger Small Molecule Drug Discovery Suite [17] was utilized for ligand docking and the program MOE developed by Chemical Computing Group [18] was used for chemical library composition and ligand-based search. We chose these programs due to their availability in our research laboratory and their popularity within the research community reflected by a myriad of cited references. The use of Glide and MOE requires the purchase of their commercial licenses from Schrödinger Inc. and Chemical Computing Group, respectively. Alternatively, other docking programs are freely available such as AutoDock [19], UCSF-DOCK [20], GOLD [21], and SwissDock [22], which have been developed and maintained by academic institutions.

2.2 Database

The virtual chemical library collection was obtained from the ZINC12 database [23, 24], which comprises around ten million commercially available compounds filtered for drug-like physical properties [25]: 150 ≤ molecular weight ≤ 500, xlogP ≤ 5, number of rotatable bonds ≤ 7, polar surface area < 150 Å^2, number of hydrogen bond donors ≤ 5, and number of hydrogen bond acceptors ≤ 10. The structures were downloaded in an SDF file format and converted to molecular database (mdb) files using the MOE program. The chemical structures were cleaned up to add/remove hydrogen atoms, remove salt and solvent molecules, and generate relevant protonation states at the physiological pH 7.4. The collection of chemical structures was then saved in an MOE database file format for future use.

3 Methods

3.1 Workflow

The virtual screening comprised five steps as outlined below (Fig. 3):

1. Ligand-based search.
2. Analysis of X-ray structures and preparation of a target structure.
3. Grid generation.
4. Glide docking.
5. Data analysis.

3.2 Ligand-Based Search

Initially, all chemical structures possessing one or more carboxylic acid functionalities, which can chelate the catalytically active Zn-ion of MMPs, were selected from the MOE database collection generated in Subheading 2.2. In the Select panel of the MOE database,

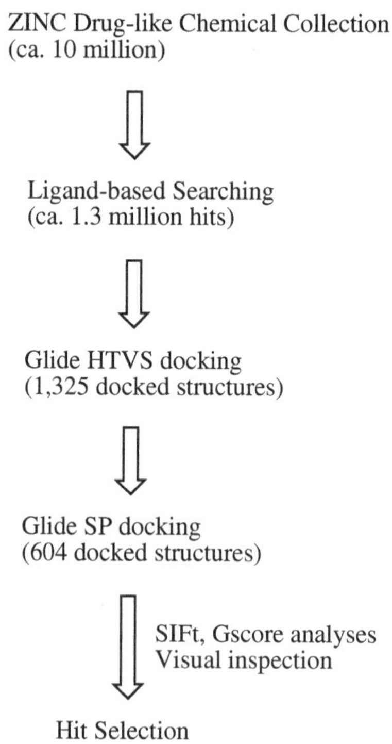

Fig. 3 Virtual screening process and outcomes

the keyword, mol="O=CO" was used to search for chemical structures containing at least one carboxylic acid moiety. This search identified 1,387,623 compounds out of ca. ten million drug-like small molecules (*see* **Note 1**). These structures were assembled and saved as SDF files for structure-based virtual screening.

3.3 Analysis of X-Ray Structures and Preparation of the Target Structure

Structure coordinates were downloaded for all MMP isozymes found in the Protein Data Bank (PDB). The sequence alignment for these MMP X-ray crystal structures showed about 50% sequence identity between each isozyme. Superimposition of the MMP structures, based on the sequence alignment, also showed that the catalytic domain of different isozymes is highly conserved, and the binding sites of co-crystallized Zn-chelating inhibitors are located around the metal ion and extended toward all subsites (S1′–S3′, S2, and S3). Furthermore, the chelating moieties of inhibitors that interact with the Zn-ion are located around the metal ion at a distance of approximately 2.1 Å (Fig. 4).

The X-ray co-crystal structure of MMP-13 (PDB ID: 3ELM) [26] was used as the target for the SBVS within this study. First, the protein structure was prepared for docking by refining the target protein structure. This step included: assigning bond orders, adding hydrogen atoms, creating zero-order bonds to metals, creating disulfide bonds, filling in missing side chains and loops, and deleting water molecules beyond 5 Å from hetero groups. Chain A of the target protein was used for further refinement that comprised assigning hydrogen bond interactions using the PROPKA module (pH = 7.4). Finally, the structure was subjected for restrained minimization using the OPLS3 force field.

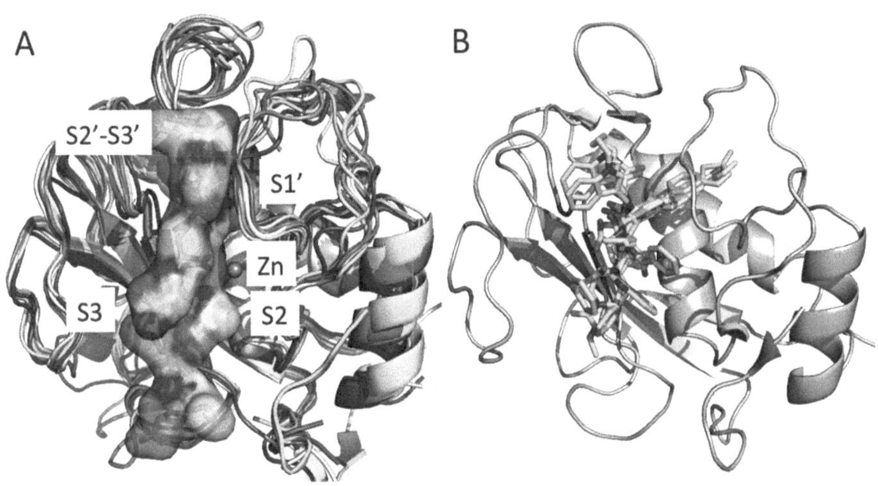

Fig. 4 Superimposed X-ray co-crystal structures of MMP isozymes (**a**) and Zn-chelating inhibitors (**b**). Substrate mimetic (decapeptide) is represented as a surface (PDB ID: 3AYU) (**a**). Each subsite in the substrate-binding cleft is annotated (**a**) [29]

3.4 Receptor Grid Generation

A receptor grid was generated for the ligand-binding site of the target protein. The grid was set to a 25 Å^3 box centered on the Zn-chelating reference ligand of the MMP-13 co-crystal structure (3ELM) and the metal coordination to the zinc ion was assigned as a constraint.

3.5 Glide Docking

The Virtual Screening Workflow module was used for Glide docking. The compound collection, containing at least one carboxylic acid moiety from the ligand-based search described in Subheading 3.2, was directly used without further ligand filtering or preparation. Ligand docking was restricted by the comparison of core patterns, which required the carboxylic acid unit of docked ligands to align with that of the reference compound with a tolerance of 0.7 Å. Initially, the ligand structures were sequentially docked into the target grid in the high-throughput virtual screening (HTVS) mode, which reduced about 1.3 million carboxylic acid-containing compounds to 1325 docked structures. This process was followed by Glide standard precision (SP) mode docking of the remaining 1325 structures, which generated 604 ligands that occupied the zinc-binding site of MMP-13. All 604 docked structures were scored and saved for data analysis and hit triage (*see* **Note 2**).

3.6 Data Analysis and Hit Triage

The 604 docked structures occupied the entire area surrounding the Zn-ion. Since the ligand interactions with the subsites S1′–S3′, S2, and S3 are known to improve selectivity within the MMP family, the goal of this virtual screen was to identify diverse sets of Zn-chelating molecules that occupied these regions centered around the Zn-containing active site of MMP-13. Docking post-processing procedures were applied in Maestro to calculate structural interaction fingerprints (SIFt) to analyze intermolecular binding interactions between the protein and the docked structures. After the analysis of the SIFt matrix, a number of structurally diverse compounds, which show interaction with the five subsites, were selected as potential MMP-13 inhibitors [27, 28].

The computed matrix map shows the clustering of SIFts correlated to the docking score of the screened compounds (Fig. 5). Various positions on the map represent ligand occupancy of different binding sites. Ligand binding poses in each cluster of SIFts were visually inspected and potential hits were selected using the following selection criteria: the Glide docking score, the conformational stability of the bound inhibitor, ligand rigidity, and uniqueness. Furthermore, compounds were triaged by comparing their binding poses to currently known inhibitors exhibiting interactions considered to be important for potency and selectivity for MMP-13, such as a π-stacking interaction with His222 in the S1′ site, hydrophobic contacts with Tyr176, His187, and Phe189 in the S3 site, and a hydrogen bond interaction with the protein backbone in the S2′ site [26, 29–31].

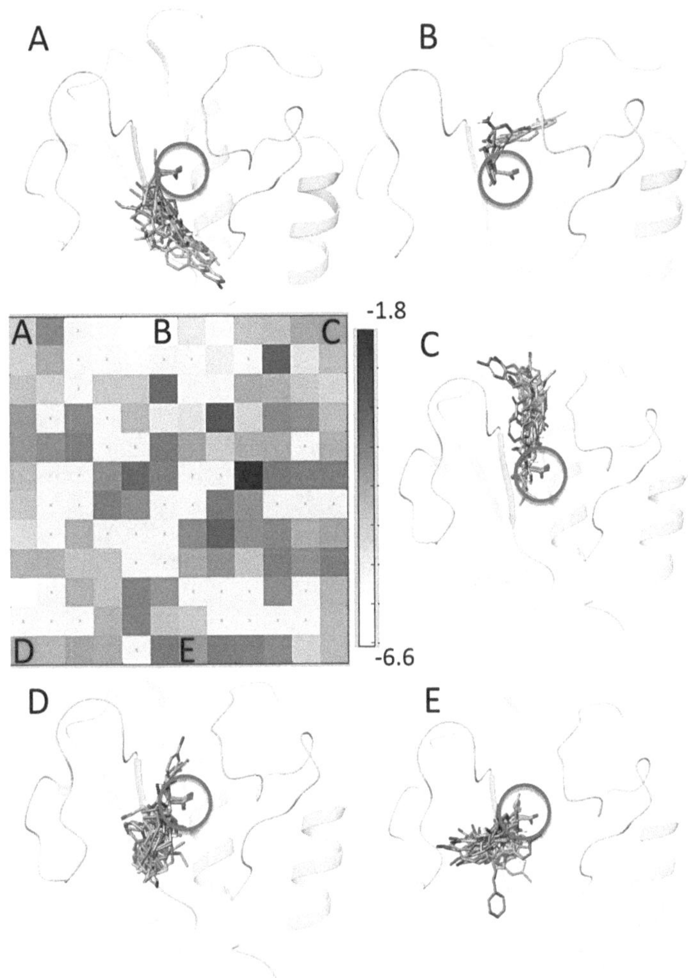

Fig. 5 Structural interaction fingerprints and binding poses of ligands clustered based on their interactions with the target structure. *Blue circle* represents the Zn-binding site. The *color code* in the map represents the average of the Glide docking score of different ligands clustered in each cell. It ranges from −6.6 (*white*, highest score) to −1.8 (*dark*, lowest score). The *top left cell* (**a**) of the map represents ligands occupying the S2 site, the *top middle cell* (**b**) represents ligands interacting with the S1′ site, the *top right cell* (**c**) represents ligands occupying the S2′ and S3′ sits, the *bottom left cell* (**d**) represents ligands binding the area between the subsites S2 and S3, and the *middle bottom cell* (**e**) represents the ligands occupying the S3 site

Finally, 23 ligands were selected as potential Zn-chelating MMP-13 inhibitors (*see* **Note 3**). The ZINC ID of the selected ligands were ZINC39475452, ZINC37417840, ZINC82555302, ZINC02860431, ZINC62430943, ZINC41498653, ZINC83306707, ZINC76441060, ZINC71632004, ZINC57989822, ZINC71694581, ZINC82401153, ZINC82867972, ZINC22145620, ZINC00229666, ZINC76186843, ZINC20231914, ZINC71726611,

ZINC33510136, ZINC82973537, ZINC82305218, ZINC19476737, and ZINC19438618. The structure and purchasing information for each of these hits can be found at the ZINC database website (www.zinc15.docking.org) using these ID numbers. The new updated version of ZINC15 was released in early 2016.

Binding poses of four representative hits are shown in Fig. 6. ZINC62430943 forms a metal-π interaction between the Zn-ion and the pyridine ring and the sulfonamide moiety can form a

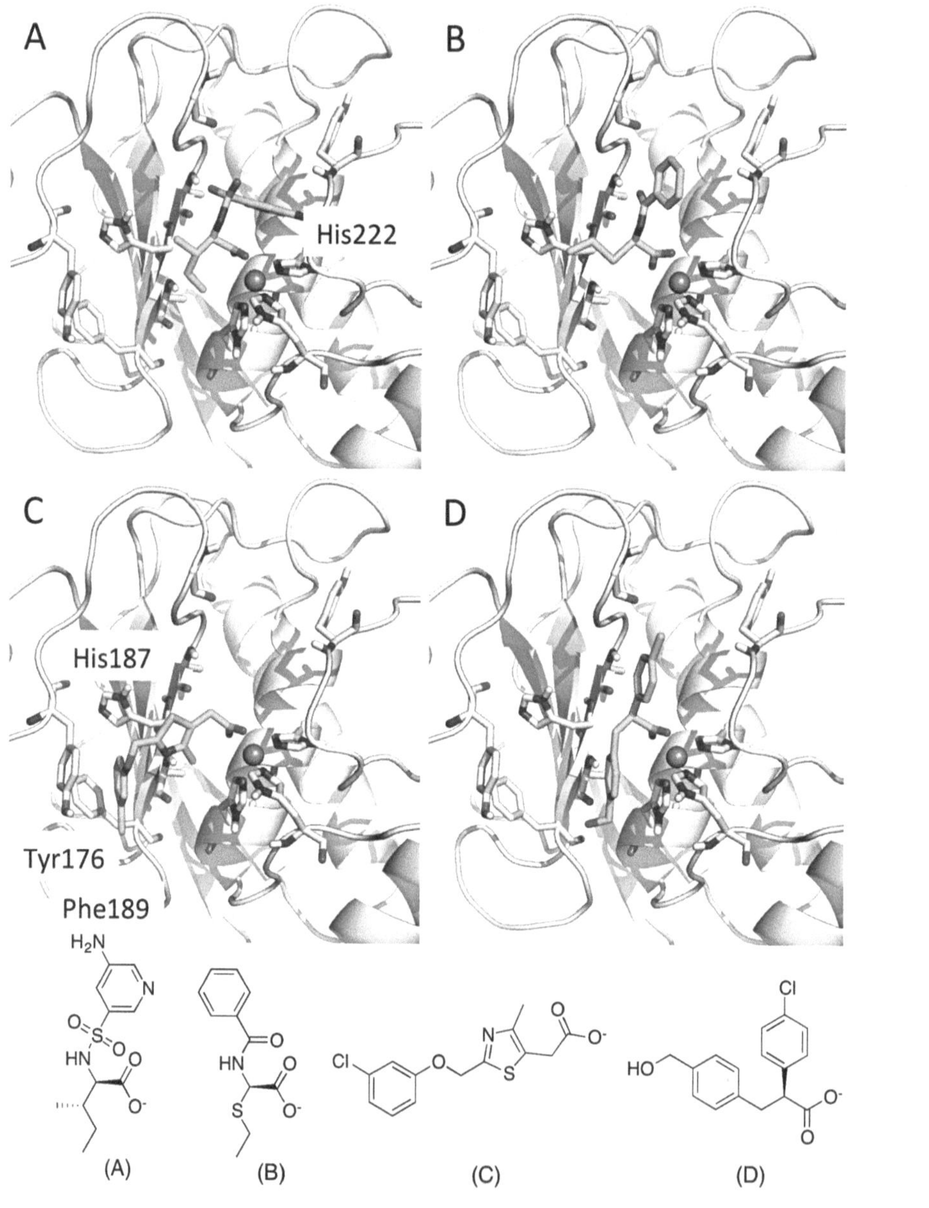

Fig. 6 Representative hits and their binding poses in MMP-13. (**a**) ZINC62430943, (**b**) ZINC82305218, (**c**) ZINC41498653, and (**d**) ZINC57989822 are shown in *green sticks*. Amino acids in the active site of MMP-13 are shown in *gray sticks*, and protein structures are shown in a *gray ribbon diagram*. Figures were generated using the program PyMol [31]

hydrogen bond interaction with the amide backbone of Leu185 (Fig. 6a). ZINC82305218 interacts with the amide backbone of Leu185 in the substrate binding site via a hydrogen bond and is oriented toward the S2′–S3′ subsite (Fig. 6b). ZINC41498653 forms hydrophobic contacts with Tyr179, His187, and Phe189 in the S3 subsite (Fig. 6c). ZINC57989822 occupies the substrate binding-site near the Zn-ion by forming a hydrogen bond interaction with the amide backbone of Ala188 (Fig. 6d).

4 Notes

1. We assembled a drug-like chemical library of structures obtained from the ZINC database and performed a ligand-based search using MOE. These procedures could also be conducted using LigPrep [32], Canvas [33], and Knime [34] in the Schrödinger program suite. In our study, the analysis of the database to search for potential ligands was faster in MOE compared to using other modules of the Schrödinger program suite.
2. We identified pan-assay interference compounds (PAINS) [35] in the 604 docked ligands obtained from Glide SP, which implies that the ZINC drug-like compound set includes PAINS. Thus, it is necessary to filter for PAINS during the chemical library composition process. Current MOE and Schrödinger programs have additional substructure filters for the removal of PAINS.
3. Coordination to the Zn-ion and the highly conserved catalytic domain of MMP isozymes (Fig. 4a) suggests that most of the identified hits will also inhibit other MMP isoforms. Currently, 199 X-ray structures of MMP isozymes are available in the PDB database: 63 for MMP-12 (macrophase metalloelastase), 43 for MMP-13 (collagenase 3), 31 for MMP-3 (Stromelysin-1), 19 for MMP-9, 18 for MMP-8 (Neutrophil collagenase), 10 for MMP-2 (72 kDa type IV collagenase), 9 for MMP-14, 3 for MMP-7, 1 for MMP-10, 1 for MMP-11, 1 for MMP-16, and a solution structure for MMP-20. Therefore, the hits identified from this virtual screen may form the basis for the structure-based design of inhibitors specifically targeting each MMP isoform.

References

1. Cheng T, Li Q, Zhou Z, Wang Y, Bryant SH (2012) Structure-based virtual screening for drug discovery: a problem-centric review. AAPS J 14(1):133–141
2. Ripphausen P, Nisius B, Peltason L, Bajorath J (2010) Quo vadis, virtual screening? A comprehensive survey of prospective applications. J Med Chem 53(24):8461–8467
3. Ghosh S, Nie A, An J, Huang Z (2006) Structure-based virtual screening of chemical libraries for drug discovery. Curr Opin Chem Biol 10(3):194–202
4. Takaishi H, Kimura T, Dalal S, Okada Y, D'Armiento J (2008) Joint diseases and matrix metalloproteinases: a role for MMP-13. Curr Pharm Biotechnol 9(1):47–54

5. Zigrino P, Kuhn I, Bauerle T, Zamek J, Fox JW, Neumann S, Licht A, Schorpp-Kistner M, Angel P, Mauch C (2009) Stromal expression of MMP-13 is required for melanoma invasion and metastasis. J Invest Dermatol 129(11):2686–2693
6. Morrison C, Mancini S, Cipollone J, Kappelhoff R, Roskelley C, Overall C (2011) Microarray and proteomic analysis of breast cancer cell and osteoblast co-cultures: role of osteoblast Matrix Metalloproteinase (MMP)-13 in bone metastasis. J Biol Chem 286(39): 34271–34285
7. Vandenbroucke RE, Libert C (2014) Is there new hope for therapeutic matrix metalloproteinase inhibition? Nat Rev Drug Discov 13(12):904–927
8. Rothenberg ML, Nelson AR, Hande KR (1999) New drugs on the horizon: matrix metalloproteinase inhibitors. Stem Cells 17(4):237–240
9. Engel CK, Pirard B, Schimanski S, Kirsch R, Habermann J, Klingler O, Schlotte V, Weithmann KU, Wendt KU (2005) Structural basis for the highly selective inhibition of MMP-13. Chem Biol 12(2):181–189
10. Li JJ, Johnson AR (2011) Selective MMP13 inhibitors. Med Res Rev 31(6):863–894
11. Dormán G, Cseh S, Hajdu I, Barna L, Konya D, Kupai K, Kovacs L, Ferdinandy P (2010) Matrix metalloproteinase inhibitors: A critical appraisal of design principles and proposed therapeutic utility. Drugs 70(8):949–964
12. Johnson AR, Pavlovsky AG, Ortwine DF, Prior F, Man C-F, Bornemeier DA, Banotai CA, Mueller WT, McConnell P, Yan C, Baragi V, Lesch C, Roark WH, Wilson M, Datta K, Guzman R, Han H-K, Dyer RD (2007) Discovery and characterization of a novel inhibitor of matrix metalloprotease-13 that reduces cartilage damage in vivo without joint fibroplasia side effects. J Biol Chem 282(38):27781–27791
13. Li JJ, Nahra J, Johnson AR, Bunker A, O'Brien P, Yue W-S, Ortwine DF, Man C-F, Baragi V, Kilgore K, Dyer RD, Han H-K (2008) Quinazolinones and pyrido[3,4-d]pyrimidin-4-ones as orally active and specific matrix metalloproteinase-13 inhibitors for the treatment of osteoarthritis. J Med Chem 51(4):835–841
14. Cai H, Agrawai AK, Putt DA, Hashim M, Reddy A, Brodfuehrer J, Surendran N, Lash LH (2009) Assessment of the renal toxicity of novel anti-inflammatory compounds using cynomolgus monkey and human kidney cells. Toxicology 258(1):56–63
15. Gege C, Bao B, Bluhm H, Boer J, Gallagher BM, Korniski B, Powers TS, Steeneck C, Taveras AG, Baragi VM (2011) Discovery and evaluation of a non-Zn chelating, selective Matrix Metalloproteinase 13 (MMP-13) inhibitor for potential intra-articular treatment of osteoarthritis. J Med Chem 55(2):709–716
16. Nara H, Sato K, Naito T, Mototani H, Oki H, Yamamoto Y, Kuno H, Santou T, Kanzaki N, Terauchi J, Uchikawa O, Kori M (2014) Discovery of novel, highly potent, and selective quinazoline-2-carboxamide-based Matrix Metalloproteinase (MMP)-13 inhibitors without a zinc binding group using a structure-based design approach. J Med Chem 57(21):8886–8902
17. Friesner RA, Banks JL, Murphy RB, Halgren TA, Klicic JJ, Mainz DT, Repasky MP, Knoll EH, Shelley M, Perry JK, Shaw DE, Francis P, Shenkin PS (2004) Glide: a new approach for rapid, accurate docking and scoring: 1. Method and assessment of docking accuracy. J Med Chem 47(7):1739–1749
18. Molecular Operating Environment (MOE), 2013.08 (2016). Chemical Computing Group Inc., 1010 Sherbooke St. West, Suite #910, Montreal, QC, Canada H3A 2R7
19. Morris GM, Huey R, Lindstrom W, Sanner MF, Belew RK, Goodsell DS, Olson AJ (2009) AutoDock4 and AutoDockTools4: automated docking with selective receptor flexibility. J Comput Chem 30(16):2785–2791
20. Allen WJ, Balius TE, Mukherjee S, Brozell SR, Moustakas DT, Lang PT, Case DA, Kuntz ID, Rizzo RC (2015) DOCK 6: impact of new features and current docking performance. J Comput Chem 36(15):1132–1156
21. Jones G, Willett P, Glen RC, Leach AR, Taylor R (1997) Development and validation of a genetic algorithm for flexible docking. J Mol Biol 267(3):727–748
22. Grosdidier A, Zoete V, Michielin O (2011) SwissDock, a protein-small molecule docking web service based on EADock DSS. Nucleic Acids Res 39(Web Server issue):W270–W277
23. Irwin JJ, Sterling T, Mysinger MM, Bolstad ES, Coleman RG (2012) ZINC: a free tool to discover chemistry for biology. J Chem Inf Model 52(7):1757–1768
24. Irwin JJ, Shoichet BK (2005) ZINC—a free database of commercially available compounds for virtual screening. J Chem Inf Model 45(1):177–182
25. Lipinski CA (2000) Drug-like properties and the causes of poor solubility and poor permeability. J Pharmacol Toxicol Methods 44(1):235–249
26. Monovich LG, Tommasi RA, Fujimoto RA, Blancuzzi V, Clark K, Cornell WD, Doti R, Doughty J, Fang J, Farley D, Fitt J, Ganu V,

Goldberg R, Goldstein R, Lavoie S, Kulathila R, Macchia W, Parker DT, Melton R, O'Byrne E, Pastor G, Pellas T, Quadros E, Reel N, Roland DM, Sakane Y, Singh H, Skiles J, Somers J, Toscano K, Wigg A, Zhou S, Zhu L, Shieh W-C, Xue S, McQuire LW (2009) Discovery of potent, selective, and orally active carboxylic acid based inhibitors of matrix metalloproteinase-13. J Med Chem 52(11):3523–3538
27. Singh J, Deng Z, Narale G, Chuaqui C (2006) Structural interaction fingerprints: a new approach to organizing, mining, analyzing, and designing protein-small molecule complexes. Chem Biol Drug Des 67(1):5–12
28. Deng Z, Chuaqui C, Singh J (2004) Structural interaction fingerprint (SIFt): a novel method for analyzing three-dimensional protein-ligand binding interactions. J Med Chem 47(2):337–344
29. Hashimoto H, Takeuchi T, Komatsu K, Miyazaki K, Sato M, Higashi S (2011) Structural basis for matrix metalloproteinase-2 (MMP-2)-selective inhibitory action of beta-amyloid precursor protein-derived inhibitor. J Biol Chem 286(38):33236–33243
30. Gall AL, Ruff M, Kannan R, Cuniasse P, Yiotakis A, Dive V, Rio MC, Basset P, Moras D (2001) Crystal structure of the stromelysin-3 (MMP-11) catalytic domain complexed with a phosphinic inhibitor mimicking the transition-state. J Mol Biol 307(2):577–586
31. Browner MF, Smith WW, Castelhano AL (1995) Matrilysin-inhibitor complexes: common themes among metalloproteases. Biochemistry 34(20):6602–6610
32. Schrödinger Release 2016-2: LigPrep, version 3.8 (2016) Schrödinger, LLC, New York, NY
33. Sastry M, Lowrie JF, Dixon SL, Sherman W (2010) Large-scale systematic analysis of 2D fingerprint methods and parameters to improve virtual screening enrichments. J Chem Inf Model 50(5):771–784
34. Berthold MR, Cebron N, Dill F, Gabriel TR, Kötter T, Meinl T, Ohl P, Sieb C, Thiel K, Wiswedel B (2007) Knime: the Konstanz information miner. In: Studies in classification, data analysis, and knowledge organization. Springer, New York
35. Baell J, Walters MA (2014) Chemistry: chemical con artists foil drug discovery. Nature 513(7519):481–483

Chapter 15

Computational Approaches to Matrix Metalloprotease Drug Design

Tanya Singh, B. Jayaram, and Olayiwola Adedotun Adekoya

Abstract

Matrix metalloproteinases (MMPs) are a family of zinc-containing enzymes required for homeostasis. These enzymes are an important class of drug targets as their over expression is associated with many disease states. Most of the inhibitors reported against this class of proteins have failed in clinical trials due to lack of specificity. In order to assist in drug design endeavors for MMP targets, a computationally tractable pathway is presented, comprising, (1) docking of small molecule inhibitors against the target MMPs, (2) derivation of quantum mechanical charges on the zinc ion in the active site and the amino acids coordinating with zinc including the inhibitor molecule, (3) molecular dynamics simulations on the docked ligand–MMP complexes, and (4) evaluation of binding affinities of the ligand–MMP complexes via an accurate scoring function for zinc containing metalloprotein–ligand complexes. The above pathway was applied to study the interaction of the inhibitor Batimastat with MMPs, which resulted in a high correlation between the predicted and experimental binding free energies, suggesting the potential applicability of the pathway.

Key words Matrix metalloprotease, Computer-aided drug design, Docking and scoring, Molecular dynamics simulations

1 Introduction

Matrix metalloproteinases constitute a family of Ca^{2+} containing and Zn^{2+} dependent endopeptidases that are involved in cleavage of extracellular matrix proteins. The family consists of more than 26 proteinases in mammals classified as collagenases (MMP-1, 8, 13, and 18), gelatinases (MMP-2, and 9), stromelysins (MMP-3, 10, 11, and 27), matrilysins (MMP-7 and 26), and membrane-type (MT-MMP) based on their substrate specificity [1]. MMPs are secreted as inactive zymogens and are activated through cleavage by other enzymes [1]. Their enzymatic activity is regulated by their natural inhibitors viz. tissue inhibitors of metalloproteinases (TIMPs) [2, 3]. These enzymes are involved in the regulation of several biological processes such as embryonic development, signal

Charles A. Galea (ed.), *Matrix Metalloproteases: Methods and Protocols*, Methods in Molecular Biology, vol. 1579, DOI 10.1007/978-1-4939-6863-3_15, © Springer Science+Business Media LLC 2017

regulation, wound healing, angiogenesis, ovulation, uterine involution, bone resorption, and nerve growth [4–11]. Dysregulation of MMP activity has been linked to a number of diseases. For example, over-expression of MMPs can result in accelerated matrix degradation mediating a number of pathologies including cancer, loss of cartilage in osteoarthritis, rheumatoid arthritis, cardiovascular diseases, acute lung injury, peridontitis, and many others [5, 12–18]. Thus, MMPs have been a pharmaceutical target for more than 20 years and many MMP inhibitors have been reported in the literature [1]. Most of these MMP inhibitors bind to the zinc at the active site, thereby blocking its activity. However, despite many research efforts, only one MMP inhibitor periostat has been approved by the FDA [1].

The number of available high-resolution X-ray crystal structures of MMP/inhibitor complexes has increased dramatically in recent years aiding in the design of potential inhibitors at an early lead generation stage [19]. Molecules exhibiting high affinity toward Zn^{2+} can effectively prevent the binding of polypeptides to MMPs, thus acting as MMP inhibitors [20]. Several Zn^{2+} binding groups (ZBGs) have been reported: the hydroxamates, reverse hydroxamates, carboxylates, hydroxyureas, hydrazides, phosphinates, sulfones, and sulfonylhydrazides of which the hydroxamates appear to be the most potent [1]. Many broad-spectrum ZBG-containing small molecule inhibitors from different pharmaceutical companies have also entered clinical trials for cancer, rheumatoid arthritis, and osteoarthritis [18, 21–26]. These broad-spectrum MMP inhibitors include hydroxamate-based Marimastat, Batimastat, Ilomastat, Prinomastat, Solimastat, Tanomastat, MMI-270, Trocade, Periomastat, and Metamastat. Nearly all of these MMP inhibitors have failed in clinical trials due to lack of specificity against a given class of MMPs [1], posing a challenge to rational design of specific MMP inhibitors. Computational design of MMP inhibitors is limited by several issues including an appropriate force field to model the metal-ligand interactions. Zinc is commonly found to be four-coordinated with a tetrahedral geometry; five and six coordinated geometries are also observed in zinc metalloproteinases and play an important role in metal/ligand binding [27]. Computational docking and prediction of binding affinities of metalloproteinase inhibitors to MMPs remains a challenge due to the multiple coordination geometries of zinc [28]. Addressing these challenges, we combine here molecular docking, quantum mechanical charge derivation of the Zn ion in the binding site followed by molecular dynamics (MD) simulations of an MMP-inhibitor complex, and post facto analyses of the MD trajectories to evaluate the binding free energies associated with MMP-inhibitor complexation. The ability to computationally predict effectively the binding modes and affinities of small molecule inhibitors [29] for these zinc containing enzymes can be of utmost importance in designing very selective clinically relevant inhibitors.

2 Methods

The over expression of various MMPs (i.e., MMP-1, MMP-2, MMP-3, MMP-8, MMP-9, and MMP-13) has been implicated in several important diseases such as breast [30–39] and gastric cancer [40–43], peripheral nerve injury [44–50], neuropathic pain [51], spinal cord damage [52–56], brain Injury [57–64], colorectal cancer [65], pathologic bone resorption [66], chronic wound [67], inflammation of skin [68] and pulmonary tract [69], hypersensitivity [70], Alzheimer's disease [71, 72], Crohn's disease [72], and peridontitis [73]. Considering the importance of MMPs in these and various other diseases, we focus here on describing computational approaches to design inhibitors against MMPs.

Batimastat, a potent, broad-spectrum MMP inhibitor was the first to enter clinical trials against cancer [1]. Experimental pIC50 values (in nM) for Batimastat against various MMPs have previously been reported [1] (Table 1). However, there are no crystal structures available for a Batimastat-MMP complex [1]. Thus, to determine the molecular basis for the binding of Batimastat to various MMPs, a docking and scoring study was performed using an in silico drug design software suite developed in house called *Sanjeevini* [74] (*see* **Note 1**). The web server incorporates several modules that include, (1) a module for the detection of binding sites for a target protein, scanning against a million compound library [75] to identify hit molecules against a target protein, (2) an all atom-based docking, and (3) a scoring module to design molecules with desired affinity and specificity against the drug target. The docking [76, 77] and scoring [28, 78, 79] modules of *Sanjeevini* have been previously validated on a large dataset of 335 protein/DNA drug targets for inhibitors with known crystal structures and known experimental binding free energies (*see* **Note 2**). The binding free energies of the top ranked docked structure predicted by *Sanjeevini* when compared with the experimental binding free energies gave a correlation coefficient of 0.83 for 335

Table 1
Experimental IC50 values in nM reported for the binding of Batimastat with MMP-1, MMP-2, MMP-3, MMP-8, MMP-9, and MMP-13 [1]

Serial number	MMPs	IC50 values (nM)
1	MMP-1	10
3	MMP-2	4
4	MMP-3	20
5	MMP-8	10
6	MMP-9	1

protein/DNA drug targets. The root mean square deviation of the top ranked docked structure against the crystal structure lies within 2 Å with 90% accuracy. Further, to test the accuracy of the docking and scoring modules of *Sanjeevini* on Matrix metalloproteinases, we analyzed a test set of known crystal structures of matrix metalloproteinase-inhibitor complexes with known experimental binding free energies. A correlation coefficient (R^2) of 0.68 was obtained between experimental and *Sanjeevini* predicted binding energies for the top ranked docked structure (Fig. 1).

Here, we demonstrate how it is possible to model the coordinates of Batimastat-MMP complexes using the docking and scoring modules integrated into *Sanjeevini* in conjunction with molecular dynamics simulations. The computational pathway designed for matrix metalloprotease drug design is illustrated in Fig. 2 and outlined in detail below.

2.1 Protein and Inhibitor (Batimastat) Preparation

Three-dimensional structures of target MMPs are downloaded from the RCSB (http://www.rcsb.org/pdb/home/home.do). PDB ids corresponding to MMP-1, MMP-2, MMP-3, MMP-8, MMP-9, and MMP-13 were 2TCL, 1HOV, 1BIW, 1MNC, 1GKC, and 456C respectively [19]. Crystallographic waters are removed from the protein [28] and hydrogen atoms are added along with the correct AMBER force field parameters [28]. The MMP catalytic site consisted of several histidine residues along with a zinc ion. The protonation state of these histidine residues is adjusted according to the catalytic site hydrogen bond network [80]. Basic amino acid residues are protonated and carboxylic groups are left deprotonated. The ionization state of the inhibitor

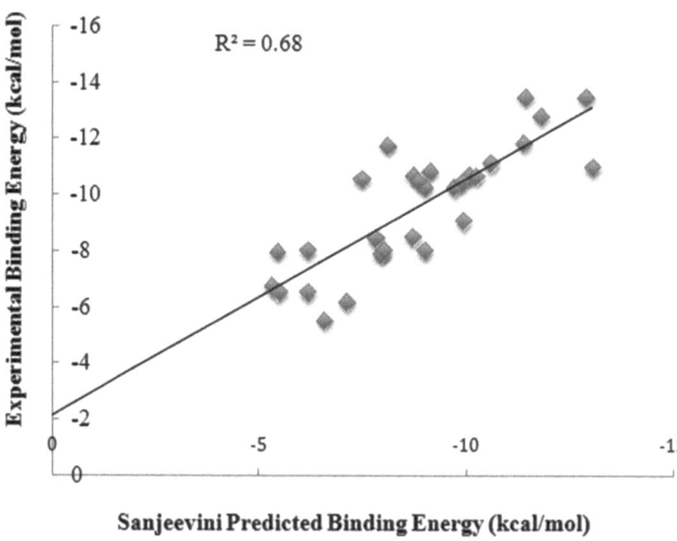

Fig. 1 Correlation plot between experimental binding energies (kcal/mol) and *Sanjeevini* predicted binding energy (in kcal/mol) of known MMP inhibitors

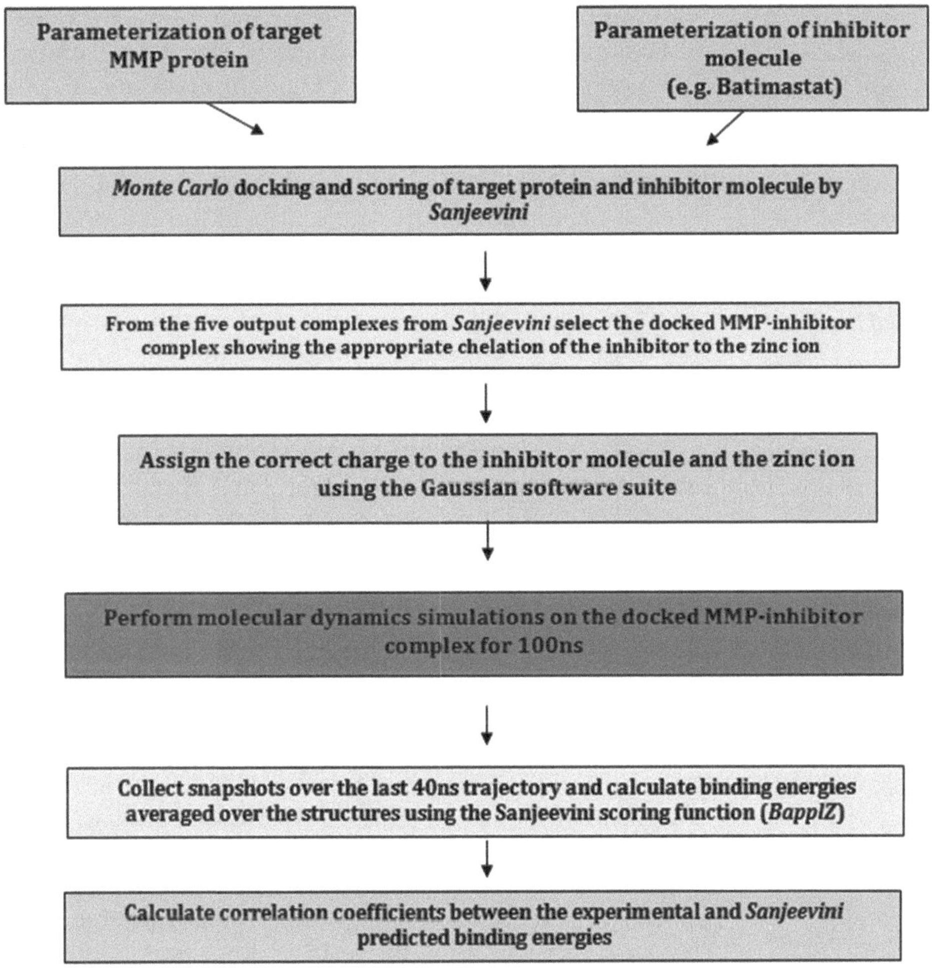

Fig. 2 Computational pathway showing the steps followed in docking and scoring study of Batimastat binding to different MMPs

molecule Batimastat is modeled as reported previously in the literature [1] and its overall formal charge [28] is maintained at −1. The molecule is further geometry optimized utilizing the AM1 procedure followed by calculation of partial charges by the AM1-BCC procedure [81]. GAFF force field parameters [82] are used to assign atom types, bond angle, dihedral, and van der Waals parameters [83] for the inhibitor molecule.

2.2 Molecular Docking and Scoring

The target protein-ligand complex and the inhibitor molecule are provided as input to *Sanjeevini* software suite for docking and scoring studies. The docking module [76, 77] of *Sanjeevini* docks the ligand molecule to the binding site and generates several (i.e., 10^3) configurations via a six-dimensional rigid body *Monte Carlo* methodology resulting in many ligand configurations that are scored based on the scoring function built into

Sanjeevini. Four docked structures representing favorable poses of the inhibitor molecule in the binding site along with the associated predicted binding free energies in kcal/mol are provided as output by the server.

2.3 Derivation of Partial Atomic Charges on the Docked Ligand, Protein and the Zinc Ion

In zinc containing metalloprotein-ligand complexes, the ligand replaces a zinc-bound water molecule and forms monodentate or bidentate coordinate bond(s) with the zinc ion. Thus, due to a charge transfer between amino acid residues and the zinc-bound ligand molecule, the total formal charge on the zinc ion is always less than +2. HF/6-31G* ab initio calculations were performed on a region encompassing the zinc ion, the ligand and zinc-bonded histidine residues using the Gaussian software [84] suite, followed by RESP fitting on the resultant electrostatic potentials to obtain partial atomic charges on the ligand and the zinc ion. For these calculations protein residues within the coordinate bond distance (<2.7 Å) of the zinc ion are deprotonated. The net charge on the zinc binding motif comprising the zinc ion, amino acid residues within a distance of 2.7 Å of the zinc ion and the ligand molecule is equal to the sum of the formal charge on each amino acid residue, the formal charge on the ligand, and the +2 charge of the zinc ion [28]. Parameters for the zinc ion are adopted from the work of Stote and Karplus [85] [σ = 1.95 Å, ε = 0.25 (kcal/mol)], while those for the ligand atoms are from the GAFF force field [82]. The AntechAMBER [86] module of AMBER is used to assign the bonded and the nonbonded parameters to the ligand atoms. Assignment of force field parameters for protein atoms (RESP derived partial atomic charges, van der Waals and bonded parameters) is carried out using the AMBER force field.

2.4 Molecular Dynamics Simulations of the Docked Complex

The docked protein-inhibitor complexes are subjected to molecular dynamics simulations to account for flexibility/dynamics of the ligand and the active site residues of the target, explicit solvent, and small ion effects [87]. These molecular dynamics simulations are performed under periodic boundary conditions within the AMBER suite of programs [86]. Prior to molecular dynamics simulations 11 Na^+ ions are added to MMP-2 and MMP-3 inhibitor complexes, 12 Na^+ ions to MMP-1, MMP-8, and MMP-9 inhibitor complexes and seven Na^+ ions to the MMP-13 inhibitor complex to ensure that there is a zero net charge on the protein-ligand complex. The complexes are solvated with an 8 Å thick layer of water modeled using TIP4PEW parameters [88]. Distances defining the zinc ion and nitrogen atoms of the zinc-bound histidine residues are restrained to prevent the zinc ion from escaping into the solvent and to maintain the orientation of the zinc-chelating histidine residues. Once the docked complexes have been prepared for molecular dynamics simulations, an initial minimization of the solvent molecules is performed followed by minimization of the solute-solvent system. Slow heating to 300 K, while keeping the volume constant, is performed over a period of 100 ps on the solute atoms using harmonic restraints of 25 kcal/mol/Å². Slow relaxation from 5 to 1 kcal/mol/Å² is applied to these restraints, in five segments consisting of 1,000 steps of energy minimization and 50 ps of equilibration, with a constant temperature of 300 K and a pressure of 1 bar via the Berendsen algorithm [89] with a coupling constant of 0.2 ps for both parameters. Further, we also apply 50 ps of equilibration with a restraint of 0.5 kcal/mol/Å² and 50 ps of unrestrained equilibration. This is followed by a molecular dynamics simulation for 100 ns under constant temperature and pressure using the Berendsen algorithm with a coupling constant of 5 ps. Molecular dynamics simulations are performed on all unbound MMPs studied and their inhibitor-bound complexes [90] and plots of RMSD versus time are used to check the stability of the docked complexes.

2.5 Post facto Analyses of Molecular Dynamics Simulations

Structures at equal intervals over the last 40 ns of the molecular dynamics simulation run trajectory are extracted (i.e., approximately 100 structures in total) for each system (*see* **Note 3**). For each structure the binding free energy is estimated using the Bappl-Z scoring function [28]. For these calculations, the system is parameterized within the additivity approximation where the net free energy change is treated as a sum of contributions from individual energy components. The equation for the estimation of the free energy change upon binding is:

$$\Delta G^{\circ} = \alpha \text{Eel} + \beta \text{Evdw} + \sum_{(\text{A}=1)}^{22} \sigma \text{A}\Delta \text{ALSA} + \lambda(\Delta \text{SCR}) + \delta \quad (1)$$

The Bappl-Z scoring function used for calculating binding free energy estimates has been thoroughly validated earlier [28]. The scoring function captures the theoretical rigor of the MMGBSA/MMPBSA [91–97] methodologies as well as the rapidity of empirical/knowledge-based methods [28, 78, 79]. The individual terms in Eq. 1 are described below.

Eel: Electrostatic interactions including interactions between protein and ligand atoms and the zinc ion with rest of the complex. These electrostatic interactions are calculated based on Coulomb's law with a sigmoidal dielectric function for solvent screening effects [28]. To model the electrostatic interactions of zinc with the rest of the complex, we have adopted the nonbonded model described by Stote and Karplus [85].

Evdw: Direct van der Waals interactions between protein and ligand atoms and the zinc ion with the rest of the complex. Van der Waals interactions are modeled using the Lennard-Jones potential [28] while interactions with the zinc ion are modeled on the lines of Stote and Karplus [85].

σAΔΔALSA: Hydrophobic contributions (nonelectrostatic components of desolvation) are captured using a modified version of the Eisenberg-Mclachlan model, where atom types for proteins and small molecules in the AMBER force field have been combined into a set of 22 atom types [28], to ensure transferability of parameters to other biological systems. In Eq. 1 ΔALSA represents the net loss in surface area for an atom type.

ΔSCR: Loss in conformational entropy of protein side chains upon binding of the ligand to the protein [28].

ΔG^{o} is calculated for each structural snapshot obtained over the last 40 ns of the trajectory file and block averaging is performed to obtain average binding free energy values (*see* **Note 4**).

2.6 Correlation between Experimental and Predicted Binding Free Energies

In zinc containing metalloprotein-ligand complexes, the ligand is bonded to the zinc ion with one or two coordinate bonds. The output of *Sanjeevini* is comprised of four docked structures representing energetically favorable poses of the ligand molecule in the binding site. The pose in which the hydroxamate group of the ligand molecule is chelated to the zinc ion is chosen for further analysis. Structures obtained from the molecular dynamics trajectory are processed using the *BapplZ* scoring function and the average binding energy is calculated for Batimastat binding against different MMPs. In Fig. 3, we present plots of the correlation between the experimental pIC50 and predicted binding free energies for Batimastat binding to different MMPs. These results show that the computational pathway outlined in this chapter, comprised

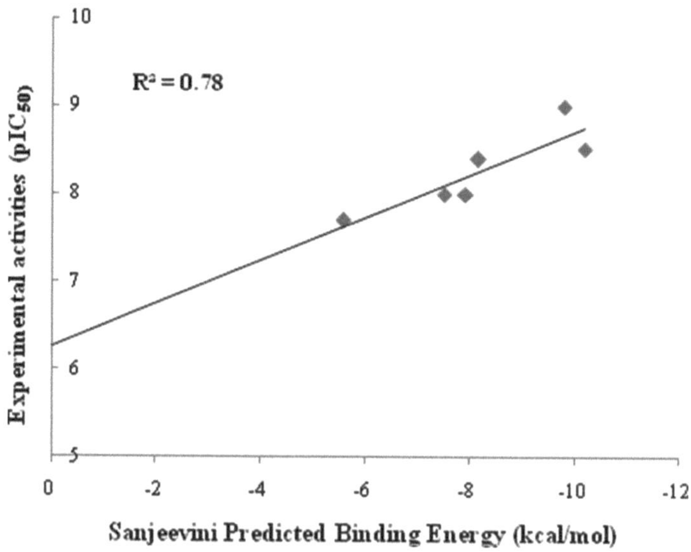

Fig. 3 Correlation between experimental activities (pIC50) and *Sanjeevini* predicted binding energies (kcal/mol) of Batimastat binding to MMP-1, MMP-2, MMP-3, MMP-8, MMP-9, and MMP-13

of (1) an efficient docking algorithm, (2) a correct charge assignment protocol, (3) a rigorous molecular dynamics simulation study, and (3) an accurate scoring function, can contribute toward structural, dynamic, and thermodynamic rationalization of the experimental inhibition data. This is evident from the high correlation coefficient [98] obtained between experimental and predicted binding free energies (Fig. 3).

3 Notes

1. *Sanjeevini* is freely available as a web server at http://www.scfbio-iitd.res.in/sanjeevini/sanjeevini.jsp for protein and DNA targeted lead molecule discovery.
2. Protein targets consisting of zinc containing metalloproteinases have also been tested earlier.
3. This approach subjects the systems to configurational averaging via molecular dynamics simulations, thereby overcoming the limitations inherent in single point calculations (i.e., studies on energy minimized structures).
4. The Bappl-Z scoring function is freely accessible as a web tool at http://www.scfbio-iitd.res.in/software/drugdesign/bapplz.jsp.

Acknowledgments

This work is supported by grants from the Department of Biotechnology, Govt. of India. Tanya Singh is a recipient of Senior Research Fellowship from Council of Scientific & Industrial Research, Govt. of India.

References

1. Skiles JW, Gonnella NC, Jeng AY (2004) The design, structure, and clinical update of small molecular weight matrix metalloproteinase inhibitors. Curr Med Chem 11:2911–2977
2. Cawston T (1998) Matrix metalloproteinases and TIMPs: properties and implications for the rheumatic diseases. Mol Med Today 4:130–137
3. Blavier L, Henriet P, Imren S, Declerck YA (1999) Tissue inhibitors of matrix metalloproteinases in cancer. Ann N Y Acad Sci 878: 108–119
4. Chang C, Werb Z (2001) The many faces of metalloproteases: cell growth, invasion, angiogenesis and metastasis. Trends Cell Biol 11:S37–S43
5. Hu J, Van den Steen PE, Sang QX, Opdenakker G (2007) Matrix metalloproteinase inhibitors as therapy for inflammatory and vascular diseases. Nat Rev Drug Discov 6:480
6. Ray JM, Stetler-Stevenson WG (1995) Gelatinase A activity directly modulates melanoma cell adhesion and spreading. EMBO J 14:908–917
7. Chambers AF, Matrisian LM (1997) Changing views of the role of matrix metalloproteinases in metastasis. J Natl Cancer Inst 89: 1260–1270
8. Kahari VM, Saarialho-Kere U (1993) Matrix metalloproteinases and their inhibitors in tumour growth and invasion. Ann Med 31:34–45

9. Kleiner DE, Stetler-Stevenson WG (1999) Matrix metalloproteinases and metastasis. Cancer Chemother Pharmacol 43:S42–S51
10. Yong VW (2005) Metalloproteinases: mediators of pathology and regeneration in the CNS. Nat Rev Neurosci 6:931–944
11. Whittaker M, Floyd CD, Brown P, Gearing AJ (1999) Design and therapeutic application of matrix metalloproteinase inhibitors. Chem Rev 99:2735
12. Weinstat-Saslow DL, Zabrenetzky VS, van Houtte K, Frazier WA, Roberts DD, Steeg PS (1994) Transfection of thrombospondin 1 complementary DNA into a human breast carcinoma cell line reduces primary tumor growth, metastatic potential, and angiogenesis. Cancer Res 54:6504–6511
13. Kessenbrock K, Plaks V, Werb Z (2010) Matrix metalloproteinases: regulators of the tumor microenviroment. Cell 141:52–67
14. Ardi VC, Kupriyanova TA, Deryugina EI, Quigley JP (2007) Human neutrophils uniquely release TIMP-free MMP-9 to provide a potent catalytic stimulator of angiogenesis. Proc Natl Acad Sci U S A 104:20262–20267
15. Egeblad M, Werb Z (2002) New functions for the matrix metalloproteinases in cancer progression. Nat Rev Cancer 2:161–174
16. Elkington P, Shiomi T, Breen R, Nuttall RK, Ugarte-Gil CA, Walker NF, Saraiva L, Pedersen B, Mauri F, Lipman M, Edwards DR, Robertson BD, D'Armiento J, Friedland JS (2011) MMP-1 drives immunopathology in human tuberculosis and transgenic mice. J Clin Invest 121:1827–1833
17. Murphy G, Knäuper V, Atkinson S, Butler G, English W, Hutton M, Stracke J, Clark I (2002) Matrix metalloproteinases in arthritic disease. Arthritis Res 4(Suppl 3):S39–S49
18. Fingleton B (2007) Matrix metalloproteinases as valid clinical targets. Curr Pharm Des 13:333–346
19. Westbrook J, Feng Z, Gilliland G, Bhat TN, Weissig H, Shindyalov IN, Bourne PE (2000) The Protein Data Bank. Nucleic Acids Res 28:235–242
20. Tu G, Xu W, Huang H, Li S (2008) Progress in the development of matrix metalloproteinase inhibitors. Curr Med Chem 15:1388–1395
21. Overall CM, López-Otín C (2002) Strategies for MMP inhibition in cancer: innovations for the post-trial era. Nat Rev Cancer 2:657–672
22. Fisher JF, Mobashery S (2006) Recent advances in MMP inhibitor design. Cancer Metastasis Rev 25:115–136
23. Dransfield DT (2011) New strategies for the next generation of matrix-metalloproteinase inhibitors: selectively targeting membrane-anchored MMPs with therapeutic antibodies. Biochem Res Int. doi:10.1155/2011/191670 Article ID 191670: 11p
24. Close DR (2001) Matrix metalloproteinase inhibitors in rheumatic diseases. Ann Rheum Dis 60:62–67
25. Baragi VM, Becher G, Bendele AM, Biesinger R, Bluhm H, Boer J, Deng H, Dodd R, Essers M, Feuerstein T, Gallagher BM, JrGege C, Hochgürtel M, Hofmann M, Jaworski A, Jin L, Kiely A, Korniski B, Kroth H, Nix D, Nolte B, Piecha D, Powers TS, Richter F, Schneider M, Steeneck C, Sucholeiki I, Taveras A, Timmermann A, Van Veldhuizen J, Weik J, Wu X, Xia B (2009) A new class of potent matrix metalloproteinase 13 inhibitors for potential treatment of osteoarthritis: evidence of histologic and clinical efficacy without musculoskeletal toxicity in rat models. Arthritis Rheum 60:2008–2018
26. Li NG, Shi ZH, Tang YP, Wang ZJ, Song SL, Qian LH, Qian DW, Duan JA (2011) New hope for the treatment of osteoarthritis through selective inhibition of MMP-13. Curr Med Chem 18:977–1001
27. Kawai K, Nagata N (2012) Metal-ligand interactions: an analysis of zinc binding groups using the Protein Data Bank. Eur J Med Chem 51:271–276
28. Jain T, Jayaram B (2007) A computational protocol for predicting the binding affinities of zinc containing metalloprotein-ligand complexes. Proteins 67:1167–1178
29. Langley DR, Walsh AW, Baldick CJ, Eggers BJ, Rose RE, Levine SM, Kapur AJ, Colonno RJ, Tenney DJ (2007) Inhibition of hepatitis B virus polymerase by entecavir. J Virol 81:3992–4001
30. Liu H, Kato Y, Erzinger SA, Kiriakova GM, Qian Y, Palmieri D, Steeg PS, Price JE (2012) The role of MMP-1 in breast cancer growth and metastasis to the brain in a xenograft model. BMC Cancer 12:583
31. Fink-Retter A, Gschwantler-Kaulich D, Hudelist G, Walter K, Czerwenka C (2007) Differential spatial expression and activation pattern of EGFR and HER2 in human breast cancer. Oncol Rep 18:299–304
32. Poola I, DeWitty RL, Marshalleck JJ, Bhatnagar R, Abraham J, Leffall LD (2005) Identification of MMP-1 as a putative breast cancer predictive marker by global gene expression analysis. Nat Med 11:481–483
33. Monteagudo C, Merino MJ, San-Juan J, Liotta LA, Stetler-Stevenson WG (1990) Immunohistochemical distribution of type IV collagenase in normal, benign, and malignant breast tissue. Am J Pathol 136:585–592

34. Onisto M, Riccio MP, Scannapieco P, Caenazzo C, Griggio L, Spina M, Stetler-Stevenson WG, Garbisa S (1995) Gelatinase A/TIMP-2 imbalance in lymph-node-positive breast carcinomas, as measured by RT-PCR. Int J Cancer 63:621–626
35. Engel G, Heselmeyer K, Auer G, Bäckdahl M, Eriksson E, Linder S (1994) Correlation between stromelysin-3 mRNA level and outcome of human breast cancer. Int J Cancer 58:830–835
36. Ahmad A, Hanby A, Dublin E, Poulsom R, Smith P, Barnes D, Rubens R, Anglard P, Hart I (1998) Stromelysin 3: an independent prognostic factor for relapse-free survival in node-positive breast cancer and demonstration of novel breast carcinoma cell expression. Am J Pathol 152:721–728
37. Freije JM, Diez-Itza I, Balbin M, Sánchez LM, Blasco R, Tolivia J, López-Otín C (1994) Molecular cloning and expression of collagenase-3, a novel human matrix metalloproteinase produced by breast carcinomas. J Biol Chem 269:16766–16773
38. Ueno H, Nakamura H, Inoue M, Imai K, Noguchi M, Sato H, Seiki M, Okada Y (1997) Expression and tissue localization of membrane-types 1, 2, and 3 matrix metalloproteinases in human invasive breast carcinomas. Cancer Res 57:2055–2060
39. Jones JL, Glynn P, Walker RA (1999) Expression of MMP-2 and MMP-9, their inhibitors, and the activator MT1-MMP in primary breast carcinomas. J Pathol 189:161–168
40. Cai QW, Li J, Li X, Wang J, Huang Y (2012) Expression of STAT3, MMP-1 and TIMP-1 in gastric cancer and correlation with pathological features. Mol Med Rep 5:1438–1442
41. Murray GI, Duncan ME, Arbuckle E, Melvin WT, Fothergill JE (1998) Matrix metalloproteinases and their inhibitors in gastric cancer. Gut 43:791–797
42. Kemik O, Kemik AS, Sümer A, Dulger AC, Adas M, Begenik H, Hasirci I, Yilmaz O, Purisa S, Kisli E, Tuzun S, Kotan C (2011) Levels of matrix metalloproteinase-1 and tissue inhibitors of metalloproteinase-1 in gastric cancer. World J Gastroenterol 17:2109–2112
43. Fujimoto D, Hirono Y, Goi T, Katayama K, Vamaguchi A (2008) Prognostic value of Protease-activated Receptor-1 (PAR-1) and Matrix Metalloproteinase-1 (MMP-1) in gastric cancer. Anticancer Res 28:847–854
44. Kawasaki Y, Xu ZZ, Wang X, Park JY, Zhuang ZY, Tan PH, Gao YJ, Roy K, Corfas G, Lo EH, Ji RR (2008) Distinct roles of matrix metalloproteases in the early- and late-phase development of neuropathic pain. Nat Med 14:331–336
45. Shubayev VI, Angert M, Dolkas J, Campana WM, Palenscar K, Myers RR (2006) TNF alpha-induced MMP-9 promotes macrophage recruitment into injured peripheral nerve. Mol Cell Neurosci 31:407–415
46. Chattopadhyay S, Myers RR, Janes J, Shubayev V (2007) Cytokine regulation of MMP-9 in peripheral glia: implications for pathological processes and pain in injured nerve. Brain Behav Immun 21:561–568
47. Liu LY, Zheng H, Xiao HL, She ZJ, Zhao SM, Chen ZL, Zhou GM (2008) Comparison of blood-nerve barrier disruption and matrix metalloprotease-9 expression in injured central and peripheral nerves in mice. Neurosci Lett 434:155–159
48. Platt CI, Krekoski CA, Ward RV, Edwards DR, Gavrilovic J (2003) Extracellular matrix and matrix metalloproteinases in sciatic nerve. J Neurosci Res 74:417–429
49. Shubayev VI, Myers RR (2008) Upregulation and interaction of TNFalpha and gelatinases A and B in painful peripheral nerve injury. Brain Res 855:83–89
50. Ramer R, Hinz B (2008) Inhibition of cancer cell invasion by cannabinoids via increased expression of tissue inhibitor of matrix metalloproteinases-1. J Natl Cancer Inst 2:59–69
51. Sommer C, Schmidt C, George A, Toyka KV (1997) A metalloprotease-inhibitor reduces pain associated behavior in mice with experimental neuropathy. Neurosci Lett 237:45–48
52. Noble LJ, Donovan F, Igarash T, Goussev S, Werb Z (2002) Matrix metalloproteinases limit functional recovery after spinal cord injury by modulation of early vascular events. J Neurosci 1:7526–7536
53. Pannu R, Christie DK, Barbosa E, Singh I, Singh AK (2007) Post-trauma Lipitor treatment prevents endothelial dysfunction, facilitates neuroprotection, and promotes locomotor recovery following spinal cord injury. J Neurochem 101:182–200
54. Fleming JC, Norenberg MD, Ramsay DA, Dekaban GA, Marcillo AE, Saenz AD, Pasqale-Styles M, Dietrich WD, Weaver LC (2008) The cellular inflammatory response in human spinal cords after injury. Brain 129:3249–3269
55. Amantea D, Corasaniti MT, Mercuri NB, Bernardi G, Bagetta G (2008) Brain regional and cellular localization of gelatinase activity in rat that have undergone transient middle cerebral artery occlusion. Neuroscience 152:8–17
56. Nagel S, Su Y, Horstmann S, Heiland S, Gardner H, Koziol J, Martinez-Torres FJ, Wagner S (2008) Minocycline and hypothermia for reperfusion injury after focal cerebral ischemia in the rat: effects on BBB breakdown

and MMP expression in the acute and subacute phase. Brain Res 1188:198–206
57. Ding YH, Li J, Rafols JA, Ding Y (2004) Reduced brain edema and matrix metalloproteinase (MMP) expression by pre-reperfusion infusion into ischemic territory in rat. Neurosci Lett 372:35–39
58. Fujimoto M, Takagi Y, Aoki T, Hayase M, Marumo T, Gomi M, Nishimura M, Kataoka H, Hashimoto N, Nozaki K (2008) Tissue inhibitor of metalloproteinases protect blood-brain barrier disruption in focal cerebral ischemia. J Cereb Blood Flow Metab 28:1674–1685
59. Vilalta A, Sahuquillo J, Rosell A, Poca MA, Riveiro M, Montaner J (2008) Moderate and severe traumatic brain injury induce early overexpression of systemic and brain gelatinases. Intensive Care Med 34:1384–1392
60. Jiang X, Namura S, Nagata I (2001) Matrix metalloproteinase inhibitor KB-R7785 attenuates brain damage resulting from permanent focal cerebral ischemia in mice. Neurosci Lett 305:41–44
61. Lee JE, Yoon YJ, Moseley ME, Yenari MA (2005) Reduction in levels of matrix metalloproteinases and increased expression of tissue inhibitor of metalloproteinase-2 in response to mild hypothermia therapy in experimental stroke. J Neurosurg 103:289–297
62. Yang Y, Estrada EY, Thompson JF, Liu W, Rosenberg GA (2007) Matrix metalloproteinase-mediated disruption of tight junction proteins in cerebral vessels is reversed by synthetic matrix metalloproteinase inhibitor in focal ischemia in rat. J Cereb Blood Flow Metab 4:697–709
63. Wang Y, Deng XL, Xiao XH, Yuan BX (2007) A non-steroidal anti-inflammatory agent provides significant protection during focal ischemic stroke with decreased expression of matrix metalloproteinases. Curr Neurovasc Res 3:176–183
64. Truettner JS, Alonso OF, Dalton DW (2005) Influence of therapeutic hypothermia on matrix metalloproteinase activity after traumatic brain injury in rats. J Cereb Blood Flow Metab 11:1505–1516
65. Sunami E, Tsuno N, Osada T, Saito S, Kitayama J, Tomozawa S, Tsuruo T, Shibata Y, Muto T, Nagawa H (2000) MMP-1 is a prognostic marker for hematogenous metastasis of colorectal cancer. Oncologist 5:108–114
66. Rodrigues WF, Madeira MF, da Silva TA, Clemente-Napimoga JT, Miguel CB, Dias-da-Silva VJ, Barbosa-Neto O, Lopes AH, Napimoga MH (2012) Low dose of propranolol down-modulates bone resorption by inhibiting inflammation and osteoclast differentiation. Br J Pharmacol 165:2140–2151
67. Wysocki AB, Kusakabe AO, Chang S, Tuan TL (1997) Temporal expression of urokinase plasminogen activator, plasminogen activator inhibitor and gelatinase-B in chronic wound fluid switches from a chronic to acute wound profile with progression to healing. Wound Repair Regen 7:154–165
68. Warner RL, Bhagavathula N, Nerusu KC, Lateef H, Younkin E, Johnson KJ, Varani J (2004) Matrix metalloproteinases in acute inflammation: induction of MMP-3 and MMP-9 in fibroblasts and epithelial cells following exposure to pro-inflammatory mediators in vitro. Exp Mol Pathol 76:189–195
69. Sagel, S.D., Kapsner, R.K, and Osberg, I. (2005) Induced sputum matrix metalloproteinase-9 correlates with lung function and airway inflammation in children with cystic fibrosis. Pediatr Pulmonol 39, 224–232.
70. Yang H, Dai Y, Dong H, Zang D, Liu Q, Duan H, Niu Y, Bin P, Zheng Y (2011) Trichloroethanol up-regulates matrix metalloproteinase-9 and tissue inhibitor of metalloproteinase-1 in HaCaT cells. Toxicol In Vitro 25:1638–1643
71. Shibataa N, Ohnumaa T, Higashia S, Usuia C, Ohkuboa T, Kitajimaa A, Uekib A, Nagaoc M, Araia H (2005) Genetic association between matrix metalloproteinase MMP-9 and MMP-3 polymorphisms and Japanese sporadic Alzheimer's disease. Neurobiol Aging 26:1011–1014
72. Kofla-Dlubacz A, Matusiewicz M, Krzystek-Korpacka M, Iwanczak B (2012) Correlation of MMP-3 and MMP-9 with Crohn's disease activity in children. Dig Dis Sci 57:706–712
73. Kumar MS, Vamsi G, Sripriya R, Sehgal PK (2006) Expression of matrix metalloproteinases (MMP-8 and -9) in chronic periodontitis patients with and without diabetes mellitus. J Periodontol 77:1803–1808
74. Jayaram B, Singh T, Mukherjee G, Mathur A, Shekhar S, Shekhar V (2012) *Sanjeevini*: a freely accessible web-server for target directed lead molecule discovery. BMC Bioinformatics 13:S7
75. Mukherjee G, Jayaram B (2013) A rapid identification of hit molecules for target proteins via physico-chemical descriptors. Phys Chem Chem Phys 15:9107–9116
76. Gupta A, Gandhimathi A, Sharma P, Jayaram B (2007) ParDOCK: an all atom energy-based Monte Carlo docking protocol for protein-ligand complexes. Protein Pept Lett 14:632–646
77. Singh T, Biswas D, Jayaram B (2011) AADS—An automated active site identification, docking and scoring protocol for protein targets based on physico-chemical descriptors. J Chem Inf Model 51:2515–2527
78. Jain T, Jayaram B (2005) An all atom energy-based computational protocol for predicting

binding affinities of protein-ligand complexes. FEBS Lett 579:6659–6666

79. Shaikh SA, Jayaram B (2007) A Swift all-atom energy-based computational protocol to predict DNA ligand binding affinity and ΔTm. J Med Chem 50:2240–2244
80. Giangreco I, Lattanzi G, Nicolotti O, Catto M, Laghezza A, Leonetti F, Stefanachi A, Carotti A (2011) Insights into the complex formed by matrix metalloproteinase-2 and alloxan inhibitors: molecular dynamics simulations and free energy calculations. PLoS One 6:1
81. Jakalian A, Bush BL, Jack DB, Bayly CI (2004) Fast, efficient generation of high-quality atomic charges. AM1-BCC model: I. Method. J Comput Chem 21:132–146
82. Wang J, Wolf RM, Caldwell JW, Kollman PA, Case DA (2004) Development and testing of a general amber force field. J Comput Chem 25:1157–1174
83. Cornell WD, Cieplak P, Bayly CI, Gould IR, Merz KM (1995) A second generation force field for the simulation of proteins, nucleic acids, and organic molecules. J Am Chem Soc 117:5179–5197
84. Frisch MJ, Trucks GW, Schlegel HB, Scuseria GE, Robb MA, Cheeseman JR, Scalmani G, Barone V, Mennucci B, Petersson GA, Nakatsuji H, Caricato M, Li X, Hratchian HP, Izmaylov AF, Bloino J, Zheng G, Sonnenberg JL, Hada M, Ehara M, Toyota K, Fukuda R, Hasegawa J, Ishida M, Nakajima T, Honda Y, Kitao O, Nakai H, Vreven T, Montgomery JA Jr, Peralta JE, Ogliaro F, Bearpark M, Heyd JJ, Brothers E, Kudin KN, Staroverov VN, Kobayashi R, Normand J, Raghavachari K, Rendell A, Burant JC, Iyengar SS, Tomasi J, Cossi M, Rega N, Millam NJ, Klene M, Knox JE, Cross JB, Bakken V, Adamo C, Jaramillo J, Gomperts R, Stratmann RE, Yazyev O, Austin AJ, Cammi R, Pomelli C, Ochterski JW, Martin RL, Morokuma K, Zakrzewski VG, Voth GA, Salvador P, Dannenberg JJ, Dapprich S, Daniels AD, Farkas Ö, Foresman JB, Ortiz JV, Cioslowski J, Fox DJ (2009) Gaussian, Inc., Wallingford, CT
85. Stote RH, Karplus M (1995) Zinc binding in proteins and solution: a simple but accurate nonbonded representation. Proteins Struct Funct Genet 23:12–31
86. Pearlman DA, Case DA, Caldwell JW, Ross WS, Cheathem JE III, DeBolt S, Ferguson D, Seibel G, Kollman P (1995) AMBER, a package of computer programs for applying molecular mechanics, normal mode analysis, molecular dynamics and free energy calculations to simulate the structural and energetic properties of molecules. Comput Phys Commun 91:1–41
87. Kalra P, Reddy TV, Jayaram B (2011) Free energy component analysis for drug design: a case study of HIV-1 protease-inhibitor binding. J Med Chem 44:4325–4338
88. Horn HW, Swope WC, Pitera JW, Madura JD, Dick TJ, Hura GL, Head-Gordon T (2004) Development of an improved four-site water model for biomolecular simulations: TIP4P-Ew. J Chem Phys 120:9665–9678
89. Berendsen HJC, Postma JPM, van Gunsteren WF, DiNola A, Haak JR (1984) Molecular dynamics with coupling to an external bath. J Chem Phys 81:3684–3690
90. Lavery R, Zakrzewska K, Beveridge DL, Bishop TC, Case DA, Cheatham T III, Dixit S, Jayaram B, Lankas F, Laughton C, Maddocks JH, Michon A, Osman R, Orozco M, Perez A, Singh T, Spackova N, Sponer J (2009) A systematic molecular dynamics study of nearest neighbor effects on base pair and base pair step conformations and fluctuations in B-DNA. Nucleic Acids Res 38:299–313
91. Rizzo RC, Toba S, Kuntz ID (2004) A molecular basis for the selectivity of thiadiazole urea inhibitors with stromelysin-1 and gelatinase-A from generalized born molecular dynamics simulations. J Med Chem 47:3065–3074
92. Chong S, Ham S (2013) Assessing the influence of solvation models on structural characteristics of intrinsically disordered protein. Comput Theor Chem 1017:194–199
93. Shaikh SA, Ahmed SR, Jayaram B (2004) A molecular thermodynamic view of DNA-drug interaction: a case study of 25 minor groove binders. Arch Biochem Biophys 429:81–99
94. Jayaram B, Mcconnell K, Dixit SB, Beveridge DL (2002) Free energy component analysis of 40 protein-DNA complexes: a consensus view on the thermodynamics of binding at the molecular level. J Comput Chem 23:1–14
95. Jayaram B, McConnell KJ, Dixit SB, Beveridge DL (1999) Free energy analysis of protein-DNA binding: the *EcoR*I endonuclease—DNA complex. J Comput Phys 151:333–357
96. Fenley MO, Harris R, Jayaram B, Boschitsch AH (2010) Revisiting the association of cationic groove-binding drugs to DNA using a Poisson-Boltzmann approach. Biophys J 99:879–886
97. Wong S, Amaro RE, McCammon JA (2009) MM-PBSA captures key role of intercalating water molecules at a protein–protein interface. J Chem Theory Comput 5:422–449
98. Singh T, Adekoya OA, Jayaram B (2015) Understanding the binding of inhibitors of matrix metalloproteinases by molecular docking, quantum mechanical calculations, molecular dynamics simulations, and a MMGBSA/MMBappl study. Mol Biosyst 11:1041–1051

Chapter 16

A Simple Adaptable Blood-Brain Barrier Cell Model for Screening Matrix Metalloproteinase Inhibitor Functionality

Jennifer S. Myers, Joan Hare, and Qing-Xiang Amy Sang

Abstract

The blood-brain barrier is a multicellular and basement membrane unit that regulates molecular transport between the blood and central nervous system. Many cerebral pathologies, such as acute stroke and chronic vascular dementia, result in a disrupted blood-brain barrier, increasing its permeability and allowing the entry of potentially neurotoxic molecules. The activation of matrix metalloproteinases mediates further blood-brain barrier damage. The inhibition of matrix metalloproteinases is a potential strategy for stroke therapy. As inhibitors are developed, efficient context-specific screening methods will be required. Models of the blood-brain barrier have been extensively used to study neuropathologies and the effect of various treatment options.

Herein, we describe a co-culture model of the blood-brain barrier composed of brain microvascular endothelial cells and astrocytes grown on an artificial basement membrane-coated membrane insert. Our cell model forms a barrier and is a simple first approximation of blood-brain barrier integrity. As currently developed, the model may be applied to testing the effect of matrix metalloproteinases and matrix metalloproteinase inhibitors on blood-brain barrier physiology and pathophysiology. The model is a quick and effective evaluation tool for generating nonclinical data in a living cell system before proceeding to animal models.

Key words Blood-brain barrier, Matrix-metalloproteinases, Matrix-metalloproteinase inhibitors, Tissue plasminogen activator, Stroke, BBB co-culture model, BBB permeability, Cell culture membrane inserts, Normal human astrocytes, Bovine brain microvascular endothelial cells

1 Introduction

Cerebral vascular endothelial cells, a basal lamina, pericytes, and astrocytes or glia combine to form the blood-brain barrier (BBB) neurovascular unit, a selectively permeable physical barrier between blood vessels and the central nervous system. Each component of the neurovascular unit plays a particular role in BBB function [1]. The BBB is a complex entity that ensures normal brain function by controlling the entry of molecules and white blood cells from

Charles A. Galea (ed.), *Matrix Metalloproteases: Methods and Protocols*, Methods in Molecular Biology, vol. 1579,
DOI 10.1007/978-1-4939-6863-3_16, © Springer Science+Business Media LLC 2017

systemic circulation and by maintaining homeostasis of the brain's extracellular fluid. Tight junctions between the cells of the BBB maintain integrity of the barrier and restrict paracellular entry. Cerebral pathologies such as strokes and chronic vascular dementia alter the structural integrity of brain tissues and blood vessels. The resulting injury eventually degrades the BBB through the release and activation of matrix metalloproteinases (MMPs) [2].

Tissue plasminogen activator (tPA) is therapeutic for dissolving clots resulting from stroke and works by activating plasminogen to plasmin that then digests the fibrin in clots. However, plasmin has a damaging secondary side effect of activating MMPs and may thus disrupt the BBB [3]. MMPs have been shown to mediate stroke injury by destabilizing the BBB and aggravating cerebral damage through stimulation of inflammatory responses [4–6]. Loss of a properly functioning BBB leads to edema, hemorrhaging, and the introduction of neurotoxic elements to the brain environment. MMP inhibition has been shown to be a promising strategy in stroke treatment and recovery [2].

To study stroke and other cerebral pathologies, a vast array of BBB models have been developed. BBB cellular models have been employed for drug discovery and development [7, 8]. The complexity of the BBB presents a plethora of options for constructing a BBB model to varying degrees of physiological accuracy. BBB models range in complexity from simple monocultures of brain endothelial cells to three-dimensional co-cultures [9–11]. Early models used only monocultures and proved to be inadequate in replicating the BBB. Most current models are a variation of brain endothelial cells grown in culture with astrocytes or grown in astrocyte-conditioned medium. More sophisticated models may even incorporate additional components of the neurovascular unit, including pericytes and/or neurons [12–14]. More complex models incorporating multiple cell types more closely mimic in vivo BBB and achieve better performance in terms of reduced monolayer permeability. However, the complexity of such systems makes them more inconvenient, especially for screening applications.

The underlying challenge of BBB model design is striking a balance between simplicity and performance- producing a model that is physiologically accurate and relevant, easy to construct and adapt, and is economically feasible. We have created a simple model that reasonably approximates BBB physiology, is suitable as a screening tool, and may be used to screen libraries of MMP inhibitors for initial phases of drug development. The model described within is formed from bovine brain microvascular endothelial cells (bBMVEC) and normal human astrocytes (NHA) grown on opposite sides of an artificial basement membrane-coated porous membrane insert. In the model presented, tPA is used to induce BBB disruption. The integrity of the BBB model was assessed by measuring the passage of the fluorescent tracer Lucifer yellow between the cell layer, building on the examples of Culot *et al.* and Megard *et al.* [15, 16]. We selected bBMVEC for relative ease of culture and included astrocytes (*see* **Note 1**) because of their significant contribution to barrier properties in endothelial cells such as BBB formation and integrity [1, 17, 18]. The economic benefit of our model is twofold. First, the permeability of cells can be simply quantified by the use of Lucifer yellow and a fluorescence plate reader, a piece of equipment that is readily available and thus avoids the use of equipment dedicated to measurement of transendothelial electric resistance. Second, our choice of using primary cell lines instead of freshly harvested cells avoids continual animal sacrifice for each experiment and the attendant difficulties and costs.

2 Materials

2.1 Cell Culture

1. Type 1 rat tail collagen.
2. 0.02 N acetic acid.
3. Tris-buffered saline (TBS): 6.0 g Trizma-base, 16 g NaCl, 0.76 g KCl, 0.2 g Na_2HPO_4, 0.6 mL phenol red (0.5% solution in H_2O) in 2 L of water. Add HCl to pH 7.4.
4. Clonetics® bovine brain microvascular endothelial cells (bBMVEC) (Lonza, Walkersville, MD, USA).
5. EBM-2 medium supplemented with ascorbic acid, endothelial cell growth factor beta, platelet-poor horse serum, heparin, and penicillin/streptomycin/fungizone.
6. Clonetics® normal human astrocytes (NHA) (Lonza, Walkersville, MD, USA).
7. ABM medium supplemented with human recombinant epidermal growth factor, insulin, ascorbic acid, l-glutamine, gentamicin/amphotericin, and fetal bovine serum (FBS).
8. Transparent, 0.4 μm pore polyethylene terephthalate (PET) or 0.4 μm pore polytetrafluoroethylene (PTFE) membrane inserts (*see* **Note 2**).
9. Sterile box with loose fitting lid.
10. Sterile multi-well culture plates corresponding to size of membrane inserts with lids.
11. Human plasma fibronectin.

2.2 Permeability Testing

1. Ringer's-HEPES buffer: 8.77 g (150 mM) NaCl, 392.9 mg (5.2 mM) KCl, 323.4 mg (2.2 mM) $CaCl_2$, 0.0406 mg (0.2 mM) $MgCl_2$, 0.504 mg (6 mM) $NaHCO_3$, 0.5045 mg (2.8 mM) glucose, 1.192 g (5 mM) HEPES in 1 L of water. Adjust pH to 7.4.

2. Lucifer yellow dye.
3. Human tPA.
4. MMP inhibitor of choice.
5. Multi-well cell culture plate.
6. 96-well fluorescence assay plate with black walls and a clear bottom.
7. Aluminum foil.
8. Microtiter plate reader.

3 Methods

3.1 Cell Culture

All experiments were performed in a biological safety cabinet at room temperature unless otherwise stated. Inserts are easily punctured by pipet tips; extreme care must be taken to avoid contact with the membrane at each handling.

1. Prepare a stock solution of Type 1 rat tail collagen dissolved in 0.02 N acetic acid. Using this stock makes a dilution of sufficient concentration to provide 15 μg collagen/cm^2 of flask surface. Add the collagen solution to the culture flask and incubate at 37°C for 1 h (*see* **Note 3**). Remove the collagen solution and rinse the flask with TBS and then with warm medium.
2. Expand bBMVEC culture by plating at 5,000 cells/cm^2 in collagen-treated flask. Culture cells until 75–80% confluence and then harvest and freeze using method of choice. This will be a stock for future experiments. Freeze in multiple small batches such that each aliquot provides enough cells for a single experiment (*see* **Note 4**).
3. Expand NHA culture by plating between 5,000 cells/cm^2 in an untreated flask. Culture cells until 75–80% confluence and then harvest and freeze using method of choice. This will be a stock for future experiments. Freeze in multiple small batches such that each aliquot provides enough cells for a single experiment (*see* **Note 4**).
4. Three to four days prior to starting the co-culture, thaw an aliquot of bBMVEC into a collagen-coated culture flask and allow to recover with regular medium changes.
5. One day prior to starting the co-culture, prepare a dilution of Type 1 rat tail collagen dissolved in sterile H_2O to a sufficient concentration to deliver 15 μg collagen/cm^2 of membrane insert surface. Apply collagen solution to the apical surface inside the well of the membrane insert and allow to dry uncovered overnight (*see* **Note 5**). After drying, rinse with TBS (*see* **Note 6**).

6. When ready to begin co-culture, prepare a stock solution of human plasma fibronectin dissolved in unsupplemented EBM medium. Make a diluted fibronectin solution in unsupplemented EBM medium of sufficient concentration to provide 5 μg fibronectin/cm^2 of membrane insert surface (*see* **Note 7**). Apply the fibronectin solution to the apical side of the membrane insert and incubate partially covered at 37°C for 6 h and then rinse the surface with TBS (*see* **Note 8**).
7. Place membrane inserts in the sterilized box so that they are inverted with the basal side exposed. Seed NHA directly from thaw on basal side of membrane insert at 50,000 cells/cm^2. Slowly add the cells in small volumes to form a droplet on the surface (Fig. 1). If the surface tension bursts, collect the cell solution and reapply. Close the lid of the box and carefully

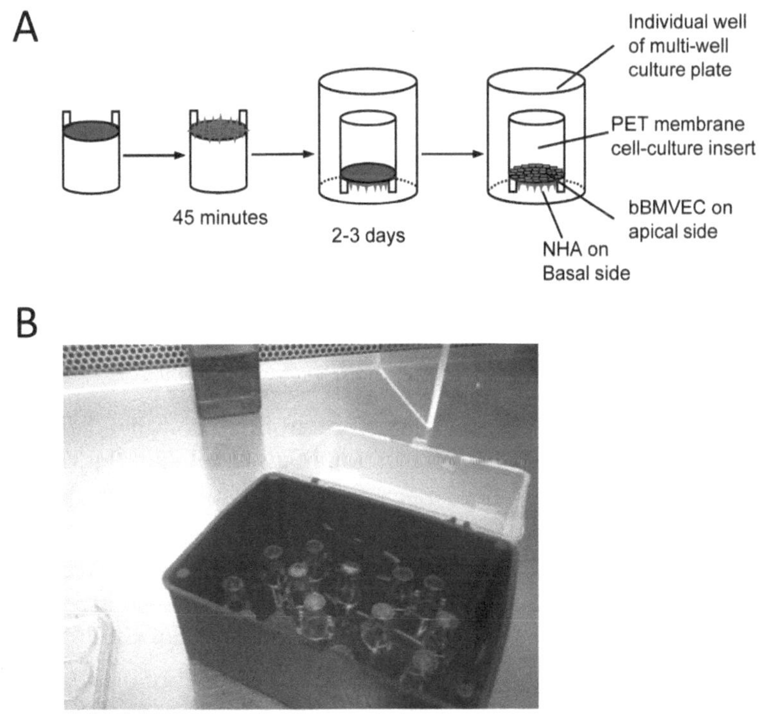

Fig. 1 Construction of co-culture on opposite sides of cell culture membrane inserts. (**a**) Schematic illustrating the application of normal human astrocytes (NHA) on the basal side and bovine brain microvascular endothelial cells (bBMVEC) on the apical side of porous polyethylene terephthalate (PET) membrane inserts. Membrane inserts were initially inverted before depositing NHA. NHA were allowed to attach to the membrane for 45 min in an incubator. The membrane inserts were then flipped into a multi-well plate. After allowing two to 3 days for NHA growth in an incubator, bBMVEC were seeded onto the apical membrane surface. (**b**) Picture showing the NHA cell solution droplets on the basal side of the membrane inserts

place it in an incubator at 37°C and 5% CO_2. Cells will attach in approximately 45 min.

8. After 45 min, use forceps to carefully flip membrane inserts into a multi-well plate containing warmed NHA-specific ABM supplemented medium. Allow to grow for 48–72 h with medium changes every 48 h (*see* **Note 9**).
9. After the prescribed NHA growing period, transfer membrane inserts to new wells containing fresh NHA-specific medium. Harvest bBMVEC from recovery flask and seed at 50,000 cells/cm^2 on the apical membrane surface in bBMVEC-specific EBM supplemented medium. Every 48 h, exchange EBM supplemented medium in the membrane insert and ABM supplemented medium in the well underneath.
10. Check the co-culture to see if it is fit for assay by observing the endothelial cells under a microscope (*see* **Note 10**). Endothelial cells display a characteristic cobblestone morphology and approach a confluent monolayer within 48–72 h of seeding.

3.2 Permeability Testing

All experiments were performed in a biological safety cabinet at room temperature unless otherwise stated. The experimental workflow consisted of treatment with tPA (with or without MMP inhibitor) followed by the addition of tracer dye to measure permeability (Fig. 2).

1. Make a stock tPA solution by diluting tPA in serum-free medium to around 1 μM. Keep in mind that the final tPA concentration in the assay will be 100-fold lower and you will need to pipet 1/100 the volume.
2. Dilute inhibitor in recommended buffer to 200 μM. Final concentration in assay will be 10 μM (*see* **Note 11**).
3. Remove medium from membrane inserts to be tested, rinse once with TBS, and replace with serum-free medium, DMSO to 2.5%, and enough stock tPA for a final tPA concentration of 10 nM. Add the appropriate amount of MMP inhibitor. Incubate at 37°C and 5% CO_2 for 6 h (*see* **Note 12**).
4. During the last 30 min of incubation, dissolve the Lucifer yellow in Ringer's-HEPES buffer at pH 7.4 to make standard solutions of 200, 100, 50, 20, 10, 1, 0.5, 0.25, 0.1, 0.05, 0.01, and 0 μM from a stock concentration of 40 mM. Make a sufficient quantity of 100 μM Lucifer yellow to use as both tracer and for the standards (*see* **Note 13**).
5. Remove the tPA solution from the membrane inserts and replace with 400 μL of 100 μM Lucifer yellow solution (*see* **Note 14**).
6. Move the co-culture membrane inserts to new, sterile multi-well plates containing a volume of transport buffer in the well

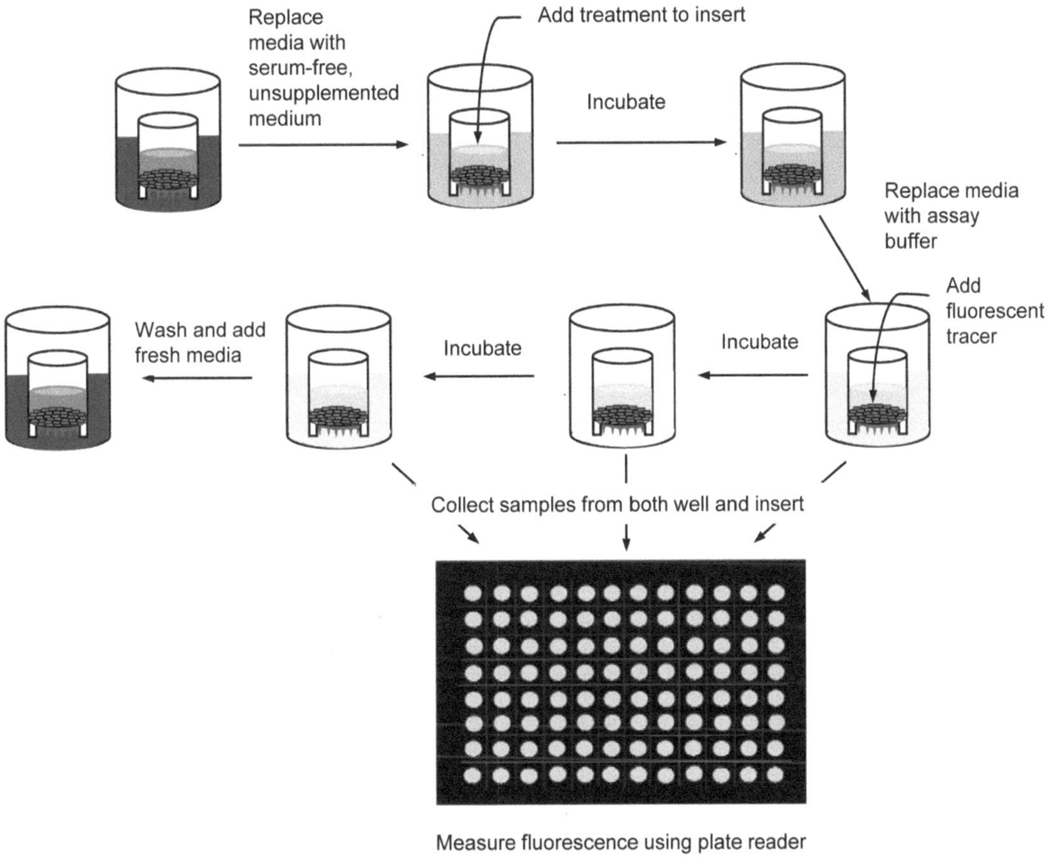

Fig. 2 Experimental workflow for testing the effect of MMP inhibition on monolayer permeability. The integrity of the monolayer formed in the co-culture was assessed by its permeability to Lucifer yellow fluorescent tracer. The effect of MMP inhibitors of the integrity of the monolayer was assessed after disrupting the monolayer with tissue plasminogen activator (tPA). First, cells were incubated with a treatment solution (tPA, tPA + inhibitor, inhibitor, etc.) for 6 h. Then, the treatment solution was replaced with a solution of Lucifer yellow. After the introduction of the tracer, the membrane insert and the well below were sampled at various time points up to 1 h. The passage of Lucifer yellow between the cells was determined by fluorescence spectroscopy

that ensures that the liquid level inside and outside the insert is roughly equal (*see* **Note 15**).

7. Immediately remove 100 μL aliquots from the well and the membrane insert and place in the 96-well assay plate. Then add 100 μL of fresh transport buffer to the membrane compartment. Repeat sampling at desired time points, keeping the culture in the incubator between time points (*see* **Note 16**).
8. Measure the fluorescence of the samples using a microtiter plate reader. Excite at 485 nm and record emission at 538 nm. Calculate the concentration of Lucifer yellow in the sample using a calibration curve.

9. Calculate the permeability of the cell model as follows using the % leakage and permeability coefficient, P_e, and mass balance (*see* **Notes 17** and **18**).

   (a) % Leakage = moles of dye in well ÷ moles of dye introduced into membrane insert.

   (b) P_e = (moles of dye in well ÷ molarity of dye introduced into membrane insert) ÷ (surface area × time).

   (c) Mass balance = (moles of dye in well + moles of dye in membrane insert) ÷ moles of dye introduced into membrane insert.

4 Notes

1. Bovine astrocytes are not commercially available therefore human astrocytes were substituted.
2. Membrane inserts can be 6-well, 12-well, or 24-well format. All have been tested, but the 24-well format was found to be easier to work with and more consistent in producing confluent monolayers. The multi-well format of this assay also allows us to scale up the experiments with relative ease so that the proportion of available membrane inserts for testing increases. Furthermore, using smaller culture membrane inserts enables experiments to be scaled down to conserve precious samples.
3. Make a sufficient volume of diluted collagen so that the entire surface of the flask is in contact with the solution. For a T75 flask, this was about 7 mL.
4. Expanding and freezing cells ensures that early passage cells (no higher than passage 4) are used for each experiment. Cells passaged more than four times are not ideal for co-culture. Only early passage bBMVEC can be seeded at this density that is lower than recommended. Cells were frozen in a 1:1 mixture of a 50% serum solution in medium with a 50% serum, 5% DMSO solution in medium, accounting for the amount of serum already added to the supplemented medium.
5. For a membrane insert with 0.3 cm^2 surface area, 50 μL of collagen solution was applied.
6. Coated inserts may be stored until needed if they have not been rinsed with TBS. Rinse stored inserts with TBS before use.
7. The importance of collagen is documented [19]. The attachment and growth of the bBMVEC on cell culture membrane inserts was optimized when both 15 μg collagen/cm^2 and 5 μg fibronectin/cm^2 were applied.
8. The fibronectin solution may be applied in advance of seeding NHA. Store the membrane inserts with the fibronectin solution at 4°C until ready for use.

9. If using a transparent membrane insert, the NHA can be seen on the basal surface with phase-contrast microscopy. If using translucent membrane inserts, the NHA will not be visible.
10. Visual observations of the attachment and proliferation of cells can be used to determine when to proceed with the co-culture. In co-culture, monolayer formation and integrity was maximal at 5–6 days with visible monolayer disruption and shedding beginning 8–11 days after seeding.
11. The inhibitor tested was dissolved in DMSO. After dilution the DMSO concentration was 2.5%. Inhibitor was delivered in 25 μL aliquots.
12. Consider the amount of solvent introduced in the diluted inhibitor so that an equivalent amount is added to wells that will not be treated with the inhibitor. If the recommended insert volume is 500 μL, then 3.25 μL of 1.54 μM tPA + 0.625 μL of solvent will be added to 496.1 μL of serum-free medium. For studies on the activation of pro-MMPs, investigators may wish to add plasminogen together with tPA [3].
13. Making the standards by serial dilution is the easiest and most precise method. The amount of each standard is based on the quantity needed during fluorescence measurements plus the amount needed to make the serial dilution. Since 100 μM Lucifer yellow is also used as the tracer, the number of membrane inserts that will be tested must also be considered.
14. 100 μM Lucifer yellow gave the most consistent results
15. For a 24-well plate, 600 μL of buffer was placed in the well and 400 μL delivered to the membrane insert. Keeping the liquid levels balanced between the well and insert reduces hydrostatic pressure.
16. When testing multiple membrane inserts it is easier to stagger the timing so that each individual membrane insert does not start its time course at the same moment. Protect the collected samples from light and keep chilled until taking the fluorescence readings.
17. The permeability coefficient, P_e, provides a way to quantify the integrity of the co-culture monolayer and the calculated mass balance provides a check on potential losses of tracer to the interior of cells or to the membrane insert itself. Inserts with a mass balance value below a user-determined threshold were discarded. The P_e equation was modified from [16].
18. Achieving uniformity in any cell culture assay is challenging. The variability in the growth and surface coverage of cells within a set of membrane inserts was a significant difficulty. This prevented the membrane inserts from reaching confluence at the same time. To compensate for this problem, co-cultures used for fitting were selected from a pool of membrane inserts based on their growth and monolayer progression.

Acknowledgments

This work was completed at the Protein Expression Facility of the Institute of Molecular Biophysics at Florida State University and supported by the James and Esther King Biomedical Research Program grant (1KF04) and a Florida State University Research Foundation GAP grant to Dr. Sang.

References

1. Abbott NJ, Ronnback L, Hansson E (2006) Astrocyte-endothelial interactions at the blood-brain barrier. Nat Rev Neurosci 7(1):41–53
2. Roycik MD, Myers JS, Newcomer RG, et al. (2013) Matrix metalloproteinase inhibition in atherosclerosis and stroke. Curr Mol Med 13(8):1299–1313
3. Niego B, Medcalf RL (2014) Plasmin-dependent modulation of the blood-brain barrier: a major consideration during tPA-induced thrombolysis? J Cereb Blood Flow Metab 34(8):1283–1296
4. Gasche Y, Soccal PM, Kanemitsu M, et al. (2006) Matrix metalloproteinases and diseases of the central nervous system with a special emphasis on ischemic brain. Front Biosci 11:1289–1301
5. Copin J, Bengualid DJ, Da Silva RF, et al. (2011) Recombinant tissue plasminogen activator induces blood–brain barrier breakdown by a matrix metalloproteinase-9-independent pathway after transient focal cerebral ischemia in mouse. Eur J Neurosci 34(7): 1085–1092
6. Candelario-Jalil E, Yang Y, Rosenberg GA (2009) Diverse roles of matrix metalloproteinases and tissue inhibitors of metalloproteinases in neuroinflammation and cerebral ischemia. Neuroscience 158(3):983–994
7. Cecchelli R, Berezowski V, Lundquist S, et al. (2007) Modelling of the blood-brain barrier in drug discovery and development. Nat Rev Drug Discov 6(8):650–661
8. Cucullo L, Hossain M, Rapp E, et al. (2007) Development of a humanized in vitro blood-brain barrier model to screen for brain penetration of antiepileptic drugs. Epilepsia 48(3):505–516
9. Deli MA, Abraham CS, Kataoka Y, et al. (2005) Permeability studies on in vitro blood-brain barrier models: physiology, pathology, and pharmacology. Cell Mol Neurobiol 25(1):59–127
10. Naik P, Cucullo L (2012) In vitro blood-brain barrier models: current and perspective technologies. J Pharm Sci 101(4):1337–1354
11. He Y, Yao Y, Tsirka SE, et al. (2014) Cell-culture models of the blood-brain barrier. Stroke 45(8):2514–2526
12. Nakagawa S, Deli MA, Kawaguchi H, et al. (2009) A new blood-brain barrier model using primary rat brain endothelial cells, pericytes and astrocytes. Neurochem Int 54(3–4):253–263
13. Hatherell K, Couraud P, Romero IA, et al. (2011) Development of a three-dimensional, all-human in vitro model of the blood–brain barrier using mono-, co-, and tri-cultivation Transwell models. J Neurosci Methods 199(2):223–229
14. Xue Q, Liu Y, Qi H, et al. (2013) A novel brain neurovascular unit model with neurons, astrocytes and microvascular endothelial cells of rat. Int J Biol Sci 9(2):174–189
15. Megard I, Garrigues A, Orlowski S, et al. (2002) A co-culture-based model of human blood-brain barrier: application to active transport of indinavir and in vivo-in vitro correlation. Brain Res 927(2):153–167
16. Culot M, Lundquist S, Vanuxeem D, et al. (2008) An in vitro blood-brain barrier model for high throughput (HTS) toxicological screening. Toxicol In Vitro 22(3):799–811
17. Siddharthan V, Kim YV, Liu S, et al. (2007) Human astrocytes/astrocyte-conditioned medium and shear stress enhance the barrier properties of human brain microvascular endothelial cells. Brain Res 1147:39–50
18. Kuo YC, Lu CH (2011) Effect of human astrocytes on the characteristics of human brain-microvascular endothelial cells in the blood-brain barrier. Colloids Surf B Biointerfaces 86(1):225–231
19. Cecchelli R, Dehouck B, Descamps L, et al. (1999) In vitro model for evaluating drug transport across the blood-brain barrier. Adv Drug Deliv Rev 36(2–3):165–178

Part VII

Matrix Metalloproteases as Biomarkers

Chapter 17

Matrix Metalloproteases as Biomarkers of Disease

Fernando Luiz Affonso Fonseca, Beatriz da Costa Aguiar Alves, Ligia Ajaime Azzalis, and Thaís Moura Gáscon Belardo

Abstract

Matrix metalloprotease play a vital role in many cellular processes. Dysfunction in activity of these enzymes has been implicated in the pathogenesis of a number of diseases. Factors that affect the balanced interaction between MMPs and their inhibitors, such as genetic mutations of extracellular matrix components or dysregulation of MMP expression, can lead to various diseases. Due to their essential role in ECM remodeling, MMPs have become targets of interest as biomarkers for the diagnosis and prognosis of diseases associated with alterations of the ECM.

Key words Matrix metalloproteases, Disease, Cancer, Inflammatory disease

1 Introduction

1.1 Definition, Classification, and Roles of MMPs

The space between the cells that form all tissues and organs is filled with a series of compounds and macromolecules that constitute the extracellular matrix (ECM). Therefore, the ECM represents the noncellular components of all tissues and organs. The ECM has tissue-specific characteristics; however, it is basically composed of water and relatively stable macromolecules, which are not only produced, exported, and bound by cells but also responsible for the modulation of their structure, physiology, biochemical interactions, and the biomechanics of tissues. Such molecules, essential for morphogenesis, differentiation, and tissue homeostasis, are classified into three categories: glycosaminoglycan polysaccharide chains, fibrous proteins with structural roles (collagens and elastins), and non-fibrous proteins with adhesion properties (laminins and fibronectins) [1].

Extracellular matrix remodeling is a dynamic and continuous process in which the cell deposits new components and degrades old ones by proteolytic action [2]. ECM degradation is essential not only for processes like embryonic development, morphogenesis, reproduction and tissue reabsorption and remodeling [3] but

Charles A. Galea (ed.), *Matrix Metalloproteases: Methods and Protocols*, Methods in Molecular Biology, vol. 1579,
DOI 10.1007/978-1-4939-6863-3_17,

also for the regulation of its quantity, composition, and structure. Moreover, ECM degradation plays an important role in the liberation of various biologically active molecules, like growth factors and hormones. The main enzymatic family that takes part in this degradation mechanism is the matrix metalloproteinases (MMPs). MMPs form part of a larger family, the metzincin enzymes, along with the adamalysins, which encompass the ADAMs disintegrin and metalloproteinase, the ADAMTS (ADAMs with thrombospondin motifs), and the astacins (meprins) [4].

Metalloproteinases are zinc-dependent enzymes secreted by cells or bound to the cell membrane. They all have many functional domains in common, such as the signal peptide domain, the propeptide domain, the hemopexin-like domain (excluding MMP-7, MMP-23, and MMP-26), and the catalytic domain [3, 4]. The catalytic domains of MMPs from a wide range of species display a high degree of sequence homology [5]. So far, a total of 25 MMPs, including 24 vertebrate MMPs, have been identified. This includes two alternatively spliced isoforms of MMP-23, namely MMP23A and MMP23B.

In the human genome, the gene cluster that contains MMP-1, -3, -7, -8, -10, -12, -13, -20, and -27 is found on chromosome 11 (11q22). The remaining MMPs are located on different chromosomes, such as chromosomes 1, 8, 12, 14, 16, 19, and 20 [5].

Based on characteristics like substrate specificity (Table 1), sequence similarity, and domain organization, MMPs can be classified into six groups: (a) collagenases (MMP-1, -8, -13, -14, and -18), which are capable of cleaving triple helical collagen types I, II, and III. Collagenases are commonly found in tissues, at cell-cell and cell-matrix contact sites, and they disappear in the event of evolving fibrosis; (b) gelatinases (MMP-2 and -9), which degrade collagen fragments released by collagenases and also cleave collagen IV and elastin; (c) stromelysins (MMP-3, -7, -10, -11, -19, and -20), which act on the proteoglycans, binding proteins, fibronectins, laminins, and collagen types III and IV; (d) matrilysins (MMP-7 and -16), which degrade versican, elastin, fibronectin, collagen type IV among others; (e) membrane-bound MMPs (MT-MMPs), exemplified by MMP-1, -3, -2, -5, -15, -16, -17, -24, -25; and (f) other MMPs [5].

Most MMPs are secreted in an inactive form (zymogens), and are subsequently activated in the extracellular space [6]. The main MMP activation mechanism is mediated by cell surface associated plasmin, which is generated from plasminogen by the activity of either tissue-type plasminogen activator (t-PA) or urokinase-type plasminogen activator (u-PA). The main MMPs activated by this pathway are pro-MMP-1, pro-MMP-3, pro-MMP-7, pro-MMP-9, pro-MMP-10, and pro-MMP-13. Once activated, these MMPs can activate other MMPs [7].

Table 1
Substrates cleaved by various MMPs and their chromosomal location

MMP	Substrate	Chromosome
1	Collagen types I, II and III	11
2	Collagen IV, V, VII, X, fibronectin, proteoglycans, elastin	16
3	Procollagen, proteoglycans, collagen X and XI, laminin, elastin, fibronectin, plasminogen, pro-MMP-9	11
7	Collagen IV, gelatin, fibronectins, laminins, elastin, transferrin, casein, plasminogen	11
8	Proteoglycans, fibronectin	11
9	Plasminogen, collagen IV, V, VII, and X, gelatin, elastin, fibronectin	20
10	Stromelysin, gelatin, casein, fibronectin	11
11	IGFPB-1, collagen, laminin, elastin, fibronectin, casein, proteoglycans	
12	Plasminogen, elastin, collagen IV, fibronectin, gelatin, casein, laminin, fibrinogen	11
13	Collagen I, plasminogen, fibronectin, proteoglycan	11
14	Collagen I, CD44, fibronectin, laminin, pro-MMP-2	14
15	Cell-surface transglutaminase, fibronectin, gelatin, laminin, pro-MMP-2	16
16	Cell-surface transglutaminase, collagen III, gelatin, casein, fibronectin, pro-MMP-2	8
17	Fibrinogen, gelatin, pro-MMP-2	12
18	Collagen I	12
19	Collagen I and IV, fibronectin, gelatin, laminin	12
20	Amelogenin	11
24	Fibronectin, pro-MMP-2, proteoglycans, gelatin	20
25	Collagen IV, gelatin, fibronectin, pro-MMP-2	
23 (A and B)	Gelatin	1
26	Collagen IV, fibronectin, fibrinogen, casein, insulin-like growth factor-binding protein 1, pro-MMP-9	11

Metalloproteinases can collectively cleave all components of the ECM [4]. The proteolytic activity of MMPs is controlled at different levels by regulation of gene expression (transcriptional regulation), activation of pro-enzymes, and inhibition by tissue inhibitors of metalloproteinases (TIMPs) or other inhibitors, such as α2-macrobulin. MMPs are not constitutively expressed but expressed after stimulation by cytokines and growth factors [3].

Normal tissue structure partly results from the balanced interaction between MMPs, α2-macrobulin, and TIMPs [8]. To date, four types of TIMPs have been identified (TIMP-1 to TIMP-4), while TIMP-3 is the main regulator of MMP activity (including ADAMs and ADAM-TSs) [3].

Genetic mutations of extracellular matrix components that affect the balanced interaction between MMPs and their inhibitors can cause various diseases, such as inflammation, ulceration, arthritis, periodontitis, vascular disease, diabetes, fibrosis, emphysema, cancer (onset, vascularization, and metastasis), apoptosis, sepsis, intestinal defense activation, and nervous system diseases like stroke and Alzheimer's [3, 4]. Due to their essential role in ECM remodeling, MMPs have become targets of interest as biomarkers for the diagnosis and prognosis of diseases associated with alterations of the ECM.

2 Metalloproteinases and Malignant Hematological Diseases

The action of metalloproteinases in various malignant hematological diseases sheds light on the role of the ECM in different blood pathologies.

Stromal cell activity plays a role in malignant transformation, the onset and progression of different kinds of malignant hematological diseases (clonal disorders resulting from the neoplastic transformation of blood progenitor cells), and solid tumors. While studies of solid and hematologic cancers focus on the analysis of gene mutations and malignant lesions due to a gain or loss of biological function it is important to study the role of MMPs in the malignant blood pathologies [9].

Two bone marrow niches play an essential role in several malignancies: the endosteal niche with its resident osteoblasts that are derived from mesenchymal stem cells, and the vascular niche, home for hematopoietic stem cells. Furthermore, the extracellular matrix is fundamentally involved in tumorigenesis not only by directing the differentiation of these pluripotent stem cells but also due to the participation of metalloproteinases [10].

Metalloproteinases are synthetized by different types of cells and tissues. Dermal fibroblasts and leukocytes are the main sources of metalloproteinases, especially MMP-2 while platelets are known to stimulate the production of MMP-1, -2, -3, and -14. MMPs are also expressed in vascular cells and tissues, such as smooth muscle cells. Alterations in the expression, distribution, or activity of MMPs may impair cell or tissue functionality, thus contributing to the development of malignant hematological diseases [11].

The bone marrow microenvironment is of paramount importance for the regulation and maintenance of hematopoietic stem cells so as to prevent the onset of hematological neoplasias. Different cell types reside within the bone marrow microenvironment, including macrophages, endothelial progenitors, and osteoblasts, and these are responsible for the maintenance of hematopoietic pluripotent cells that are located close to endothelial cells and the vascularized area. During hematopoietic stem cell transplantation, proteolytic enzymes released from neutrophils, including several metalloproteinases, cleave granulocyte colony stimulating factor (G-CSF) that then stimulates $CD34^+$ cells. This also results in metalloprotease-mediated extensive remodeling of the bone marrow environment [12].

Adult stem cells are characterized by their capacity for self-renewal, maintaining bone marrow homeostasis through cellular proliferation and differentiation via processes that are under strict control. Deletion of genes involved in the maintenance of quiescent cell cycle status and dysregulated proliferation of hematopoietic stem cells results in the loss of cell functionality. It is important to point out that quiescent and proliferative cells can be found in the same niche [13].

Angiogenesis comprises the formation of new vascularized networks that contribute to the growth of tumors through the transport of neoplastic cells to secondary sites via lymph and blood vessels, which is of great importance in various onco-hematological diseases. Angiogenesis involves the continuous proteolytic modification of the ECM and cleavage of various cell-surface molecules that are released into the microenvironment of the tumor through tumor infiltrating leukocytes. This process is mediated by various proteases including several metalloproteinases, predominantly MMP-9 [14].

Precursor cells of megakaryocytes, platelets, and their microparticules (MP) are involved in various physiological and pathological events that contribute to the aggressiveness and persistence of tumors. This leads to complications in the treatment of inflammation, vascular complications, and various malignant hematological diseases. Platelet apoptosis exceeding normal physiological levels promotes the formation of microparticules (MP). Platelet-derived MPs (PMPs), which constitute around 80% of circulating MPs, express adhesion molecules and are responsible for the promotion of angiogenesis and dissemination of neoplasias. These PMPs along with growth factor and metalloproteinases show a combined action with vascular endothelial growth factor (VEGF) [15].

Acute lymphoblastic leukemias are the most common (99%) form of lymphoid leukemia and for this reason are commonly diagnosed as acute lymphoid leukemia. In lymphoblastic leukemias there is an increase in the density of microvessels in the bone marrow. While in acute leukemias, VEGF-A is considered the main dominant inducer of blood vessel growth. Matrix-bound VEGF-A is cleaved and released by several MMPs, especially gelatinases A and B (denoted as MMP-2 and MMP-9, respectively), which hydrolyze basement membrane collagens of blood vessels, stimulating extramedullary infiltration by lymphoblasts. VEGF-A stimulated secretion of MMPs in Pre-B (B) and pre-T (T) lineage cells is currently used for disease prognosis in patients [16].

Metalloproteinases also play an important role in chronic lymphocytic leukemias, where there is an increase of CD5 B lymphocytes in the peripheral blood. Gelatinases B/(pro-) MMP-9, which are involved in the migration and survival of these malignant hematologic cells, have been shown to speed up disease progression. (Pro-) MMP-9, whose expression is regulated by cytokines and other factors, is detected on the surface of LLC cells and accumulates in these cells [17].

Mutations in Notch 1–4 genes may cause malignancy in mature B cells (via inhibition of the maturation of common lymphoid progenitor cells), chronic lymphocytic leukemia, Hodgkin's lymphoma, Burkitt's lymphoma, and multiple myeloma. The latter is a peculiar onco-hematologic disease that depends on interaction with the bone marrow microenvironment, where metalloproteinases are responsible for the interrelation of the stromal tumor, tumor invasion, and metastasis. This interrelation is characterized by the production of high levels of MMP-1 and MMP-2, which in turn generate active MMP-9 and MMP-2 that degrade collagen IV, the major structural component of the basement membrane [18].

Chronic myeloid leukemia (CML) is a myeloproliferative disorder that is characterized by the Philadelphia (Ph) chromosome, a BCR-ABL fusion gene resulting from a specific translocation t [19, 20], which leads to the expansion of immature leukemic cells. MMP-9 is known for its ability to cleave specific components of the extracellular matrix, including collagen (types IV, V, and XI) and elastin. MMP-9 is also involved in angiogenesis, demonstrating that MMPs and TIMPs play a key role in hematopoiesis once they are secreted by stromal and tumor cells [9, 20, 21].

Chronic myeloproliferative disorders are characterized by the progressive remodeling of the bone marrow stroma, highlighted by an increase in the deposition of proteins in the extracellular matrix, neoangiogenesis and the substitution of hematopoietic cells for fibrous tissue [22]. However, in CML the participation of MMPs is not limited to the extracellular matrix. They are also involved in the stimulation of various cytokines in different cell types and tissues through a specific MMP9-induced pathway [21].

MMPs are involved in coagulation in sepsis and inflammatory responses, and in malignant hematological diseases, like multiple myeloma and may play a role in disseminated intravascular coagulation (DIC). It has been suggested that impairment of the interaction of MMP-8 with TIMP-1, which are involved in the activation of the tissue factor pathway inhibitor (TFPI), is essential for triggering DIC. MMP-8, also known as collagenase-2, is a protein stored in neutrophil granules (as well as in smaller quantities in activated macrophages and endothelial cells) that acts via neutrophil chemotaxis in the inflammation process [23].

Studies are currently underway to better understand the relationship between MMP expression and polymorphisms associated with various malignant hematological diseases, to properly target medications for the treatment of these diseases. Detection of the deregulation of MMPs is of utmost importance from the clinical perspective, since it can be used as a diagnostic and prognostic tool, or as a new oncologic therapy. In addition, monitoring MMP regulation may be useful in hematopoietic stem cell transplantation where MMPs induce hematopoietic cell mobilization [9].

3 Metalloproteinases and Inflammatory Disorders

Whenever tissue injury occurs, modifications ensue which eliminate the offending agent, limit tissue damage, and regenerate the site of the lesion. These modifications depend on the regulation of proteins (inflammatory markers) involved in the inflammatory response [24, 25]. Matrix metalloproteinases (MMPs) figure among these many types of inflammatory markers. MMPs are a family of zinc dependent endopeptidases capable of degrading extracellular matrix components and proteins of the basement membrane, which actively participate in tissue remodeling, both physiologically and pathologically [26].

The analysis of MMPs as inflammatory biomarkers can help not only in the diagnosis but also in the treatment of some inflammatory diseases [27]. Some of the diseases in which MMPs could be used as biomarkers are described below.

3.1 Ankylosing Spondylitis (AS) and Rheumatoid Arthritis (RA)

Ankylosing Spondylitis (AS) is a chronic rheumatic disease that first affects the spine and sacroiliac joints. Rheumatoid arthritis (RA) is an autoimmune disease that most commonly affects joints; however, it is actually a systemic disease that may involve other organs [28].

MMP-2 and MMP-9 levels are increased in the synovial liquid of patients with RA [6, 7]. It has lately been observed that high serum levels of active MMP-3, both in animal and human models, is also a good marker for AS and RA owing to the fact that MMP-3 is more specifically localized in joints and suffers less from systemic factors [28–30].

3.2 Sepsis

Some studies show a direct correlation between increased levels of MMPs and low survival rates of patients with sepsis [31]. Aerts *et al.* [32] have recently demonstrated that a hydroxypyrone-based inhibitor selective for MMP-12 displays anti-inflammatory activity in mice.

3.3 Chronic Obstructive Pulmonary Disease (COPD)

COPD is characterized by progressive obstruction of the airway associated with chronic inflammation. MMP-1, MMP-8, MMP-12, TIMP-1 (tissue inhibitor of metalloproteinase 1), and especially MMP-9 have been shown to be associated with COPD [33].

3.4 Conjunctivochalasis (CCh)

Despite the fact the causes of CCh are still uncertain, it is very likely that the accumulation of MMPs in the tears of these patients may contribute to the progression of the disease [34]. In addition, the effectiveness of surgery to treat CCh has been shown to be associated with a postoperative decrease in MMP-9 levels in tears. In fact, many authors suggest that the quantification of MMP-9 could be important for the evaluation of many eye diseases [34].

3.5 Coronary Heart Disease (CHD)

The inflammatory process seems to play a role in the pathogenesis of vascular diseases. Markers of inflammation like C-reactive protein, fibrinogen, and interleukin-6 are associated with the risk of developing CHD [35]. However, other markers, such as MMP-9, also need to be accessed [36, 37].

3.6 Inflammatory Bowel Disease (IBD)

Ulcerative colitis (UC) and Crohn's disease (CD) are autoimmune diseases characterized by inflammation of the gastrointestinal system. Although MMPs are present in IBD, their role is not clear and further studies are required [38]. The colon of patients with IBD exhibit high levels of MMP-1, -3, -7, -9, -10, and -12 [39], and intestinal biopsies of patients with UC demonstrate that vascular endothelial cells and leukocytic infiltrate produce MMP-7 and -13 [40]. MMP-9 serum levels are correlated to both CD and UC disease, with higher levels detected in the latter. This difference in levels between these MMPs could help to differentiate between UC disease and CD [41].

Future studies should elucidate the roles of MMPs and tissue inhibitors of metalloproteinases (TIMPs), thus contributing to the diagnosis and treatment of inflammatory diseases [42, 43].

4 Metalloproteinases and Solid Tumors

Interactions between tumor cells and their microenvironment reveal the important role played by matrix metalloproteinases (MMPs) during carcinogenesis. Tumor growth and metastasis depend on cell-cell and cell-matrix interactions, as well as on tissue modifications through the action of proteolytic enzymes. The MMP family is composed of 25 endopeptidases in humans and these enzymes have an important role in cancer invasion, metastasis, and angiogenesis. They also exert significant impact on cell behavior, like the growth of metastatic tumor cells and the increase in motility of epithelial cells. MMPs not only act on the extracellular matrix (ECM) via protein degradation, cleavage of cell surface receptors, and the release of apoptotic signals but they are also associated with advanced stages and clinical prognosis of many types of cancer [44].

Various methods can be used to study the relation between MMPs and cancer. Studies examining various aspects of tumor biology will shed light on which methodologies should be used to examine this relationship. Some examples of potential methods that may be used are the enzyme-linked immunosorbent assay (ELISA), immunohistochemical (IHC) staining, and various molecular biology methods, which range from the detection of genetic polymorphisms to measuring gene expression. The importance of MMPs as potential biomarkers is highlighted by their known involvement in many types of cancers.

4.1 Breast Cancer

Overexpression of MMPs in the serum of breast cancer patients is associated with a poor prognosis. For example, overexpression of MMP-9 and MMP-2 correlates with a higher risk and poor prognosis for this type of cancer [45].

MMP-13 has also been studied and characterized with tumor progression. Overexpression of MMP-13 has been correlated with the expression of HER2/neu, more aggressive tumor phenotypes and decreased life expectancy [46]. A recent study examining the relationship between MMP and TIMP expression in tumors (i.e., tumor center or tumor invasive front) and distant metastasis in mononuclear inflammatory cells (MICs) and cancer associated fibroblasts (CAFs) of patients with primary breast cancer, showed that the expression of MMP-11 by MICs and TIMP-2 by CAFs in the tumor center or tumor invasive front were the most potent independent prognostic factors for predicting clinical outcome for patients [47].

4.2 Prostate Cancer

It has recently been shown that genotypic variations of MMP-3 and TIMP-3 (polymorphisms) may affect the balance between these proteins, thus increasing the risk of developing prostate cancer [48].

The relationship between testosterone and MMP-13 levels in prostate cancer patients has been previously reported [8]. Low levels of testosterone have been linked to a poor prognosis [49]. This relationship suggests an important biochemical interaction between this hormone and these endopeptidases. Immunohistochemical analysis of MMP-3, MMP-13 and TIMP-3, MMP-9, and TIMP-1 levels in prostate cancer also suggests that an imbalance between MMP-9 and TIMP-1 may explain the invasive potential of prostatic adenocarcinoma [50].

4.3 Lung Cancer

MMP-13 is not only associated with breast and prostate cancers but may also be the most important target among the MMPs in lung cancer. In a recent study, MMP-2, -3, -9, and -13 were analyzed by fluorescence molecular tomography (FMT) and only MMP-13 was detected early in the development of pulmonary invasive adenocarcinomas [51]. It has also been reported that MMP-2 polymorphisms may be a potential prognostic marker for inoperable non-small cell lung cancers (NSCLC) treated with radio- or chemotherapy [52].

4.4 Colorectal Cancer

Studies of advanced stage colorectal adenocarcinoma found that MMP-9 was expressed at high but inconstant levels in tumor cells. Highest expression levels were found in little or moderately differentiated carcinomas, and surprisingly, the lowest expression levels were observed in well-differentiated colorectal cancers. MMP-9 expression has also been observed in peritumoral macrophages and stroma cells. While high levels of expression were shown in metastasis-free lymphnodes in both macrophages and lymphocytes in areas of tumor necrosis, angiogenesis was in general found to be correlated with the intensity of MMP-9 expression [53, 54].

Besides MMP-9, MMP-7 also appears to have an important role in the development and progression of colorectal cancer. A recent meta-analysis involving 1631 colorectal cancer patients showed that overexpression of MMP-7 unfavorably impacted on the overall survival rate. Moreover, this study demonstrated the potential of MMP-7 as a prognostic predictor. In summary, overexpression of MMP-7, detected by immunohistochemical staining, indicated a poor prognosis for colorectal cancer patients, and could be used as a guide for the clinical-therapeutic management of this neoplasia [55].

Studies have also reported that the expression of MMP-13 is related to pathogenesis, metastasis, and tumor relapse, and for this reason it may be a valuable predictor of metastasis and relapse of colorectal cancer [56].

4.5 Bladder Cancer

Studies have investigated the role of MMPs in bladder cancer by identifying metalloproteinase polymorphisms and gene expression levels as well as measuring MMP levels in urine. These analyses of neoplasia have looked at various biological samples, including urine, and the results establish MMPs as potential diagnostic biomarkers for this neoplasia.

Both MMP-3 and MMP-9 have recently been proposed as possible early markers for bladder cancer. Using ELISA-based assays the authors were able to show that levels of both MMPs were elevated in the urine of patients with bladder cancer when compared with a control group. Furthermore, MMP-3 levels were evaluated in urine samples from patients with early stage schistosomal and non-schistosomal bladder cancer. Remarkably, MMP-3 levels remained elevated at advanced stages of these two histological types of cancer. On the other hand, MMP-9 levels were elevated only at advanced stages of schistosomal and non-schistosomal bladder cancer. Assays displayed improved sensitivity and specificity for MMP-3 compared to MMP-9 at diagnosis [57].

A more recent study has examined the role of gene polymorphisms of MMP-1, -2, -3, -7, and -9 in bladder cancer. The presence of MMP-1 gene polymorphisms was detected in all studied populations. However, MMP-2 and MMP-7 polymorphisms were only detected in Asian populations, and it was proposed that these polymorphisms may be protective for the development of bladder cancer [58].

In conclusion, many studies have shown the involvement of MMPs at all stages of carcinogenesis. Nevertheless, the diagnostic and prognostic roles of MMPs are difficult to define not only because most studies do not examine the same MMPs or use methods with similar sensitivity and specificity, but also because of the complexity of carcinogenesis. Nevertheless, MMP-3 and -13 have been identified as potential biomarkers in a range of cancers including breast, prostate, lung, colorectal and bladder cancers, regardless of the biological matrix examined or the method used.

References

1. Frantz C, Stewart KM, Weaver VM (2010) The extracellular matrix at a glance. J Cell Sci 123:4195–4200
2. Harjanto D, Zaman MH (2013) Modeling extracellular matrix reorganization in 3D environments. PLoS One 8:e52509
3. Tallant C, Marrero A, Gomis-Rüth FX (2010) Matrix metalloproteinases: fold and function of their catalytic domains. Biochim Biophys Acta 1803:20–28
4. Bonnans C, Chou J, Werb Z (2014) Remodelling the extracellular matrix in development and disease. Nat Rev Mol Cell Biol 15:786–801
5. Theocharis AD, Skandalis SS, Gialeli C, Karamanos NK (2015) Extracellular matrix structure. Adv Drug Deliv Rev 97:4–27
6. Giannandrea M, Parks WC (2014) Diverse functions of matrix metalloproteinases during fibrosis. Dis Model Mech 7:193–203
7. Tenório PP, Duque MAA, Araújo MM, Pontes Filho NT, Melo MR Jr (2012) The role of matrix metalloproteinases in the development of aortic aneurysm. Avaliable at: http://files.bvs.br/upload/S/0101-5907/2012/v26n4/a3488.pdf [Portuguese]
8. Bonaldi CM, Azzalis LA, Junqueira VBC, Oliveira CGB, Vilas Boas VA, Gáscon TM, Gehrke FS, Kuniyoshi RK, Alves BCA, Fonseca FLA (2015) Plasmatic levels of E-Cadherin and MMP-13 in prostate cancer patients: correlation with PSA, testosterone and pathological parameters. Tumori 101:185–188
9. Chaudhary KA, Pandya S, Ghosh K, Nadkarni A (2013) Matrix metalloproteinase and its drug targets therapy in solid and hematological malignances: an overview. Mutat Res 753:7–23
10. Klein G, Schmal O, Aicher KW (2015) Matrix metalloproteinases in stem cell mobilization. Matrix Biol 44–46:175–183
11. MacColl E, Khalil RA (2015) Matrix metalloproteinases as regulators of vein structure and function: implications in chronic venous disease. J Pharmacol Exp Ther 355:410–428
12. Szmigielska-Kaplon A, Szemraj J, Hamara K, et al. (2014) Polymorphism of CD44 influences the efficacy of CD34(+) cells mobilization in patients with hematological malignancies. Biol Blood Marrow Transplant 20:986–991
13. Kunisaki Y, Bruns I, Scheiermann C (2013) Arteriolar niches maintain haematopoietic stem cell quiescence. Nature 502:637–643
14. Deryugina IE, Quigley PJ (2015) Tumor angiogenesis: MMP-mediated induction of intravasation- and metastasis-sustaining neovasculature. Matrix Biol 44–46:94–112
15. Thushara MR, Hemshekhar M, Kemparaju B, et al. (2015) Biologicals, platelet apoptosis and human diseases: an outlook. Crit Rev Oncol Hematol 93:149–158
16. Poyer F, Coquerel B, Pegahi R, et al. (2009) Secretion of MMP-2 and MMP-9 induced by VEGF autocrine loop correlates with clinical features in childhood acute lymphoblastic leukemia. Leuk Res 33:407–417

17. Bailón E, Berzal-Ugarte E, Amigo-Jimenes I (2014) Overexpression of progelatinase B/ proMMP-9 affects migration regulatory pathways and impairs chronic lymphocytic leukemia cell homing to bone marrow and spleen. J Leukoc Biol 96:185–199
18. Mirandola L, Comi P, Cobos E, et al. (2011) Notch-ing from T-cell to B-cell lymphoid malignancies. Cancer Lett 308:1–13
19. Lu P, Takai K, Weaver VM, Werb Z (2011) Extracellular matrix degradation and remodeling in development and disease. Cold Spring Harb Perspect Biol 3:a005058
20. Arpino V, Brock M, Gill ES (2015) The role of TIMPs in regulation of extracellular matrix proteolysis. Matrix Biol 44–46:247–254
21. Nakahara F, Kitaura J, Uchida T, et al. (2014) Hes1 promotes blast crisis in chronic myelogenous leukemia through MMP-9 upregulation in leukemic cells. Blood 123:3932–3942
22. Maral S, Acar M, Balcik SO, et al. (2015) Matrix metalloproteinase 2 and 9 polymorphism in patients with myeloproliferative diseases. Medicine 94:e732
23. Sivula M, Hästbacka J, Kuitunen A, et al. (2015) Systemic matrix metalloproteinase-8 and tissue inhibitor ofmetalloproteinases-1 levels in severe sepsis-associated coagulopathy. Acta Anaesthesiol Scand 59:176–184
24. McGuire JK, Manicorne AM (2008) Matrix metalloproteinases as modulators of inflammation. Semin Cell Dev Biol 19:34–41
25. Rosa Neto NS, de Carvalho JF (2009) O uso de provas de atividade inflamatória em reumatologia. Rev Bras Reumatol 49:413–430
26. Amar S, Fields GB (2015) Potential clinical implications of recent matrix metalloproteinase inhibitor design strategies. Expert Rev Proteomics 12:445–447
27. Peng WJ, Yan JW, Wan YN, Wang BX, Tao JH, Yang GJ, Pan HF, Wang J (2012) Matrix metalloproteinases: a review of their structure and role in systemic sclerosis. J Clin Immunol 32:1409–1414
28. Sun S, Bay-Jensen AC, Karsdal MA, Siebuhr AS, Zheng Q, Maksymowych WP, Christiansen TG, Henriksen K (2014) The active form of MMP-3 is a marker of synovial inflammation and cartilage turnover in inflammatory joint diseases. BMC Musculoskelet Disord 15:93
29. Ateş A, Türkçapar N, Olmez U, Tiryaki O, Düzgün N, Uğuz E, et al. (2007) Serum pro-matrix metalloproteinase-3 as an Indicator of disease activity and severity in rheumatoid arthritis: comparison with traditional markers. Rheumatol Int 27:715–722
30. Young-Min S, Cawston T, Marshall N, Coady D, Christgau S, Saxne T, et al. (2007) Biomarkers predict radiographic progression in early rheumatoid arthritis and perform well compared with traditional markers. Arthritis Rheum 56:3236–3247
31. Vandenbroucke RE, Dejager L, Libert C (2011) The first MMP in sepsis. EMBO Mol Med 3:367–369
32. Aerts J, Vandenbroucke RE, Dera R, Balusu S, Vanwonterghem E, Moons L, Libert C, Dehaen W, Arckens L (2015) Synthesis and validation of a hydroxypyrone-based, potent, and specific matrix metalloproteinase-12 inhibitor with anti-inflammatory activity in vitro and in vivo. Mediators Inflamm 2015:510679
33. Navratilova Z, Kolek V, Petrek M (2016) Matrix metalloproteinases and their inhibitors in chronic obstructive pulmonary disease. Arch Immunol Ther Exp (Warsz) 64:177–193
34. Acera A, Vecino E, Duran JA (2013) Tear MMP-9 levels as a marker of ocular surface inflammation in conjunctivochalasis. Invest Ophthalmol Vis Sci 54:8285–8291
35. Libby P, Ridker PM, Hansson GK (2011) Progress and challenges in translating the biology of atherosclerosis. Nature 473:317–325
36. Kaptoge S, Seshasai SRK, Gao P, Freitag DF, Butterworth AS, Borglykke A, Angelantonio ED, Gudnason V, Rumley A, Lowe GDO, Jørgensen T, Danesh J (2014) Inflammatory cytokines and risk of coronary heart disease: new prospective study and updated meta-analysis. Eur Heart J 35:578–589
37. Alfakry H, Malle E, Koyani CN, Pussinen PJ, Sorsa T (2015) Neutrophil proteolytic activation cascades: a possible mechanistic link between chronic periodontitis and coronary heart disease. Innate Immun 22:85–99
38. O'Sullivan S, Gilmer JF, Medina C (2015) Matrix metalloproteinases in inflammatory bowel disease: an update. Mediators Inflamm. doi:10.1155/2015/964131 Article ID 964131
39. Pedersen G, Saermark T, Kirkegaard T, Brynskov J (2009) Spontaneous and cytokine induced expression and activity of matrix metalloproteinases in human colonic epithelium. Clin Exp Immunol 155:257–265
40. Rath T, Roderfeld M, Halwe JM, Tschuschner A, Roeb E, Graf J (2010) Cellular sources of MMP-7, MMP-13 and MMP-28 in ulcerative colitis. Scand J Gastroenterol 45:1186–1196
41. Matusiewicz M, Neubauer K, Mierzchala-Pasierb M, Gamian A, Krzystek-Korpacka M (2014) Matrix metalloproteinase-9: its interplay with angiogenic factors in inflammatory bowel diseases. Dis Markers doi:10.1155/2014/643645 Article ID 643645
42. Khokha R, Murthy A, Weiss A (2013) Metalloproteinases and their natural inhibitors

in inflammation and immunity. Nat Rev Immunol 13:649–665
43. Vandenbroucke RE, Libert C (2014) Is there new hope for therapeutic matrix metalloproteinase inhibition? Nat Rev Drug Discov 13:904–927
44. Wieczorek E, Jablonska E, Wasowicz W, Reszka E (2015) Matrix metalloproteinases and genetic mouse models in cancer research: a mini-review. Tumour Biol 36:163–175
45. Ren F, Tang R, Zhang X, Madushi WM, Luo D, Dang Y, Li Z, Wei K, Chen G (2015) Overexpression of MMP family members functions as prognostic biomarker for breast cancer patients: a systematic review and meta-analysis. PLoS One 10(8):e0135544
46. Rosseti C, Reis B, Delgado PO, et al. (2015) Adhesion molecules in breast carcinoma: a challenge to the pathologist. Rev Assoc Med Bras 61:81–85
47. Eiró N, Fernandez-Garcia B, Vázquez J, Del Casar JM, González LO, Vizoso FJ (2015) A phenotype from tumor stroma based on the expression of metalloproteases and their inhibitors, associated with prognosis in breast cancer. Oncoimmunology 4(7):e992222 eCollection 2015
48. Srivastava P, Kapoor R, Mittal RD (2013) Impact of MMP-3 and TIMP-3 gene polymorphisms on prostate cancer susceptibility in North Indian cohort. Gene 530:273–277
49. García-Cruz E, Piqueras M, Huguet J, Peri L, Izquierdo L, Musquera M, Franco A, Alvarez-Vijande R, Ribal MJ, Alcaraz A (2012) Low testosterone levels are related to poor prognosis factors in men with prostate cancer prior to treatment. BJU Int 110(11 Pt B):E541–E546
50. Babichenko II, Andriukhin MI, Pulbere S, Loktev A (2014) Immunohistochemical expression of matrix metalloproteinase-9 and inhibitor of matrix metalloproteinase-1 in prostate adenocarcinoma. Int J Clin Exp Pathol 7:9090–9098
51. Salaün M, Peng J, Hensley HH, Roder N, Flieder DB, Houlle-Crépin S, Abramovici-Roels O, Sabourin JC, Thiberville L, Clapper ML (2015) MMP-13 in-vivo molecular imaging reveals early expression in lung adenocarcinoma. PLoS One 10:e0132960
52. Butkiewicz D, Krześniak M, Drosik A, Giglok M, Gdowicz-Kłosok A, Kosarewicz A, Rusin M, Masłyk B, Gawkowska-Suwińska M, Suwiński R (2015) The VEGFR2, COX-2 and MMP-2 polymorphisms are associated with clinical outcome of patients with inoperable non-small cell lung cancer. Int J Cancer 137:2332–2342
53. Georgescu EF, Mogoantă SŞ, Costache A, Pârvănescu V, Totolici BD, Pătraşcu Ş, Stănescu C (2015) The assessment of matrix metalloproteinase-9 expression and angiogenesis in colorectal cancer. Rom J Morphol Embryol 56:1137–1144
54. Zheng CG, Chen R, Xie JB, Liu CB, Jin Z, Jin C (2015) Immunohistochemical expression of Notch1, Jagged1, NF-κB and MMP-9 in colorectal cancer patients and the relationship to clinicopathological parameters. Cancer Biomark 15:889–897
55. Chen H, Hu Y, Xiang W, Cai Y, Wang Z, Xiao Q, Liu Y, Li Q, Ding K (2015) Prognostic significance of matrix metalloproteinase 7 immunohistochemical expression in colorectal cancer: a meta-analysis. Int J Clin Exp Med 8:3281–3290
56. Yan Q, Yuan Y, Yankui L, Jingjie F, Linfang J, Yong P, Dong H, Xiaowei Q (2015) The expression and significance of CXCR5 and MMP-13 in colorectal cancer. Cell Biochem Biophys 73:253–259
57. El-Sharkawi F, El Sabah M, Hassan Z, Khaled H (2014) The biochemical value of urinary metalloproteinases 3 and 9 in diagnosis and prognosis of bladder cancer in Egypt. J Biomed Sci 21:72
58. Tao L, Li Z, Lin L, Lei Y, Hongyuan Y, Hongwei J, Yang L, Chuize K (2015) MMP1, 2, 3, 7, and 9 gene polymorphisms and urinary cancer risk: a meta-analysis. Genet Test Mol Biomarkers 19:548–555

Index

Charles A. Galea (ed.), *Matrix Metalloproteases: Methods and Protocols*, Methods in Molecular Biology, vol. 1579, DOI 10.1007/978-1-4939-6863-3, © Springer Science+Business Media LLC 2017